Routledge Revivals

Earth and Man

Originally published in 1982, this book offered a new approach to an ever-popular subject. Throughout the text, the relevance of physical geography to the human side is stressed. The book uses material drawn from the events of the real world and each chapter begins with a 'story' revealing a geographical message and the treatment is developed through examples, topics and analysis to provide factual and conceptual information. The chapters are written largely in a narrative style which is complemented by special topics, data-response exercises, chapter summaries to aid revision and exam questions. Part 1 covers the global geological foundations of the subject; Part 2 deals with atmosphere and ocean, including meteorology and the world's climates; Part 3 is concerned with soils and vegetation; and Part 4 introduces geomorphology, with an emphasis on processes. There are sections on power and mineral resources, fishing, types of agriculture, environmental conservation, communications and urban location.

Earth and Man

B.J. Knapp

Routledge
Taylor & Francis Group

First published in 1982 by George Allen & Unwin (Publishers) Ltd

This edition first published in 2026 by Routledge
4 Park Square, Milton Park, Abingdon, Oxon, OX14 4RN
and by Routledge
605 Third Avenue, New York, NY 10158.

Routledge is an imprint of the Taylor & Francis Group, an informa business

ISBN 13: 978-1-041-15643-7 (hbk)
ISBN 13: 978-1-003-68051-2 (ebk)
ISBN 13: 978-1-041-15649-9 (pbk)
Book DOI 10.4324/9781003680512

Contents

Preface	*page* v
Acknowledgements	v
Starting points	1
Part 1 – THE ORIGIN OF LAND AND SEA	5
1 The origin of the Earth's rocks	6
2 Rocks made from sediments	23
3 Earthquakes, faults and folds	33
4 Continental drift and mountain building	40
Part 2 – ATMOSPHERE AND OCEAN	49
5 The weather machine	50
6 The world's weather	71
7 The waters of the oceans	81
8 The water cycle	89
Part 3 – SOILS, VEGETATION AND MAN	107
9 Weathering	108
10 Soil formation	114
11 Clothing the surface	121
12 Man-made ecosystems	138
Part 4 – SHAPING THE LAND	157
13 Mass movement processes	158
14 Rivers and valleys in humid regions	170
15 Running water and limestones	191
16 Deserts	195
17 Glacial landforms	203
18 Coastal landforms	219
Appendix 1 Latitude and longitude	234
Appendix 2 Time	235
Appendix 3 Instruments to measure things	235
Index	238

Preface

Geography is the study of man and his environment. Whenever you walk from home to school, take a train, plane, ship or go on holiday by car, the opportunity for appreciating geography is all around you. Geography is also a field study. In fact the approach of a geographer is, in many ways, like that of a detective – a modern-day Sherlock Holmes. Like the investigator who attempts to solve a crime, the geographer is rarely able to witness the events in progress. Rather, he goes to the 'scene of the crime' and begins to assemble the evidence: the plan of a river, broken rock fragments, the shape of slopes, the location of a town, and so on. As the clues are assembled and arranged, many possible explanations are developed. Ideas are tested. Some are rejected and others held for further testing as new evidence continues to accumulate. The results of generations of detective work make up the contents of this book.

One of the main purposes of the book is to help you to prepare for an examination. But if the accounts and explanations help you to develop an interest in your environment, it will have achieved something just as valuable. And if you become interested in the work then, in any case, the final examination will become, as Sherlock Holmes said, 'Elementary, my dear Watson'.

B. J. Knapp
PANAMINT SPRINGS, 1980

Acknowledgements

This book covers a wide range of subjects and it has benefited greatly from the advice and guidance of a large number of people. However, in particular I would like to thank John Whittow and Antoinette Mannion from the Geography Department, Roland Goldring from the Geology Department, Ross Reynolds of the Meteorology Department and Ray Jessop of the Education Department – all in Reading University; Ian Simmons (Bristol University); Simon Burn and Michael Merrett (students of Lanchester Polytechnic) for their help with fieldwork; Gordon Grierson and Barry Stanley (Laker Airways); and Jack Brettle (Plessey). I am also very grateful to the following schoolteachers who read and commented constructively upon early drafts of the book: Nigel Bates (Magdalen College School, Oxford), Tony Wibberley (Plymouth), John Rolfe (Haberdashers' Aske's School for Boys, Elstree), Graham Agnew (Forest Hill), Mrs E. J. Walters (Cardiff), Jim Hanwell (The Blue School, Wells), Mr D. Agerskow (Longcroft School, Beverley), Mr J. A. Williamson (Ilkley); Mrs Butcher (Cleveland).

I am grateful to the following individuals and organisations who have given permission for the reproduction of copyright material:

USGS (Fig. C); Darrell Weyman (Figs 8.7 & 8); A. Collinson (Fig. 11.21); R. Jessop (Fig. 12.12); The Geological Society of London (Fig. 13.13); Figure 18.12 based on sheets 1 and 2, Scotland, reproduced by permission of the Director, Institute of Geological Sciences, NERC copyright reserved.

Starting points

The study of geography

Dying for a drink

JOHN MACK

Figure A Sudanese men collecting water from a shallow well dug in a river bed

Did you know that each year more people die from drinking water than alcohol? As you read this page hundreds of thousands of people will have to trudge several kilometres to their nearest water supply (Fig. A); it will be a hole in the ground. They will fill old petrol cans and leather buckets, trudge back home and then drink the dirty water they have collected. The water will make many ill and some will die. But why should some places have so little water and other resources, and how does this affect the way their people live?

Did you know that, one morning a few years ago, a whole junior school in a Welsh valley was buried under a landslide and 145 children were killed? Why should such accidents happen? What are the forces that cause landslides? Can they be predicted or prevented?

Why is it unsafe to build near some rivers? Why does a road have to go down a slope and what created the shape of the distant hills? How does the shape of land beneath your feet affect your life?

If you have ever asked questions like these about man and his environment you will have been asking questions about geography. This book tries to answer some of the common questions and to help to put the pieces of the environmental puzzle into some sort of order. But geography is a wide subject, and here the emphasis is on the natural environment – air, sea, plants, soil and rock. Using our understanding of the natural environment we then try to show some of the ways in which the environment has affected the way man has lived in the past and lives today. Today, more than ever, we need to understand the natural processes at work on the Earth's surface so that we can help make good use of our environment, not destroy it or be destroyed by it.

Often, starting points for interest in the environment are the stories of human tragedy that make the headlines. These are events where man and environment are in dramatic conflict. However, in general, more subtle processes are at work, going unnoticed because they operate slowly. Below we show how a real landscape can be a mixture of such subtle processes.

ROGER JONES

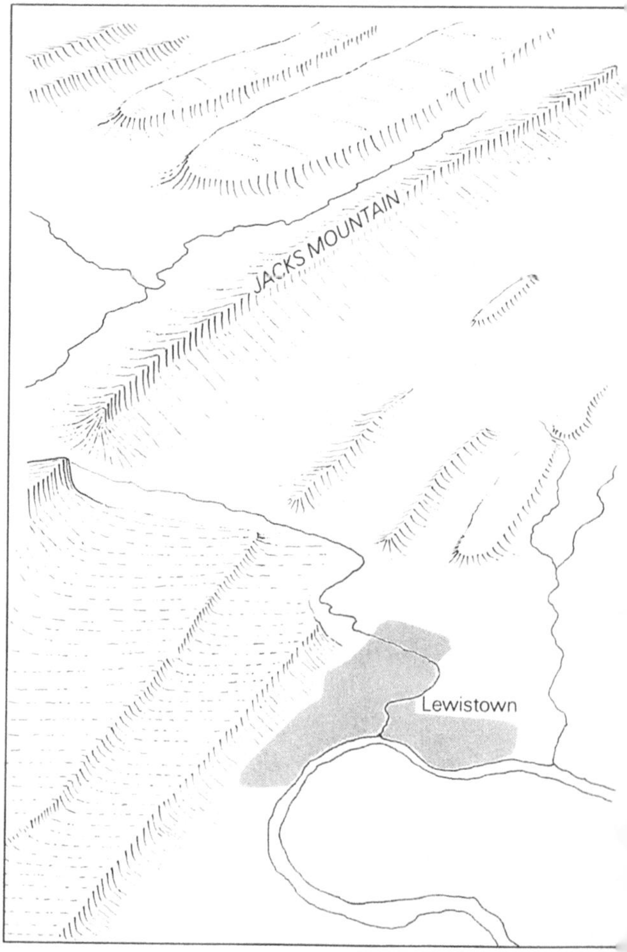

Figure B   An area of scarp and vale scenery, Appalachians (USA)

Topic   Air photos, maps and sections

There are many photographs in this book, some taken from the ground and others from the air. Some are taken from satellites and yet others from manned space flights. All of these photographs gives us different views of the Earth's surface and they are very valuable in helping to build an accurate picture of the landscape. However, photographs can be deceptive, and the geographer must make extensive use of maps, cross sections and detailed field work to complete the picture.

Figure C is a map of the area shown in Figure B. Here the height of the landscape is shown by contour lines. The spacing of the contours tells us about the steepness of the slopes. Because the ridges, for example, have contours spaced more closely on one side than the other, we know the steeper slopes face south-east. The cross section (Fig. D) also helps to show the main features of the landscape in a view from the side. In the cross section the steeper south-east slopes show up very clearly and we get a better impression of the relative steepness of the slopes and the shape of the valleys. We can also see how landscape and rock types are closely related.

Figure C A contour map of the area of Figure B

Figure D A cross section of Figure B

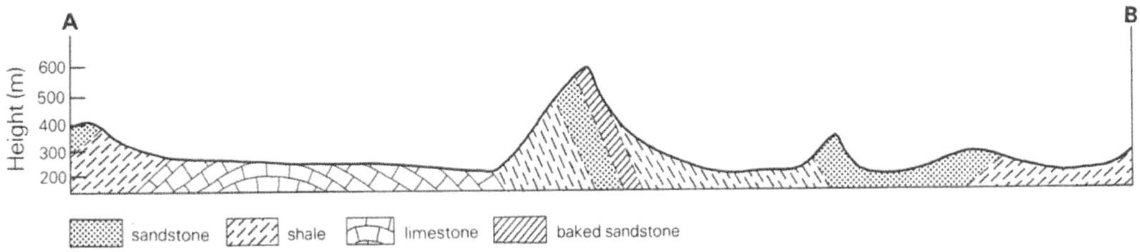

Earth and man

Look at Figure B. It was taken from an aeroplane and it shows some common relationships between Earth and man. A long, dark, forested ridge cuts across the landscape like a knife, separating two areas of snow-covered lowland. A small river flows through a deep cleft in the ridge and then twists in its path to flow through two more gaps in parallel ridges. Eventually it joins the large river in the foreground.

This is the natural landscape. It owes its shape to the rocks below the surface and the delicate etching of water and ice which have emphasised differences in the resistance of the rocks. But what causes some rocks to be more resistant than others? How did the rocks form, why were they tilted and exactly how were they attacked by water and ice?

Man has settled in the natural landscape. He must have been influenced by the gaps in the ridges when he first made roads, laid out railways and built towns. Clearly the gaps serve to focus railways, roads and people, as well as the river system. People have not settled on the upland ridges which remain forested and uninhabited. Instead cultivation of the more fertile lowland areas has produced a pattern of fields where once trees grew. But how dependent is the agricultural system on the climate of the area and the soils in which the crops grow? What restrictions do they impose?

The large river has also played its part. It has influenced the site of the town because it provided a means of carrying goods. As the town grew away from the river its streets were laid out in straight, unnatural lines, but the direction of development has still been strikingly influenced by the ridge and the river, forcing it to squeeze in between them.

The area in the photograph is typical of the way in which man has taken advantage of the landscape. As technology advances, so man has a greater impact on the environment and is less at its mercy. He can build tunnels through mountains where before he had to find a natural pass between them; he can harness rivers to supply energy, prevent floods and irrigate vast areas of desert. Nevertheless nature cannot be ignored, nor can each part of the environment be looked at in isolation and without a plan. We must study the Earth in a systematic way so that, at the end, we can see how all of the parts are related. In the past too many important environmental decisions have been taken with little understanding of their wider effects. Some will be described in later chapters. Examples are wide-ranging and they include a dam built for hydro-electric power that caused an earthquake, a landslide and a flood; and aid from developed countries to help irrigate parts of the Indian subcontinent that killed the very plants it was intended to save. Geography can make an important contribution to man's welfare by helping us to see the complex way in which the environment works. So in our study of the Earth's patterns and the processes that cause them we have very good reasons for starting geography with our feet on the ground – and with the rocks beneath our feet.

Part one **The origin of land and sea**

The materials of the Earth's surface are continually being eroded, renewed and rearranged.

USGS

Figure 1.0 Mt St Helens volcano erupting

1 The origin of the Earth's rocks

Evidence of Earth history

Ancient rocks

In 1966 a young New Zealand scientist working in Greenland found some unusual light and dark blotchy rock. When he sent samples of the rock for analysis they were found to be the most ancient rocks known to man – 3800 million years old!

There are also small areas of very old rocks elsewhere. Rocks in Minnesota, USA, for example, have recently been found to be of similar age to those in Greenland, and in South Africa there are rocks that formed 3500 million years ago. The South African rocks are also very interesting because they are made from muddy debris eroded by rivers and deposited on the floor of the ocean. This tells us that we have had land and oceans, clouds, rain and rivers for at least 3500 million years. Indeed, the Earth's surface must have looked much as it does today, except that there was no vegetation or animal life of any kind on the land.

If the Earth's surface has looked familiar for so long, how old is our planet? Most scientists agree that the Earth developed out of a cloud of 'cosmic dust' about 4500 million years ago – about the same time as our neighbouring planets and the Sun. This has been confirmed by dating the lunar rocks (brought back by astronauts) and the meteorites that occasionally come crashing through our atmosphere.

Nevertheless, although dating the birth of the planet is becoming possible, we are still far from understanding how it formed. Scientists are still arguing whether the space dust suddenly crashed together or whether it came together more slowly, picking up space debris such as meteorites as it spun through our part of the galaxy. Whichever process was responsible, it is clear from the outpourings of volcanoes that deep inside the Earth the rocks become very hot. Some would argue that an initial crash of dust could have generated enough heat to melt the Earth's interior, but, as that was over 4500 million years ago, it can hardly be important today. However, we may get a clue to present processes by looking at the processes operating in a nuclear power station. Nuclear power stations use the heat given out by radioactive substances such as uranium when they change from one form to another. There are many such substances contained within the Earth's rocks whose chemical changes could continuously fuel the Earth's internal cauldron. But their significance goes much farther than just providing heat because, as we shall see, volcanoes have provided us with the foundation of the Earth's surface rocks.

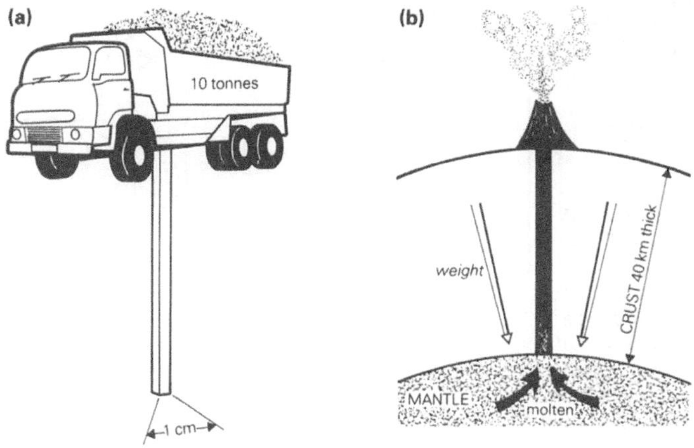

Figure 1.1 Volcanic eruption
(a) At 40 km below the Earth's surface the pressure is nearly 10 tonnes/cm^2 as a result of the weight of the overlying rock.
(b) The immense pressure deep in the Earth soon forces out molten rock if a line of crustal weakness develops.

Rock under pressure

Solid→liquid→gas

With a cool surface and very hot interior, there must be enormous changes in rock temperatures with depth. Indeed, near the surface, temperatures often increase by 30°C every kilometre. This change is even noticeable in deep coal mines where miners often have to work in 'tropical' temperatures.

Although this high rate of temperature increase is not maintained deep into the Earth, it would still mean the rock should become hot enough to melt at just a few tens of kilometres below the surface. We cannot prove this directly, but the evidence indicates that molten conditions only occur very close to the centre of the Earth, thousands of kilometres below the surface. Above this the great weight of overlying rock provides enough pressure to keep most of the Earth solid. But, if the surface rocks crack and the pressure is eased, there is an opportunity for the greatly heated rock to become liquid (Fig. 1.1). Any subsurface rock that becomes liquid in this way is called **magma.** As magma reaches the surfaces it pours out as spectacular flows of **lava**. At the same time some parts of the liquid boil off as gas and help form the huge plumes of cloud that tower over a volcanic eruption (Fig. 1.0).

Convection within the Earth

Fire burn and cauldron bubble

Most probably at its beginning the Earth did not have a separate solid surface of low density rocks and a molten core of much heavier materials. However, with time, heat built up and caused the centre to melt, but the outside remained solid and much colder, constantly losing heat to space. As a result a temperature difference developed between the surface and the core sufficient to cause a slow, circulating movement (**convection**) of the molten rock in the Earth. In some ways the process may have been similar to heating a saucepan full of dark, thick treacle (Fig. 1.2). Although extremely slow, over many hundreds of millions of years these convection currents may have heaved the whole fabric of the Earth's interior round and round, helping to separate out heavy and light materials: the light materials would be inclined to float to the surface as a sort of scum, and the heavy materials would tend to sink, thereby dividing the Earth's interior into a series of shells, one inside the other (Fig. 1.3).

Figure 1.2 Earth convection When a saucepan of treacle is heated, convection currents develop and the treacle is slowly churned around in a manner similar to that within the Earth.

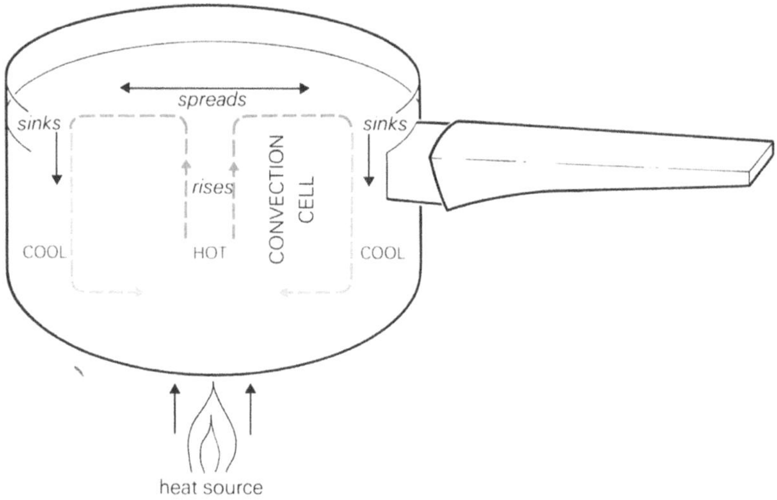

Figure 1.3 Structure of the Earth (a) The Earth is composed of a series of shells and rock: the outer shell is the rigid *crust* a few tens of kilometres thick; the *core* (about 3300 km in radius) is molten; the *mantle* is probably semi-molten. (b) The Earth's crust is therefore relatively thinner than the shell of an egg.

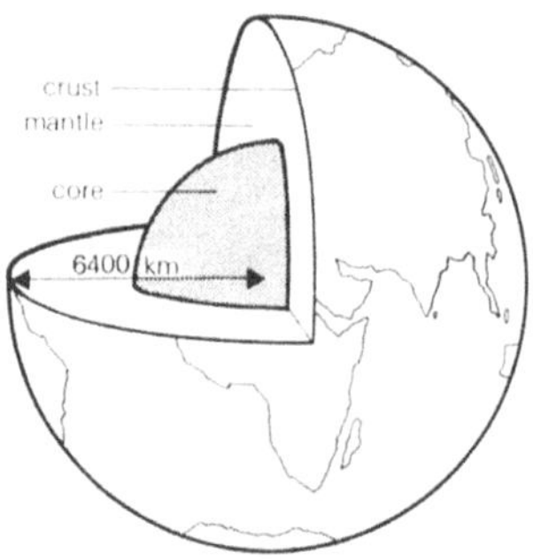

(a)

(b)

At present this 'scum' layer is up to 100 km thick. We call it the **crust** of the Earth and it is the solid rock on which we live. It is thickest under mountains and thinnest under oceans. However, compared with the radius of the Earth – nearly 6400 km – the crust is thinner than the shell of an egg (Fig. 1.3), and just like an egg shell the crust is strong but brittle and can be cracked. As we shall see, it is along the cracks that some of the interior rock escapes and volcanoes are born.

Once the crust has formed, it may have acted as a skin of insulating material, because solid rock is a very poor conductor of heat. If this were not the case the Earth's surface would still be burning hot, or alternatively the heat of the Earth might have been lost entirely. In either case, life on Earth would be impossible today. As it was, with only a quarter of Earth history gone, the surface seems to have cooled enough for oceans, clouds and rain to form and even for the first primitive life forms to develop.

The crust has remained floating on the **mantle** of denser semi-molten rocks below like polystyrene floats in a swimming bath. Nevertheless, the vast outpourings of lava from volcanoes that occur quite frequently even today tell us that the 'cauldron' remains beneath our feet, its molten rocks awaiting any opportunity to escape.

Volcanoes on Iceland

Island 'fireworks'

The oceans of the world are dotted with thousands of islands, many of which have one very surprising feature in common: they are dominated by volcanoes. For example, Japan's major city, Tokyo, is overshadowed by Mt Fuji, and Lanzarote (part of the Canary Islands just off the coast of North Africa) has a mountain whose slopes are still covered with cindery volcanic boulders (Fig. 1.4). Similarly all of the islands of the Caribbean Sea owe their existence to volcanic outpourings.

A volcanic eruption is one of the most spectacular and frightening sights on Earth. One recently active volcano was on the Icelandic island of Heimaey in 1973. A reporter who saw the first stages of the eruption gave this vivid account:

'A new vent has now appeared on the north-west side of the long dormant Helgafell volcano. Great clouds of debris, hot lava and black ash have been set free and now cover much of the island of Heimaey. The eruption created a rift nearly 3 km long across the eastern edge of the island, throwing up a torrent of boiling lava. Tonight the eruption is growing in intensity and the rift is still developing, northwards now, beneath the sea.

The refugees, many of whom were ferried to the mainland by air, say they have given up all hope, they fear they may never be able to return to their homes. Tonight Helgafell is angrier than ever and geologists are aware that it could be weeks, even months, before they can judge if the island can ever be re-inhabited . . .

(*and three months' later*) . . . For the people of this island this eruption was totally unexpected. Though only a tenth of the town was destroyed by lava, the remaining buildings are now collapsing under the weight of the volcanic dust that constantly falls. The eruption has continued for more than three months and it could be more than three years before it ends, no one knows . . .' (Based on BBC News Reports)

In the Atlantic Ocean most of the active volcanoes (such as those on Iceland) form a line down the centre of the ocean (Fig. 1.5). But although the eruptions in the Atlantic can be violent, in general the materials that flow from these volcanoes do so relatively quietly. The lavas are mostly very dark in colour and they flow very freely out of fissures. Lavas of this kind are called **basalts.** Only rarely do they build up large steep-sided cones.

Free-flowing material can often force its way in between layers of rock as well as pushing to the surface. When this happens sheets of magma are injected. Although the result is not immediately visible, the long-term results on the landscape may be important. These sheets often make much harder rock than the layers they pushed between. The Fair Head Sill in Northern Ireland shows the edge of just such a sheet. Sheets of cooled magma that push between rock layers are called **sills**; if they cut across rock layers they are called **dykes.**

ROGER JONES

Figure 1.4 Cinder cones When solid rock is ejected from a volcano it builds up into cinder cones. These lumps of material are fist sized.

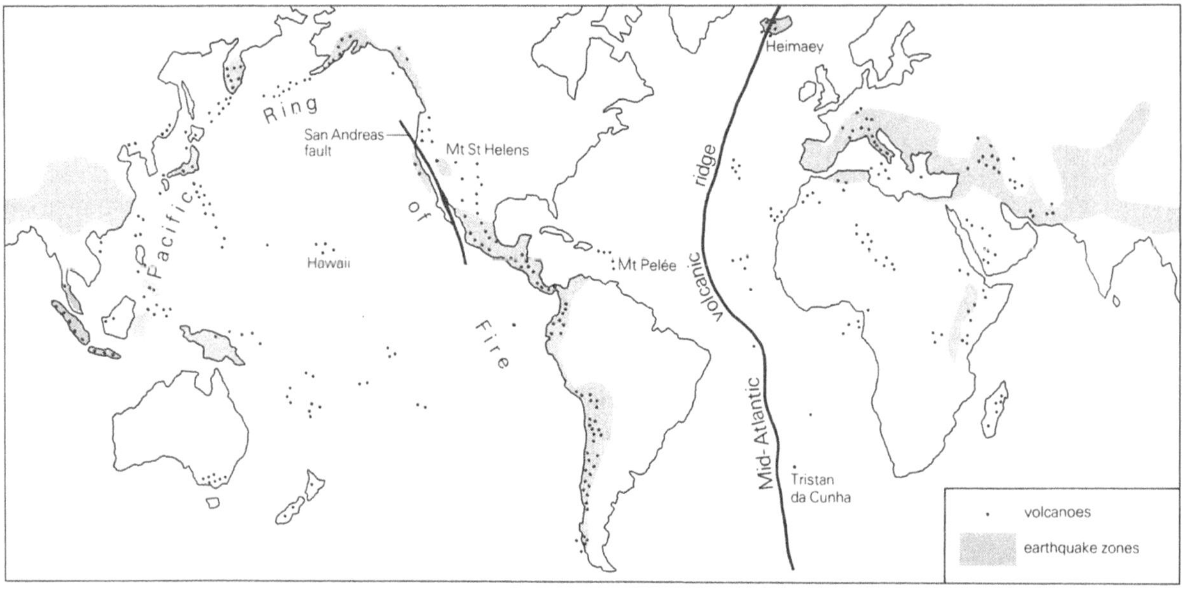

Figure 1.5 The distribution of volcanoes and earthquakes around the world

The magma that fed the Helgafell volcano has now cooled in its fissure, producing a new dyke. Indeed, sills and dykes are most commonly made of basaltic material because it flows so readily. For example, Iceland is made up of many thousands of sills, dykes and lava flows, all cutting through one another and piled up in a giant mound some 6000 m high. What we see of Iceland is only the tiny exposed tip of this huge volcanic mountain.

Topic What are rocks and minerals?

The whole outer shell of the Earth is made from solid rock. Sometimes we see it exposed in cliffs or mountains, although it usually lies hidden below soil and vegetation. The first rocks to form on the Earth's surface were cooling magmas. Rocks which have solidified from magma are called **igneous** rocks (the word igneous comes from the latin word for fire). Granite is an example of an igneous rock. Here the chemicals in the magma formed into substances of fixed chemical content called **minerals.** Quartz, feldspar and mica are examples of minerals found in granite. These minerals then solidified in the form of interlocking crystals (Fig. 1.6).

Eventually some of the igneous rocks were eroded by water, wind and ice, and the small particles of eroded rock (called **sediment**), were carried away and deposited in the sea. These particles were eventually buried by more sediment and were slowly compacted into new rocks. These are **sedimentary** rocks. For example, sand particles on a beach can be compacted to form sandstone rock (Fig. 1.6b). Similarly, clay particles eroded from land by rivers settle out as mud on the sea floor are compacted to form shale rock.

Sometimes the burial is so deep that the rocks are crushed and partly melted until they take on a new form. These (changed) rocks are called **metamorphic** rocks (Fig. 1.6c). (See also p. 44)

There are about 2000 different minerals on the Earth and in various combinations, they make up a wide variety of rocks. We use such names as granite and sandstone to describe groups of rocks that are similar, but each rock within a group has its own special features, thereby helping to give variety to our landscape.

Figure 1.6 Rock types
Through a microscope we can see (a) a piece of granite made from crystals that interlock, (b) a piece of sandstone made from rounded sand grains, (c) a piece of metamorphic gneiss with all of its crystals melted into each other.

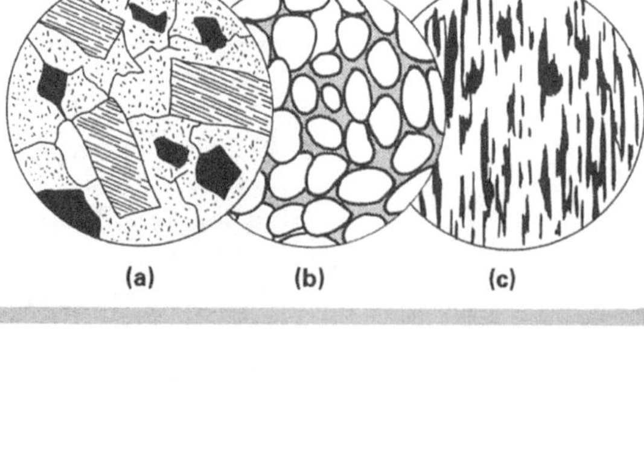

Volcanoes and mountains

The ring of fire

In contrast to Iceland the many volcanoes that form a 'ring of fire' around the Pacific Ocean (Fig. 1.5) are made from types of lava that flow much less readily. One main type is called **andesite** after the Andes mountains where so many of the volcanoes are found. These volcanoes produce steep-sided cones which are mostly composed of ash rather than lava. Although they are usually smaller than those that form in the oceans (Fig. 1.7), their violence, coupled with their closeness to many centres of population, makes them some of the most dangerous natural features on Earth.

Because this magma is more sticky than the basaltic type, dykes and sills are less common, but under each volcano is a vast chamber of magma deep in the crust. The pressure builds up in this chamber before a new eruption. But when the volcano finally ceases to erupt and there is no more pressure in the chamber, the magma solidifies into a rock called **granite.** Dartmoor in Devon is the top of a solidifed magma chamber called a **batholith** (Fig. 1.17).

There are many other examples in Africa (e.g. much of the Jos Plateau, Nigeria or the area between Bulawayo and Salisbury, Zimbabwe) and America (e.g. the Sierra Nevada mountains of California).

The Mt St Helens eruption

Terror, 20 km high!

Quite recently we have been able to observe a violent volcanic event similar in size to the famous eruption of Vesuvius in Italy in AD 79.

In May 1980, after two months of rumbling, the beautiful ice-clad peak of

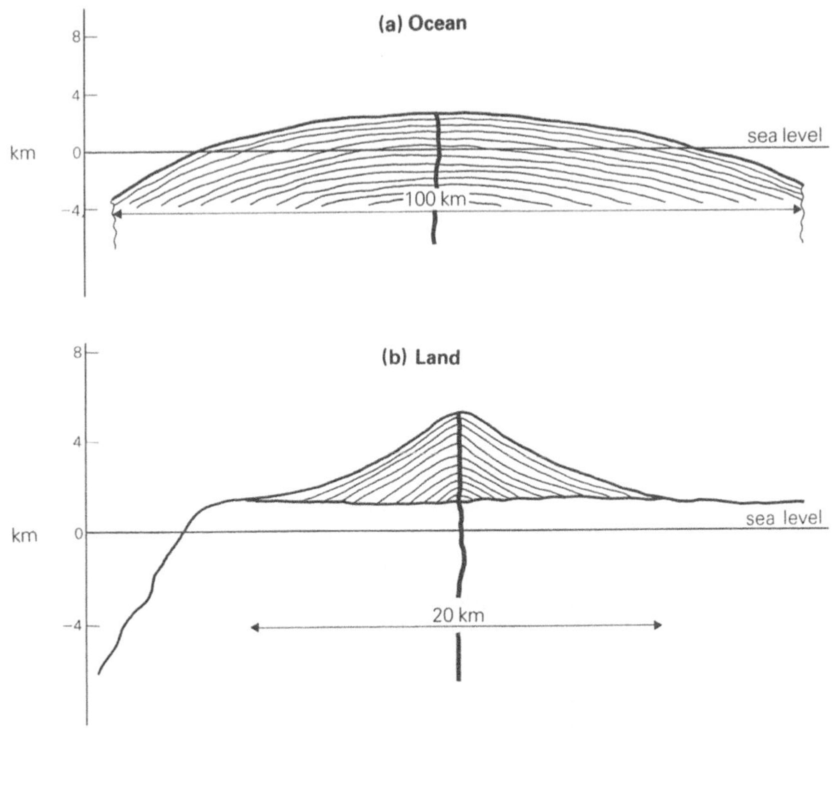

Figure 1.7 The shapes of volcanoes
Volcanoes vary in shape and size depending on their location and how easily the lava flows. (a) Mobile basaltic lava, (b) sticky andesitic lava.

KEITH RONNHOLM

Figure 1.8 The eruption of Mt St Helens volcano

Mount St Helens in Washington state, USA, suddenly erupted (Fig. 1.8). Early on a Sunday morning an earthquake shook the area and this was followed immediately by an explosion inside the volcano equivalent to 10 million tonnes of TNT. A nuclear bomb of this capacity would destroy London.

Over six cubic kilometres of rock were blasted away, lopping 400 m from the top of the mountain (a piece of rock about a third the size of Ben Nevis!). 400 km² of forest (about the size of the New Forest in Hampshire) were destroyed. A helicopter pilot sent out to look for survivors reported 'all the trees were lying parallel on the ground like big spokes emanating from the centre of the mountain . . . the whole area was burned over and I just couldn't believe it. It looked like the surface of the moon' (Fig. 1.9).

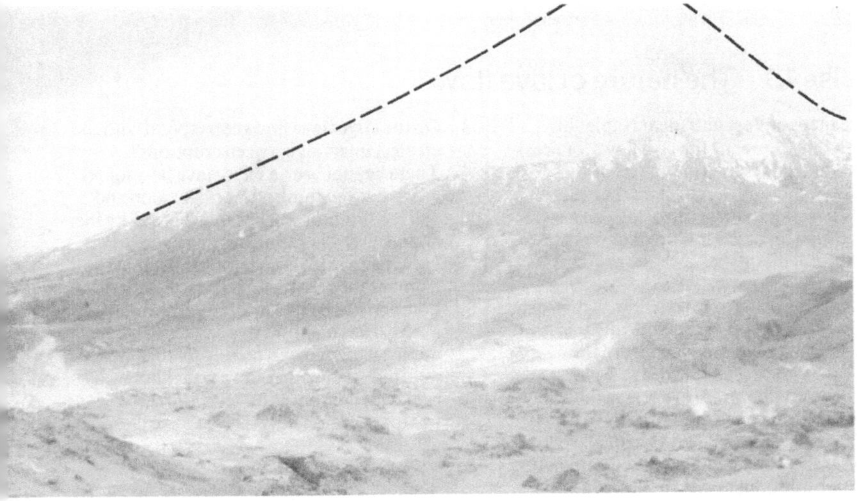

STEPHEN PORTER

Figure 1.9 Mt St Helens *Left.* The eruption blew off a whole section of the mountain and, *right*, the blast also destroyed a vast area of forest and stripped the branches from trees. The tree trunks shown here are over 30 m long.

Next a column of sulphurous ash belched 20 km into the sky, turning day into night as it blocked out the Sun. Mudflows were triggered by millions of tonnes of ice and snow which thawed instantly in the tremendous heat. The mudflows, moving at 60 km/hr tore out bridges, and buried cabins. Walls of water 3 m high hurtled down the nearby rivers carrying homes, cars and hundreds of logs in their path (Fig. 1.10).

Soon the whole air became foul for hundreds of kilometres around, carrying the acrid stench like a burned-out building. On the same scale the smell and ash cloud of a volcano erupting in South Wales would reach London. There was five hundred times more dust and grit in the air than normal, so that face masks became a necessity. One bank in a nearby town even put up a sign: 'For security reasons please remove face masks before entering the bank!' The Columbia River was choked with mud and ash, and 33 ocean going ships were trapped in the harbour at Portland. Indeed, the prevailing westerly winds slowly sent a huge cloud of ash right across the United States and out over the Atlantic Ocean (Fig. 1.11).

Ash is an important part of a violent volcanic eruption and it is often much greater in volume than any lava. In the 1980 eruption, Mount St Helens produced hardly any lava. Notice also that the eruption triggered off other natural disasters such as mudflows, which reminds us that the effects of a volcanic eruption are not just confined to the ejection of lava and ash.

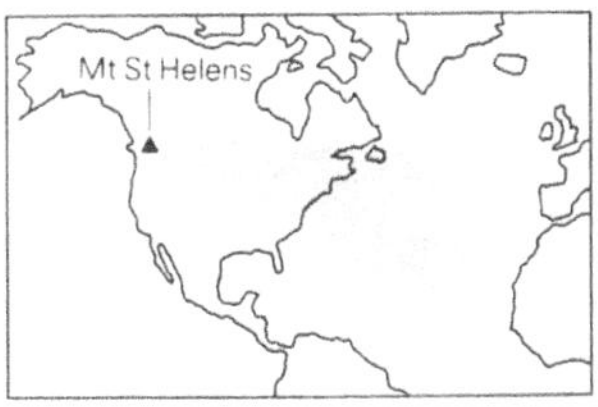

Figure 1.11 Mt St Helens The extent of the ash cloud one week after the eruption.

STEPHEN PORTER

Figure 1.10 Mt St Helens Mudflows and melting ice caused severe flooding in the nearby valleys.

Exercise 1.1 The nature of lava flows

We have maps of Vesuvius (near Naples in Southern Italy) showing the lava flows for nearly 300 years and this provides an excellent opportunity to study the effects of the eruptions (Fig. 1.12). Notice that the top of the cone is 'lopsided' and for many hundreds of years lava has flowed out mainly towards the south.

1 Choose one of the long lava flows with a date on and make a copy of it. Note that this lava is of the sticky type and it does not flow very far. What shape is a typical lava flow?
2 Write down the date of every lava flow in order, putting the earliest first. Put a cross next to each date that had a particularly large flow. Is there any regularity or pattern in the dates of eruptions?
3 Do the larger lava flows correspond with the greatest intervals between eruptions?
4 There has not been a major lava flow for 50 years. Do you think one is likely soon and will it be large or small? Note that before the famous AD 79 eruption the volcano had been inactive (**dormant**) for so long that the Romans thought it was extinct. Also Pompeii was buried in falling ash while nearby Herculanium was destroyed by a mudflow caused by a mixture of torrential rain and ash. Neither town was engulfed by lava. After the eruption of AD 79 the volcano was quiet again until AD 172..

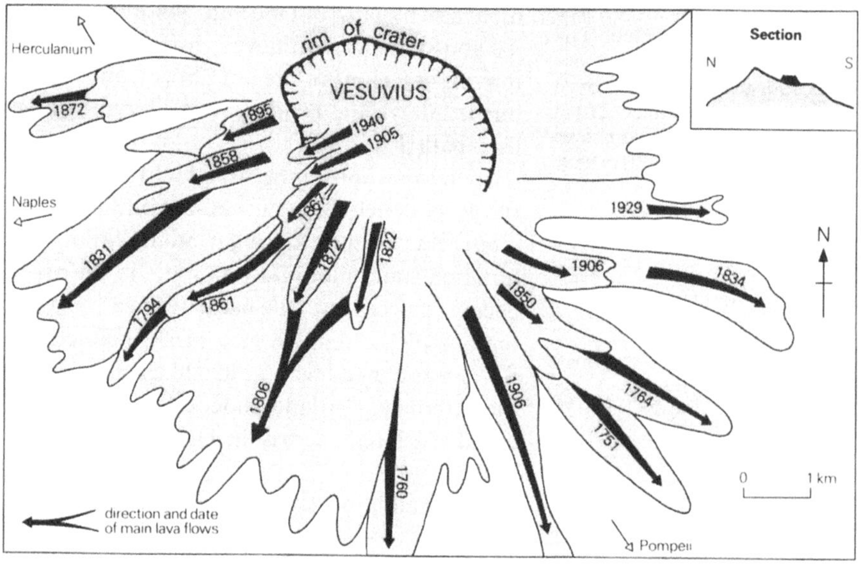

Figure 1.12 The eruptions of Vesuvius

Exercise 1.2 The deposits of an active volcano

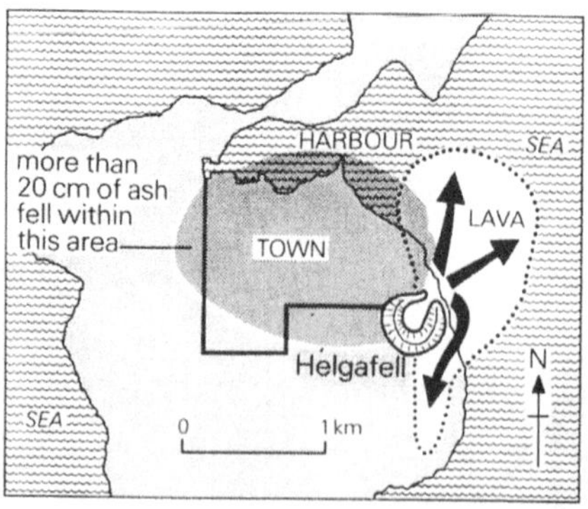

Figure 1.13 The Helgafell volcano
The lava and ash deposits that partially covered Heimaey, Iceland.

1 Turn back to page 9 and read the description of the eruption of Helgafell, a volcano on an island near Iceland. Also study the map (Fig. 1.13). What were the two main products of this eruption? Which was the most extensive?
2 The map shows the areas affected by the eruption of the Helgafell volcano. The eruption was partly from a long fissure. How did this affect the distribution of lava?
3 At the time of the eruption the wind was blowing from the east. How does the distribution of ash deposits tell us this?
4 What would have been the least hazardous direction for the wind to have been blowing from at the time? Why?
5 The description tells us that a volcanic eruption produces more than one type of material. Describe the differences between the distributions of lava and ash.

Topic Dykes, sills and batholiths

Dykes are sheets of igneous rock which cut across other rocks. Very often they are harder than the rocks they cut, and erosion leaves them standing up as ridges (Fig. 1.14). This can be seen clearly along the southern shore of the Isle of Arran, Scotland (Fig. 1.15). Sills are made from the same materials as dykes and are sheets of igneous rock. However, they were injected between, rather than across, other rocks. As with dykes they can be more resistant to erosion than are other rocks. Hadrian's Wall was built on the Great Whin Sill because it stood proud as a long, natural ridge in the landscape. High Force, a famous waterfall in Durham, England, has been formed by the same sill (Fig. 1.16). Batholiths are large magma chambers that once fed volcanoes but are now filled with igneous rock. They are usually made from granite and they provide the largest intrusive landscape features. Much of the Cairngorm mountains in Scotland and the Sierra Nevada mountains in California are granite batholiths. The top of another batholith is just emerging to give the moors of Devon and Cornwall (Fig. 1.17).

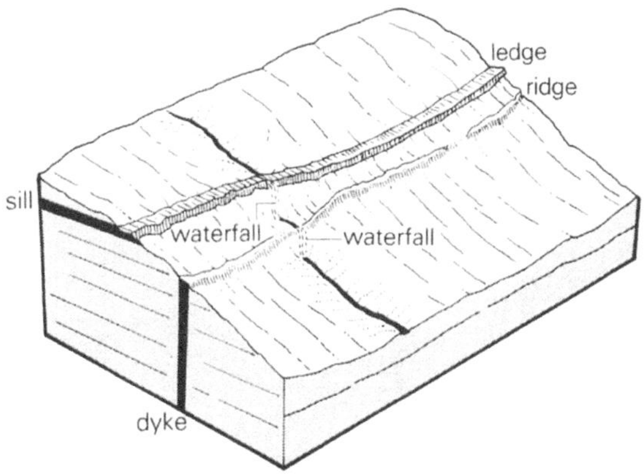

Figure 1.14 Intrusions
Many small dykes and sills form ridges and benches in the landscape.

Figure 1.16 Sills
The River Tees in Durham (England) flows over the Great Whin Sill, forming a spectacular waterfall.

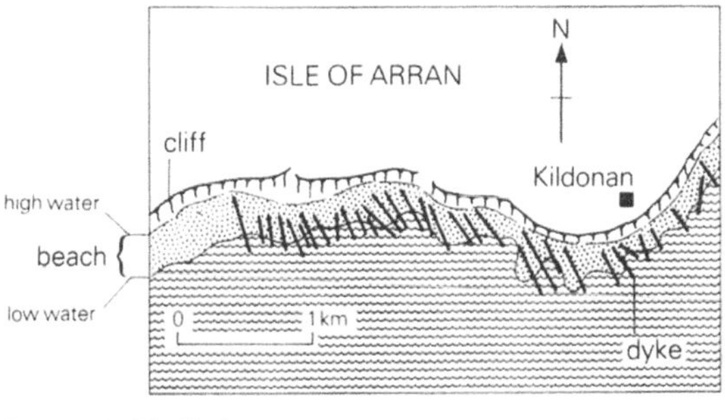

Figure 1.15 Dykes
The southern shore of the Isle of Arran (Scotland) has dozens of thin dykes jutting out to sea.

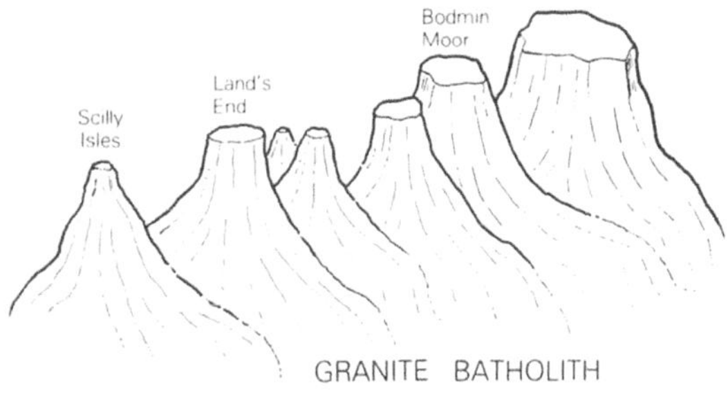

Figure 1.17 A batholith
The granite of southwest England is shown here with the cover rocks stripped away for clarity. These nobbly tops of the batholith are also the exposed cores of a mountain chain now showing through after a long period of erosion.

SUMMARY: IGNEOUS ROCKS

Igneous rocks are those that have solidified from **magma** as it cooled on or near the Earth's surface. Beneath the Earth's rigid **crust** there is a mixture of solid and partly molten rock. When a line of weakness forms in the rigid outer crust, the magma nearer the surface could possibly erupt as a **volcano** (Fig. 1.18). If it remains in liquid form the magma is called **lava,** but often the magma boils off as gas or is blown out of the volcano in small solid lumps called **ash** or in larger lumps called cinders. Much of the time magma cannot push its way to the surface either because it is too sticky or because there is very little pressure from below. In these cases the magma will solidify within the crustal rocks.

Magma that reaches the surface cools quickly in air and forms only tiny crystals, whereas buried magma cools much more slowly and forms larger crystals. Lava is called an **extrusive** rock because it is extruded (forced out) onto the Earth's surface. Basalts are lavas that are very runny; andesites and rhyolites are sticky lavas that help make steep volcanic cones (Fig. 1.0). Some of the magma is usually trapped underground and may then be intruded (forced between) sheets of cold rock. These sheets of **intrusive** rocks are called **dykes,** if they cut across the cold rocks; or **sills,** if they form parallel to the sheets of cold rock. Eventually even the magma chamber that supplied the materials for the volcanoes, dykes and sills may cool down. These magma chambers can be very large – sometimes hundreds of cubic kilometres in volume. When they cool they form granite rocks. A solidified magma chamber is called a **batholith.**

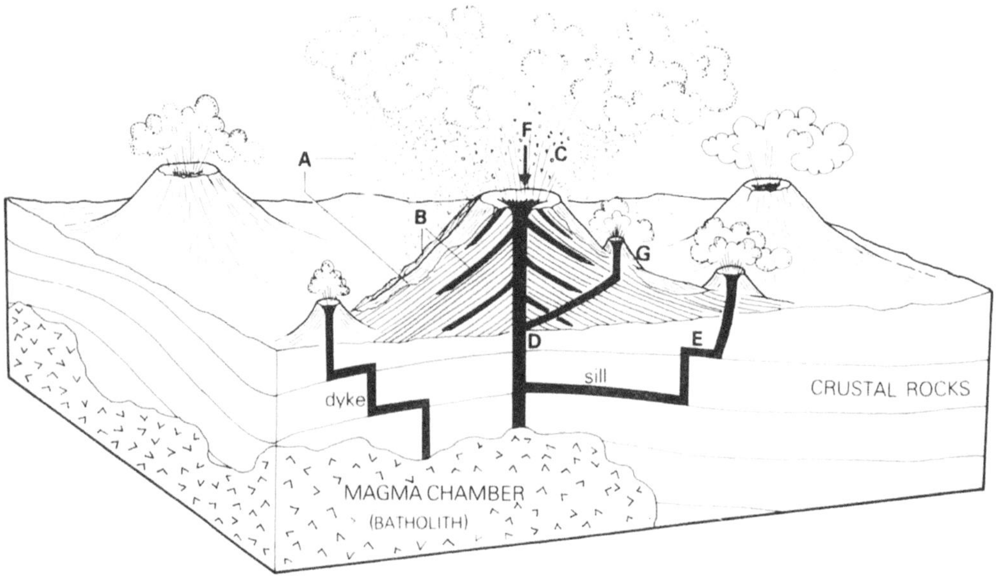

Figure 1.18 Characteristic forms of igneous activity
Write down the names for the features labelled A-G.

The Cascades mountains

Fiery Cascades

Mount St Helens is just one of a whole series of volcanic peaks that form the Cascade range of the Western USA (Fig. 1.19). It is the most active, but nearby there are volcanoes that must have been even more stupendous in their day. One of the largest of these 'inactive' volcanoes contains the famous Crater Lake (Fig. 1.20). About 6000 years ago this volcano was over 4000 m high and it erupted in what must have been a terrifying explosion. It blew 60 km^3 of rock into the air – ten times the volume blown from Mount St Helens! But not only did it blow a lot of ash into the air and send streams of lava pouring down its slopes, but the whole top of the cone collapsed inwards, so that at the end of the eruption the high ice-clad peak had vanished and in its place was a vast pit about 8 km in diameter and

1.5 km deep. Later a very small cone was built up by a small eruption, while rain-water gradually filled the crater until the cone was only a small island in a lake. This very violent explosion probably removed the magma from the magma chamber faster than it could seep in from the mantle below. With the chamber partly emptied, the support for the volcano was largely removed, allowing the weight of the cone to cause the spectacular collapse (Fig. 1.21). Collapsed volcanoes are known as **calderas.**

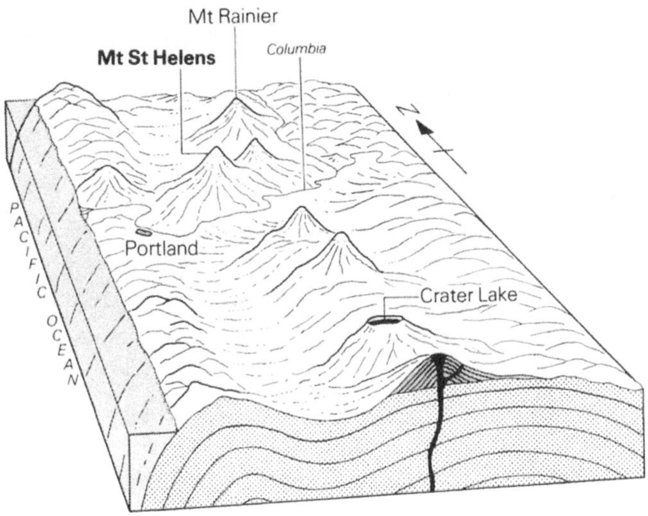

Figure 1.19 The Cascade Mountains in the western USA
The photograph shows the region near Mt St Helens. Notice that the volcanoes rise above a plateau. Compare with Figure 1.7.

STEPHEN PORTER

Figure 1.20 Crater Lake This lake now occupies the centre of a volcano that blew its top only 6000 years ago. The small secondary ash cone that forms Wizard Island in the lake tells us that there might be more violence to come.

Exercise 1.3 Caldera formation

The development of Crater Lake can be summarised by the diagrams in Figure 1.21. The diagrams and captions have, however, not been put in the correct order. Match the following captions with the diagrams, and write them down in the correct order.

(x) Very violent explosions remove magma from the upper chamber faster than it can be supplied from below. Support for the volcano is reduced and cracks develop in the rocks over the magma chamber.

(y) The cone of the volcano collapses into the magma chamber leaving a huge crater within the outer parts of the cone (caldera). Activity is much reduced and a lake forms in the crater. The start of a new period of building produces a new small cone.

(z) Violently active volcano throwing out ash and spilling lava from the cone.

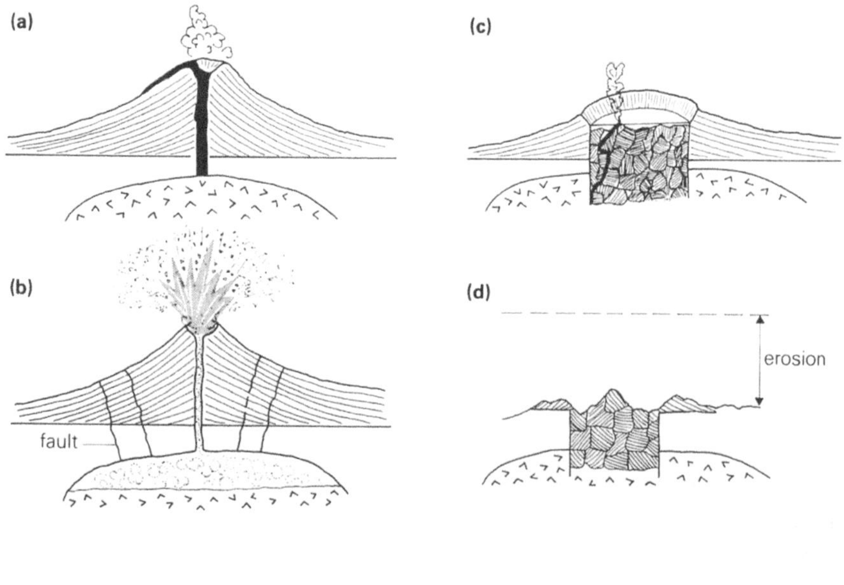

Figure 1.21 Calderas (a–c) The stages leading to the formation of a caldera. (d) The subsequent change in the shape of a caldera due to 300 million years of erosion around Ben Nevis, Scotland. The top of Ben Nevis is now 1400 m above sea level. When it looked like (a) it may have been 4000 m high. At least 2500 m of rock has been eroded away.

Topic British volcanic landscapes

How can we apply knowledge of the formation of relatively new volcanic landscapes to, say, Britain? None of the mountains appear to look like Helgafell, Mt St Helens, or Crater Lake. Nevetheless Britain has been the scene of massive volcanic activity in the past. The difference is that long periods of erosion have changed the appearance of our volcanoes so that they are largely camouflaged. Some of the biggest mountains are the remains of volcanoes – Ben Nevis is a remnant of a volcano whose history is similar to that which produced the caldera of Crater Lake, but all the events took place 400 million years ago (Fig. 1.21d). Similarly, Snowdon and the central parts of the Lake District all have volcanic origins. In Edinburgh, the hill called Arthur's Seat and the nearby smaller hill, on which Edinburgh Castle stands, are also the remains of volcanoes, although only the lava that once filled the vent is preserved. The newest volcanoes – some 25–100 million years old – still form some of the most dramatic scenery: the Cuillin hills of Skye and the mountains of other Scottish islands owe much of their rugged character to the resistance of some volcanic rocks to erosion. Nevertheless, these high mountains are still only the 'stumps' of old volcanoes or magma chambers. Much of Northern Ireland is covered by the eroded remains of lava sheets, sometimes 800 m thick, although at places such as the famous Giant's Causeway only the edge of it can be seen.

Beneath the surface igneous rocks are no less widespread. The Cairngorm mountains of Scotland and much of Devon and Cornwall in south-west England are made from giant batholiths, and everywhere dykes and sills create minor ridges in the landscape (e.g. Figs. 1.15, 1.16).

The dangers of volcanicity

Baked alive

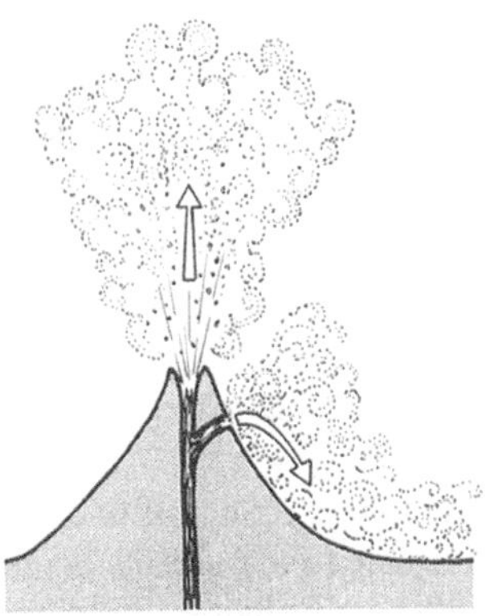

Figure 1.22 Volcanic disasters
The danger from a volcanic eruption depends on where the magma explodes into gas and ash. The Mt Pelée disaster and Mt St Helens eruption occurred because of high-level sideways explosions.

Although the weight of ash from the Helgafell volcano described earlier caused many houses to collapse, loss of life was avoided because there was time for evacuation to be effective. However, this is not always the case. When a natural catastrophe involves loss of life, it is termed a disaster. Without doubt one of the worst disasters happened to the people of St Pierre on the Caribbean Island of Martinique in 1902.

Mount Pelée, which overlooks the town of St Pierre, began erupting on 23 April. Nobody was particularly alarmed by this because in 1792 and 1851 (the last occasions the volcano had erupted) the lava flows had been small. This time a small amount of ash was thrown up into the air and a small cone started to form. But at 7.50 on the morning of 8 May a series of violent explosions suddenly shook the mountain and a great black plume of ash rose many kilometres into the sky. Simultaneously another great cloud burst down the side of the mountain (Fig. 1.22). It was a spectacular sight, but few of those who saw it were to survive for long because the cloud hid a massive glowing avalanche of frothy lava, gas and boiling liquid. Moving at 160 km/hr, the tongue of flaming gas ahead of the lava flow had raced down the mountainside and engulfed the town within two minutes leaving 30 000 people dead. Mount St Helens erupted in the same way but the surrounding area was fortunately almost uninhabited.

The benefits of volcanicity

Volcanic gifts

On the Caribbean island of St Vincent – near Martinique – a similar glowing avalanche swept down from the mountain top of La Soufrière, killing 1350 people. There have been many small eruptions of this volcano since 1902, including those of 1971, 1976 and 1979, but none caused much destruction. On all occasions many people were evacuated, just in case. With the volcano still erupting, this was the experience of one journalist in 1979:

'... there were plantations of bananas and coconuts. It was near harvest time but the fruit trees had been left and, in any case, the bananas had been con-

TIMOTHY GOULTER

ROGER JONES

Figure 1.23 Volcanic soil (a) These slopes of a volcano in Java are being used to produce rice despite the danger of a new eruption. (b) Plants can be grown directly in ash and lava only a short time after the eruption has ceased.

siderably damaged by the ash. The road, which was just a dirt track, was covered with coconuts. (Near the top of the cone) the terrain changed quickly. After climbing through grey ash for ten minutes the vegetation . . . disappeared completely – the gases from the eruption had burned it all . . . The climb was hard. It was difficult to find footholds and the further we climbed the worse it became. What some weeks ago had been a tropical forest was now a heap of ashes . . . ' (Geographical Magazine)

As a result of these experiences you may wonder why people still live near volcanoes. However, a visit to the slopes of La Soufrière just a few years after any one of the eruptions would have shown the lava and ash weathered into a deep and fertile soil by the tropical rain and high temperatures. Equally quickly a luxuriant forest begins to reclothe the mountain top in a rich green. And what suits the recolonisation of trees, suits the growth of crops such as coconuts and bananas. The soils that develop on the volcano sides are the means of life on these islands and, in many ways, the people rely on periodic lava deposits and ash to renew the soils and provide fresh nutrients for the crops to use. The volcano is a blessing as well as a danger, and many people are prepared to brave the danger for the benefits (Fig. 1.23).

Hot springs and geysers

Hubble, bubble, toil and trouble

When water moving in water-bearing rocks flows to great depths, it becomes hotter (Fig. 1.24) especially where molten magma is closer to the surface than is usual. In these areas the surrounding rock – and any water it contains – become very hot. Just as rock can be heated to great temperatures but remain solid if under pressure, when water is under pressure from its own weight, it can be heated to temperatures well over 100°C. Sometimes there may also be faults which provide a direct escape route to the surface for the heated water. The hot springs and geysers that are major tourist attractions are often at places where

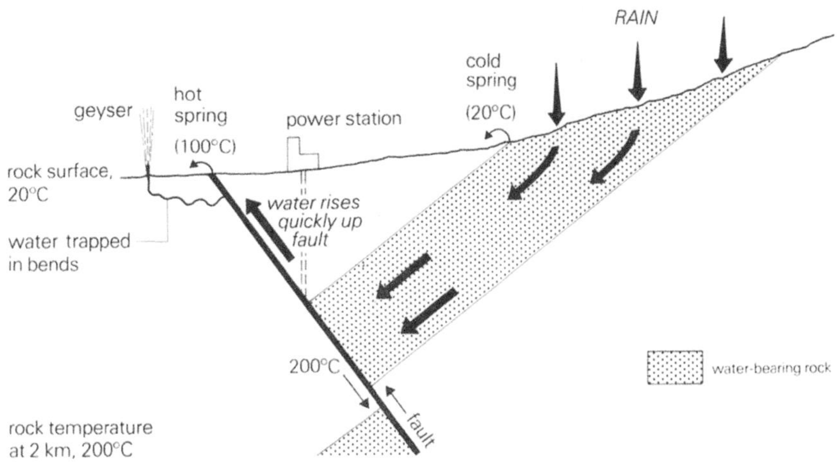

Figure 1.24 Hot springs and geyser formation Water moves down through water-bearing rock. The fault provides an easy escape route for hot water which cools only slightly as it rises quickly.

these faults reach the surface; for example 'Bumpass Hell' (Fig. 1.25) and the 'Old Faithful' geyser in Yellowstone Park, USA, and 'Geysir' in western Iceland from which the name geyser comes. At a hot spring, water bubbles to the surface constantly. By contrast a geyser forms when the hot water is temporarily trapped below ground in fissures that are often full of cold water near the surface. The build-up of pressure eventually becomes sufficient to eject both hot and cold water as a jet that may rise tens of metres into the air. After this there is a period of inactivity while sufficient pressure builds up again to cause a new spurt of boiling water. Mudpools usually occur where the supply of water is too restricted to produce hot springs. In these grey pools of boiling mud, gas rises continually and forms 'chewing gum' bubbles which grow and burst every few seconds.

These features show that magma is still quite close to the surface. For example, 'Bumpass Hell' in the Cascades (Fig. 1.25) is a region of hot water in a crater that poured out lava only 70 years ago. In this crater mud spatters and steam belches out of holes in the ground or rushes up into the air with a pulsating motion. At the same time continual release of sulphur gases gives an overpowering rotten-eggs smell. Together, steam and sulphur gas make breathing difficult within the crater.

Figure 1.25 Bumpass Hell Geysers, boiling water and mudpools show that magma is still quite close to the surface. Good examples are found at Bumpass Hell, an aptly named crater on a mountain in the Cascades range of western United States. In this crater, steam belches out of holes in the ground or rushes up into the air with a pulsating motion. At Bumpass Hell water is provided by melting snow in the spring which seeps down cracks in the crater rocks. After the snow has melted away each summer there is less water available and many hot springs turn into mudpools. In these grey pools of boiling mud, gas continuously rises and forms 'chewing gum' bubbles in the mud. These grow and burst every few seconds, splattering the mud out to the sides of the pools.

Topic The origin of metal ores

Today metals are a vital part of any economy, so it is very important to understand how they may have been formed. The sites of high-quality **ores** (rocks with a high proportion of valuable minerals) are often near to previous igneous activity, especially to granite formations. In Devon and Cornwall, for example, we sometimes see the tops of these granite intrusions exposed to the air; but when they cooled hundreds of millions of years ago, they were still deep within the Earth. The intruding magma, with temperatures well over 1000°C, 'boiled off' its lightest materials, much as oil is heated today in a chemical works in order to separate its different components. These lighter fluids pushed up through fissures in the overlying rock. Here they cooled, gradually depositing the metal ores. Subsequently erosion removed much of these ore bodies and exposed the granite in patches. As a result most of the major mining activity has been in areas where rocks still cover the granite.

We call different types of fissure filled with high-grade metal ores **veins** and **lodes** (Fig. 1.26). Because the ores occupy fissures, mines are often narrow and crack-like in appearance. In some mines more than one metal ore is found and many famous Cornish lead and copper mines gradually became tin or silver mines as miners explored deeper down the fissures. In other areas, such as the Mendip Hills and the Pennines, it appears that fissures in limestone were particularly useful in allowing the rise of hot fluids. In these cases the main mines were for lead.

Modern exploration for metal continues to investigate possible sites over old magma chambers. Even diamonds are found in some plugged up vents of old volcanoes (e.g. at Kimberley in South Africa). However, with some exceptions such as Kiruna in Northern Sweden, these ore bodies are not large, and most have been worked out. Today most metals come from sedimentary rocks (p. 23).

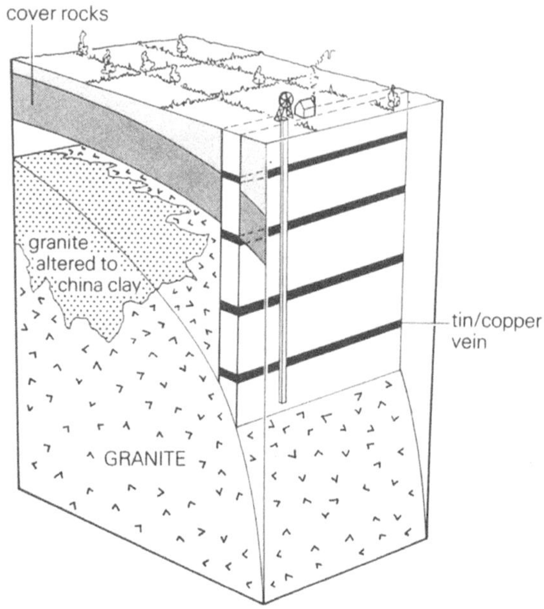

Figure 1.26 A section of a tin mine similar to those found in Cornwall
You can see that the thin sheet of tin (the vein) had to be mined at many levels.

Exercise 1.4 Hot springs and geothermal power

1. Figure 1.24 shows how a hot spring forms. Rainwater moves slowly downwards through cracks in the rock, heating up as it goes. If the water moved as slowly to the surface would it still emerge hot? What, then, is the importance of having a quick return route such as a fault (a long, continuous crack)?
2. Hot water provides a useful source of power in some areas where the water temperature is high enough (e.g. parts of Italy, New Zealand and Iceland). When the water is to be used to generate power, a large-diameter hole is drilled to cut the fault at depth. What are the advantages of by-passing the fault and bringing the water to the surface more quickly?
3. When the hot water reaches the surface it is no longer under pressure and it boils off as steam. The steam is used to drive turbines in the same way as conventional power stations. This type of energy is called **geothermal energy**. Why might geothermal energy be of limited use in solving Britain's energy problems?
4. There are hot springs in Britain at Bath and hot wells near Bristol. In this region all of the conditions described above are fulfilled except that the water does not circulate very deeply. As a result the water reaches a maximum temperature of 80°C at depth, and by the time it reaches the surface it is 46°C. Explain why it is not possible to use the hot springs at Bath to generate electricity.
5. The hot springs at Bath are not wasted and they have been important as a tourist attraction for several hundred years. To what other use do you think they could be put?
6. A geothermal power station has recently been completed in Iceland. Called the Krafla power station, it is located beside a volcano from which lava still frequently flows. Suggest some of the problems that such a site faces. (Remember the lava on Iceland flows from fissures).

2 Rocks from sediments

Sedimentary rocks

New land for old!

Originally, the Earth's crust consisted only of igneous rocks made from cooling magma. However, rain and running water soon broke down the igneous rocks into small particles and carried them away to the sea. In this way material for the first **sedimentary rocks** was formed. Sedimentary rocks (mainly those made from the eroded remains of earlier rocks) now make up a major part of the surface rocks of continents, and igneous rock exposures are less extensive. Because most sediment is brought to the sea by rivers, it builds up in the shallow seas near to continents. There is only a thin layer of sediment in the deeper parts of the ocean basins, and igneous rocks still dominate the ocean crust.

The way in which sedimentary rocks form and build up into layers is a very important influence on landscape development. Variations in the compaction of sediment and the types of rock debris produce rocks that show much variation in hardness and resistance to erosion, characteristics which are seen very clearly in the Grand Canyon (Arizona, USA) because here the rocks are not covered by soils or vegetation.

The Grand Canyon story

Back to base

In the south-west of the state of Arizona you can walk a few steps from a modern road and find yourself standing on the edge of the greatest valley in the world (Fig. 2.1). This is the Grand Canyon, which has been cut through 1000 million years of Earth's history to reach down a breathtaking 2 km to the Colorado river below.

Figure 2.1 The Grand Canyon (Arizona)
This canyon has been eroded through over 2 km of horizontally bedded sedimentary rock.

One of the most impressive features of the Grand Canyon is its many-coloured walls which descend from the rim like a giant staircase. The 'treads' of the staircase have been formed as each rock layer has been eroded back by a different amount, depending on its hardness.

From the rim of the canyon steep hiking trails snake downwards. Following one such trail, at first you cross over 100 m of white limestone – a mere 200 million years old – then comes 400 m of coarse deep-red sandstone. This in turn is followed by more limestone and then thick beds of shale. The final gorge reaches down through yet more layers of rock, finally exposing the top of a granite batholith.

All of the rock layers above the granite are made from the eroded remains of earlier rocks, that is they are sedimentary rocks. In the Grand Canyon the rocks are piled up horizontally, one upon the other, with the youngest rocks on top and the oldest ones below. The upper ones contain **fossils** – the remains of animals and plants that have been turned into rock and preserved with the other sediments. Fossils are particularly useful because they tell us of the environment that existed when the rock was being laid down. For example, fossils in the limestone near the rim show that the Grand Canyon was once part of the sea bed, even though today it is more than 2 km above sea level!

Here, without doubt, is proof that sedimentary rocks are part of a cycle of events that takes eroded material out to the oceans, forms it into new rocks and then lifts it up into land once more.

The Grand Canyon is spectacular partly because we can still see the rocks in the horizontal layers as they were first laid down. In many other parts of the world the same types of rocks have been laid down, but they are not as easily seen. For example, a cross section through the rocks below the North Sea clearly shows a build-up of rocks far greater than in the Grand Canyon (Fig. 2.2). The same sediments are also found above sea level in Eastern Britain, but here they have been tilted and then planed off by erosion, making the sequence difficult to see.

Today sedimentary rocks on land are far more widespread than igneous rocks, even though igneous rocks may dominate the landscape locally. The cycle of erosion from the land, deposition in the sea and compaction into new rock is now the main process of rock formation.

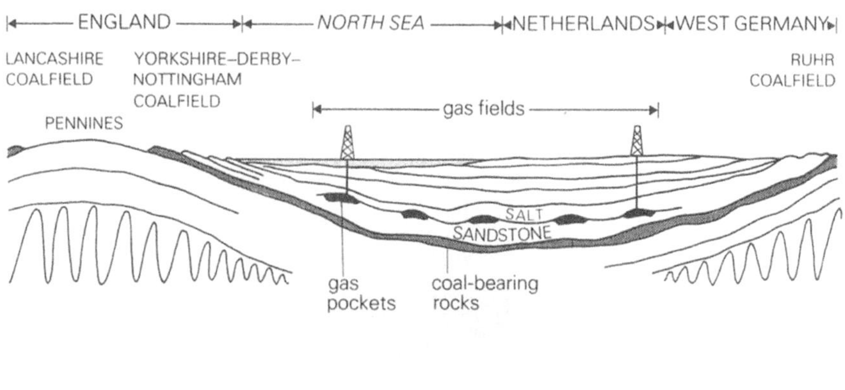

Figure 2.2 A generalised cross section of the North Sea basin and surrounding land
Almost all of the rocks and minerals discussed in Chapter 2 are found here.

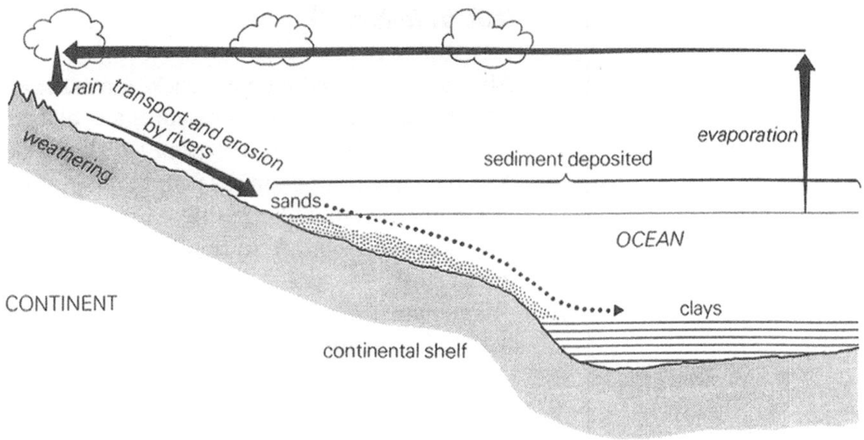

Figure 2.3 Sedimentary rocks
Sedimentary rocks form as the sediments are compacted by the weight of new debris. The shells and skeletons of sea animals are also present.

Sedimentary processes

From rivers to rocks

Sedimentary rocks are made in many ways. Mostly rivers carry eroded debris down to the sea where it settles out in the relatively still water. Coarse particles such as sand settle out first near the shore and eventually become compacted into sandstone; fine materials such as clay settle out in deeper water and eventually become compacted into shale (Fig. 2.3). However, in the past other forms of transport have been important (Fig. 2.4). Glaciers, volcanoes, wind and coastal waves have all played their part in carrying materials away from the land and out to sea. Remember that sediment builds up on the sea bed very slowly – perhaps less than a millimetre a year. In the past, climates have changed considerably and the sea floor has been moved up and down many times. This is why very different rocks have built up one upon another. Figure 2.2. shows that, although sandstones and shales are being formed in the North Sea today, before this sediments have included materials of tropical origin such as limestones, coal and salt.

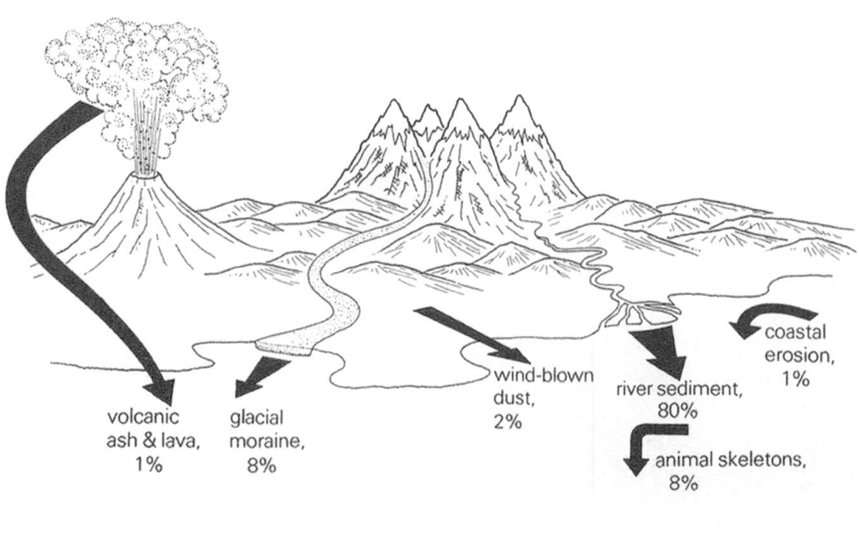

Figure 2.4 The sources of sediments in the oceans
The proportions shown here are based on recent estimates; in the past these proportions have varied considerably.

Alive, alive-o!

Although many sedimentary rocks are made of tiny particles of rock that settle on the sea floor, some rocks are made almost entirely from the shells of sea creatures. These creatures produce their skeletons or shells by extracting dissolved materials from sea water. The rocks they form are called **limestones.** The remains of shells are easy to see in some limestones although in rocks such as chalk the skeletons are so small that you would need a microscope to identify them.

Some creatures build shells into rock as they grow. Coral reefs form in this way. Other creatures, which move about in the sea, help to provide material for limestone rock after they die when their shells settle on the sea floor.

Warm seas often provide the best conditions for marine creatures to grow, because food supplies are abundant. But the shells of the creatures can still be swamped and lost if a large amount of sand or clay is also settling on the sea floor. So even in warm seas limestones are still uncommon, and in general shells make up no more than the occasional fossil in most rocks.

Because of these restrictions shells build up into limestones most effectively when there is little sediment produced on the land. Suitable conditions occur off the coasts of low-lying desert regions, or on remote islands in the middle of oceans.

Coral reefs are made of limestone (Fig. 2.5). Often they grow around islands or in the form of rings called atolls (Fig. 2.6) but their extent is very limited. Many

Figure 2.5 Fringing reefs This fringing reef has formed around a new volcano. It shows as the light blue band in the azure blue sea.

BRUCE COLEMAN LTD

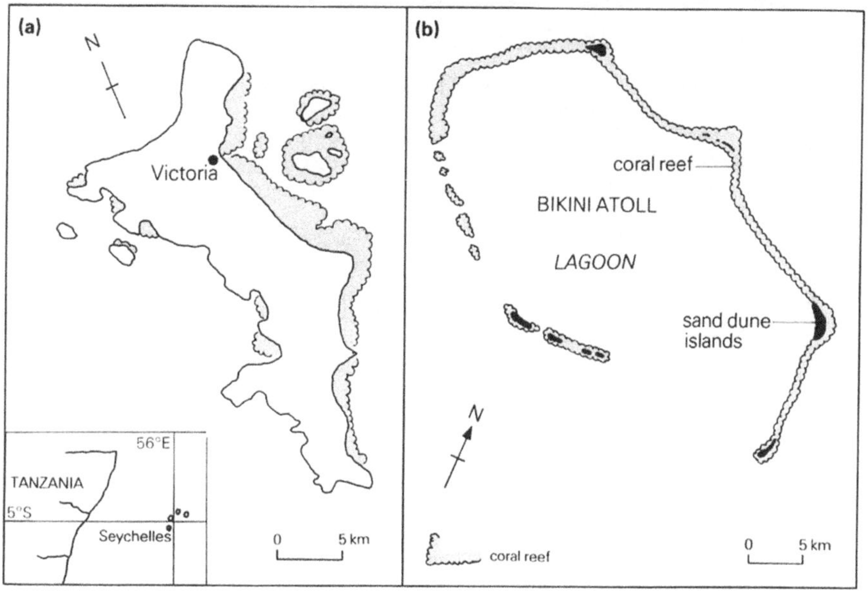

Figure 2.6 Coral reefs (a) Mahé, Seychelles and (b) Bikini atoll.

limestones do not contain any coral at all. But whatever the source of the limestone material it is soon formed into rock by burial under other sediments.

Some limestones are forming today. For example coral reefs occur off the northeast coast of Australia (the Great Barrier Reef) (Fig. 2.7), while finer-grained limestones are building up off the Florida coast of the USA. However, in the past climates have been warmer and limestones have been much more extensive. Limestones have formed over areas that now make up the Cotswold Hills and Mendip Hills in western England, parts of South Wales, the Dales and Peak District of the Pennines in northern England, the chalk downs of south-eastern England, and many other places. But although all of these areas contain limestone rock, its hardness is very variable. For example, although many limestones have been eroded to form rugged scenery (the Yorkshire Dales are a good example) by contrast chalk is soft and erosion has produced landscapes that are of 'rolling' hills. Cliffs occur in chalk only when it is undercut by the sea.

Figure 2.7 The Great Barrier Reef, Australia

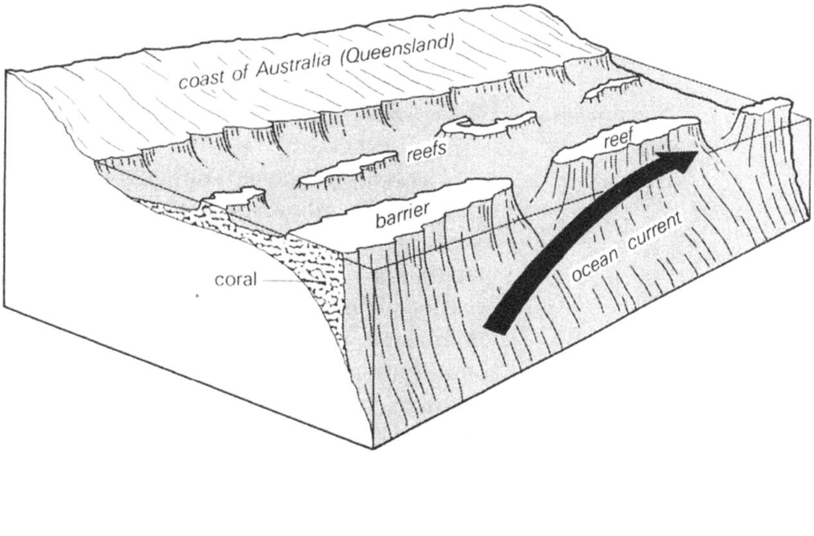

Topic Coral reefs and atolls

Most reef-building corals live in warm, clear water with plenty of sunlight. They grow fastest in the top 10 m and rarely grow in water deeper than 70 m. A coral reef is made only partly from coral animal skeletons. A large part of the reef is made from tiny plants called algae that help fill in the spaces between corals.

All coral animals need a constant food supply and, because they are fixed to the reef, they take the food that drifts past them in ocean currents. Consequently, coral reefs grow best on their seaward margins; they often cease to grow on the side of the reef facing the land. The **lagoons** between the reef and land never fill with coral because there is little food in water cut off from ocean currents.

Coral reefs are found in several distinct forms:

1 *On continental margins.* Here reefs are very extensive, e.g. the Great Barrier Reef, Queensland, Australia (Fig. 2.7) which is 2000 km long.
2 *On oceanic islands.* Here reefs occur in much smaller, isolated forms. There are three main subgroups: (a) **fringing reefs** (e.g. Mahé, Seychelles in the Indian Ocean, Fig. 2.6) where the reef grows right up to the beach, (b) **barrier reefs** where the volcanoes around which they form have been sinking for millions of years, and (c) **atolls** (e.g. Bikini, Marshall Islands, Pacific Ocean, Fig. 2.6) where the volcanic island has sunk out of sight and a lagoon is left within the ring of coral. The small atoll islands are made from broken coral 'sand' that forms dunes 3–5 m above sea level. Even coconut palms have colonised the larger dune islands, making them classic 'tropical paradise' islands. The sequence of development of these islands is shown in Figure 2.8.

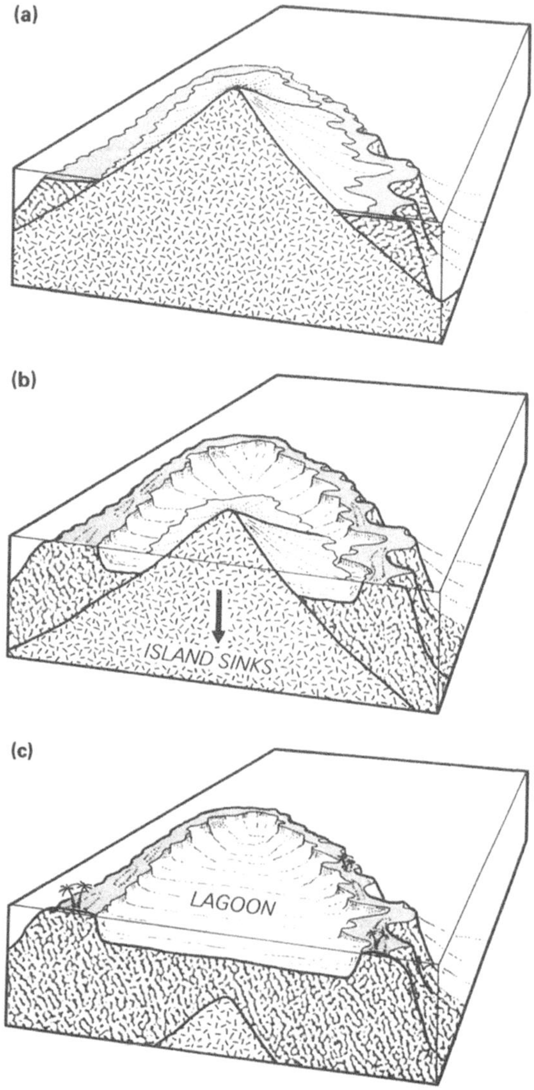

Figure 2.8 Types of reef
(a) With a reasonably constant sea level a **fringing reef** develops. (b) If sea level rises or the land sinks, the coral has to grow upwards to keep within the zone of sunlight and plankton food supply. A channel develops behind the new **barrier reef** because coral cannot grow away from the open-sea food supply. (c) An atoll is usually a volcanic island which used to have a fringing reef; but as the weight of the volcano caused it to sink, the coral grew upwards, enclosing a lagoon.

Oceanic minerals

Oceans of plenty

In oceans, eroded materials finally come to rest. We might be tempted to call them nature's 'rubbish dumps' and yet, along with the sand and clay, millions of tonnes of valuable minerals are continuously poured in. In many cases these materials are so scattered that it is too expensive to recover them, but occasionally natural processes concentrate these resources. So when we search for minerals we use our knowledge of nature to find places where natural concentration has occurred. Possibly the most important minerals for industry are gas, oil, coal, iron and salt. Coal in particular has influenced the location of much of the world's industry.

The North Sea is now known to contain many valuable minerals. This is because materials have accumulated in it more or less constantly for over 300 million years. But many of the rocks and minerals that occur in the North Sea also occur on land and they are sometimes recovered in the most surprising places. Recently, outside Basingstoke on the A32 in Hampshire a new kind of feature joined the fruit trees, wheat fields and telegraph poles of southern Britain. Just off this road there is an oil derrick – a sign that exploration and development is big business on land as well as in the sea. Nowadays it would be hard to overestimate the importance of oil and gas in every aspect of our daily lives, not just for heating and lighting but also in plastics, drugs, transport and artificial fibres. Oil recovery on land has taken place for over 40 years in southern England near Kimmeridge in Dorset, but the scale has been small. Now the situation seems likely to change. But while further exploration and development on land awaits the future, the present large-scale exploitation is concentrated in the North Sea and the shallow regions of the Atlantic.

However, because there is no real geological difference between the parts of Europe we call land (which happen to be a few feet above present sea level) and the North Sea (which happen to be a few feet below), the same rocks occur under Basingstoke as occur in the North Sea or Germany. This is why coal, natural gas and oil can be found in a zone stretching across from the British Isles, through the Netherlands to Germany (Fig. 2.2).

Building materials

An Englishman's home . . .

Igneous, sedimentary and metamorphic rocks provide the raw material from which the landscape has been fashioned. However, these rocks are also basic to the livelihood of man. He has used rocks to build roads, houses, walls and bridges . . . and of course castles.

Rock is a heavy and bulky material and it is transported more than a few kilometres only for special purposes. For example, granite from Dartmoor has been used all over southern England for curb stones because it is so hard. Similarly Shap granite from the Lake District is taken all over Britain and used to make fireplace surrounds, and facings to offices and shops and gravestones. Marble is another rock that is transported extensively. However, these rocks are exceptional; they cut well, are attractive to look at and can be polished. But most rocks are quarried and used locally. Indeed it is often possible to make a rough guess at the geology of an area by looking at the local buildings. Limestone houses (grey-white or creamy-white) are mostly found close to their rock outcrops in the Pennines, Cotswold hills and elsewhere. Central Wales has dark grey shales; Cornwall has brown-grey shales; and the Midlands of England and the Pennine borders have red and yellow sandstones.

Many areas do not have good building stones, and here clay bricks have traditionally been used. It is for this reason that East Anglia and South-East England have very few stone buildings.

Unfortunately stone quarrying does not lend itself to mass production, and today the great majority of houses are built from brick or concrete even in areas where stone has been used in the past. Nevertheless, bricks can be made only of suitable types of clay. Many of London's bricks come from Peterborough in Northamptonshire where there are large supplies of clay.

Topic The basis of industry – coal, gas, oil, iron and salt

Coal

Over 200 million years ago coal deposits were formed from dense forests of large trees. These trees grew in tropical swamps and as they died they fell into the stagnant swamp water and thus, without air, were protected from rotting. However, from time to time the land sank quite quickly, the sea flooded in and it covered the fallen trees with sand and mud. Eventually the land rose, and new trees grew on the sand and mud until there was a new period of sinking. Mostly there was only enough time for thin tree layers to build up between deposits of sand and mud. As a result of the weight of the overlying sediment, the coal seams were squashed much thinner than the original tree layers. Today it is the thin seams scattered amongst sandstone and mudstone that the coal miners seek, so it is no surprise to find that coal is often mined at a number of different levels.

Because coal was formed a long time ago there has been plenty of time for it to be buried under hundreds of metres of rock and to be affected by later movements of the Earth. One particularly important Earth movement slowly formed the Pennine range of hills in Northern England. Rivers subsequently eroded the centre of the arch, taking the coal in the process. This is why most of England's coalfields now occur either side of the Pennines. By contrast in South Wales and Central Scotland the coal has been preserved in basins. It used to be too costly to move coal over long distances, so factories were built on top of the coalfields. People were drawn close to their places of work and many large settlements are found on and near coalfields (Fig. 2.9).

Gas

Coal provides us not only with a fuel to burn, but, during its compaction from peat, it also gave off gas. In many regions the gas has subsequently risen through the overlying rocks, reached the surface and been lost. However, some rocks, especially salt beds, are 'gas tight' and when gas rises from beneath them it becomes trapped. Despite past losses, therefore, huge volumes of gas are still trapped in the rocks above coalfields, particularly under the North Sea (Fig. 2.2). A belt of coal stretches from Britain to Germany, although it is too deep to be mined. The gas has risen and been trapped in upward folds of the rocks, so that it now occurs in isolated pockets. Each of these pockets of gas has to be found and then a well drilled before the gas can be piped ashore. It is an expensive and difficult process but the rewards are huge.

Figure 2.9 The distribution of coalfields, oil and gas in NW Europe, and the major centres of population

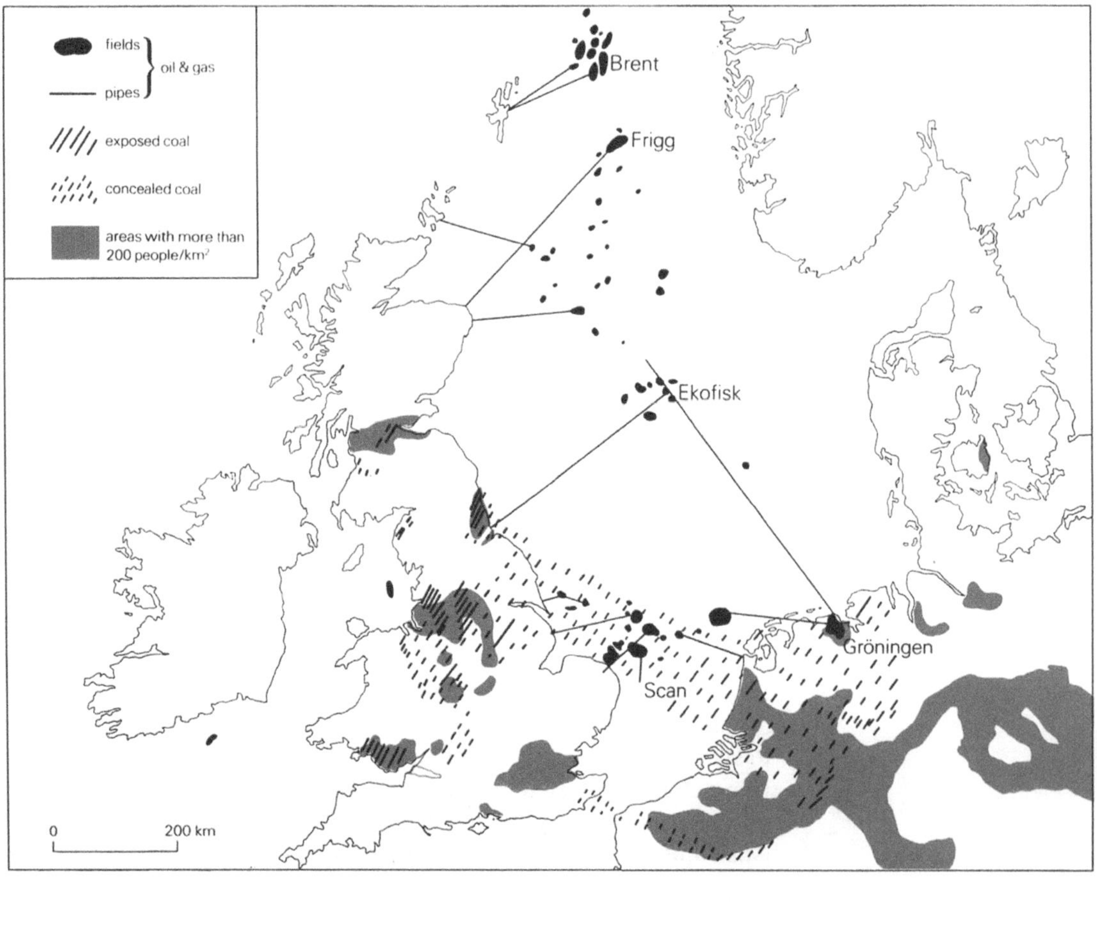

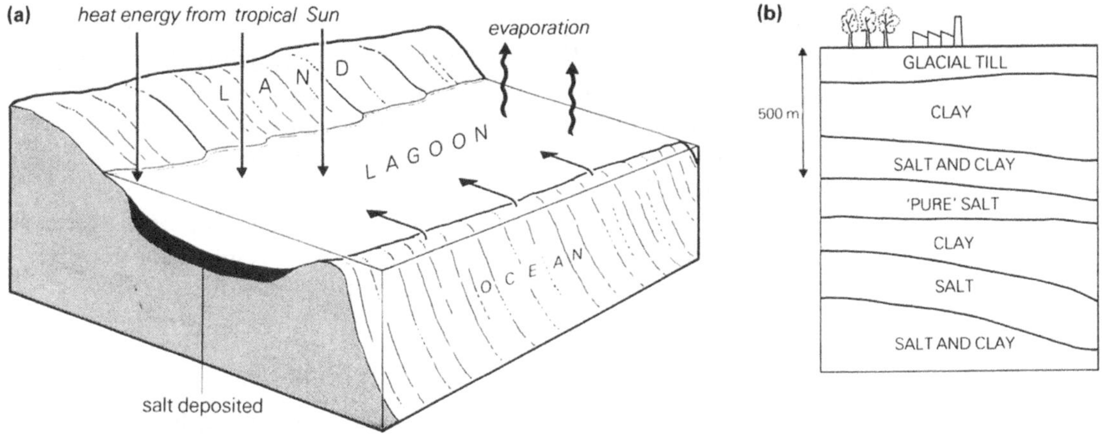

Figure 2.10 Salt formation
(a) Salt is formed in tropical coastal lagoons or inland lakes. (b) Such lagoons once formed in places that are now Cheshire and Cleveland.

Oil
Oil is formed by the slow decay of the bodies of microscopic sea creatures that were buried in the muds on a sea floor. Although these muds contain only a tiny proportion of dead bodies, their slow decay releases oil and gas which then seeps upwards through the rocks above. It is trapped below upward folds of salt beds or mudstones just like the gas from coal. Some of these oil and gas pockets lie under Britain, but most seem to be under our seas. At the moment this has forced oil companies into costly exploration a long way from land. (Fig. 2.9).

Oil and gas are found in many parts of the world. The largest reserves are in the Middle East where extraction is easy. There are other large oilfields in Texas, Venezuela, Mexico and Central Russia. There are also large reserves of oil in less hospitable places such as Arctic Alaska. This has forced man to work in areas where the climate makes life very difficult. However, once the initial drilling was over, a pipeline was built across Alaska so that the oil could be taken to the main centres of population. Being more easily transported than coal, oil and gas travel to the cities in contrast with the factories and settlements on coalfields in the 18th and 19th centuries.

Iron ore
Some iron ore is found near to magma areas as described on page 22. The 'mountain of iron ore' at Kiruna in Northern Sweden is probably an example of molten iron cooling out into a very rich vein. Iron ore in Kiruna is 68 % iron and is so rich as to be mined profitably even in such a remote location.

Most iron ore is of considerably lower quality. In Europe the large ore fields contain less than 30 % iron. Nevertheless these ore fields are another example of the way natural processes concentrate minerals in the sea. Rich ore veins like Kiruna have been eroded from mountains and then transported to the sea in solution. However, sometimes this eroded ore can be concentrated in sedimentary rocks. Limestones and sandstones often contain precipitated iron in sufficient concentrations for mining to be worthwhile. The Northamptonshire iron ore fields of central England contain this sort of iron. Very occasionally bands of iron-rich sandstone also occur intermixed with coal seams; this is called 'black band' iron ore. Most of Britain's early industrial areas were based on these chance occurrences of two raw materials together. South Wales and the Black Country of the West Midlands were both developed on local iron and coal reserves.

Salt
Our last example of raw materials for industry is salt, a precipitate from sea water. Salt is formed in coastal lagoons. The sun's heat causes the lagoon water to evaporate, concentrating the salt water until some salt is precipitated on the lagoon bottom (Fig. 2.10). New sea water continually flows in from the sea to replace the water evaporated and this is evaporated in its turn. By this process salt beds hundreds of metres thick have been built up only to be buried by later sediment. The major salt reserves in Cheshire and Cleveland in Northern England were once tropical lagoons. In Cheshire a typical salt bore hole may now have to pass down through 500 m of solid rock before the salt beds are reached but then there might be a layer of salt rock 200 m thick! There are thousands of millions of tonnes of salt remaining in these two areas alone and they will continue to provide a basis for industry for a very long time to come. It is no accident that Britain's biggest chemical industry is on the River Mersey, half way between the Cheshire salt fields and the Lancashire coalfield.

Salt is also deposited in inland lakes in desert regions. The Great Salt Lake, Utah, USA; Bonneville salt flats, Australia; and the Dead Sea on the Israel/Jordan border are examples of land-locked lakes where river water is continually evaporating and leaving its salt content behind. These areas, and many more, provide a valuable resource wherever they occur. For example both Israel and Jordan have recently built chemical works beside the Dead Sea.

Broken into small pieces, rock is also widely used as a foundation for roads. Some limestone is hard enough to be broken into chips for road bases, the huge quarries near Cheddar (Somerset) and Buxton (Derbyshire) being signs of this demand; elsewhere granite or other hard rocks are used – some dykes and sills have been so heavily quarried that many a prominent landscape ridge has been turned into a trench. The Cleveland dyke crossing the North York Moors in Northern England is a very good example.

Finally, limestone can be burned to make lime which is used as a fertiliser in agriculture and as an ingredient in cement. As a result many fertiliser and cement works are located very close to outcrops of limestone.

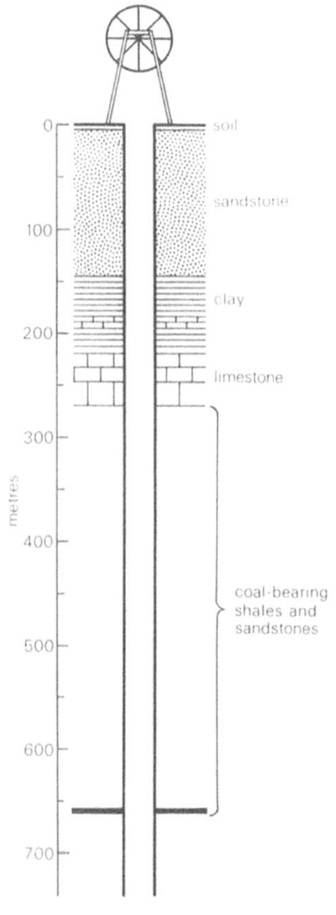

Figure 2.11 Part of the Markham Colliery log

Exercise 2.1 Coal mining

The Markham colliery shaft was the first sunk into the concealed coalfield of Derbyshire Northern England in 1922. The 'log' of the shaft is partly shown in Figure 2.11. In order to reach the coal-bearing rock the mine shaft had to be dug through nearly 300 m of unproductive rock.

1. Coal seams were found at 30 depths but only nine are more than a few centimetres thick. These nine are found at the following depths: 365 m (0.5 m thick); 407 m (1.2 m thick); 453 m (0.7 m thick); 526 m (1 m thick); 543 m (1.3 m thick); 533 m (1.6 m thick); 589 m (0.5 m thick; 619 m (3.2 m thick); 668 m (3.2 m thick – the Barnsley seam). Copy Figure 2.11 and add the other coal seams in their correct positions (a single pencil line will be sufficient for each seam).
2. Explain why it is common to find lots of coal seams in coal-bearing rock.
3. What problems are there in mining these seams with machinery? Which seams do you think are worth exploiting?

Exercise 2.2 Problems of oil and gas exploration

1. Figure 2.9 shows the location and size of known gas and oil fields. Why are there so many small oil and gas fields rather than one big one?
2. What are some of the problems of drilling in the North Sea that you know of from newspaper reports (e.g. financial, weather, depth of water, length of pipeline)?
3. Measure the distance of the oil and gas pipelines needed to bring the materials on shore. What problems does this cause?
4. How has the distribution of oil and gas affected the distribution of the new oil refineries in Britain?
5. Because the oil and gas occur in small pockets, do you think the chances of finding more pockets is high or low, and why?

3 Earthquakes, faults and folds

Faulting and earthquakes

All fall down

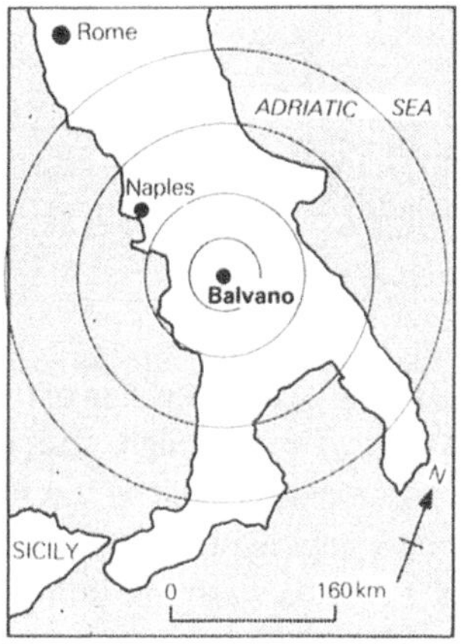

Figure 3.1 Earthquakes The spread of shock waves from the disaster area of the Italian earthquake of 1980.

Being at the scene of an earthquake is a terrifying experience. Just try to imagine the real events by reading between the lines of a typical report given in the press. We might read:

> '**AT least 112 people were killed last night as six violent earthquakes rocked Italy from the deep south to the Alps. And the death toll was mounting by the hour.**
>
> Some reports spoke of "hundreds dead." In one southern village, Balvano, near Potenza, almost 100 people were killed. '...

However, the newspaper report gives little indication of the real impact of the event (Fig. 3.1). In a village only 10 km from the earthquake focus the people were awakened to find the ground shaking violently and their houses falling apart around them. Walls bulged, cracked and then crashed to the ground. Roofs tumbled down and added to the piles of rubble that had once been the homes of hundreds of people. No one could do anything to help except to cling to the ground and hope that they would not be crushed by falling masonry.

Earthquakes occur deep in the rigid crust of the Earth. We see the results of ancient earthquakes every time we look at a section of cliff and notice some rocks are displaced with respect to others (Fig. 3.2). A **fault** (displacement) can

Figure 3.2 A small normal fault in a cliff

Figure 3.3 The Great Glen Fault
Erosion has etched out the fault into a trough that cuts right across Scotland. Faults often cause a rock to shatter along the line of movement and this allows erosion to proceed more quickly than elsewhere. Note there has been no vertical displacement.

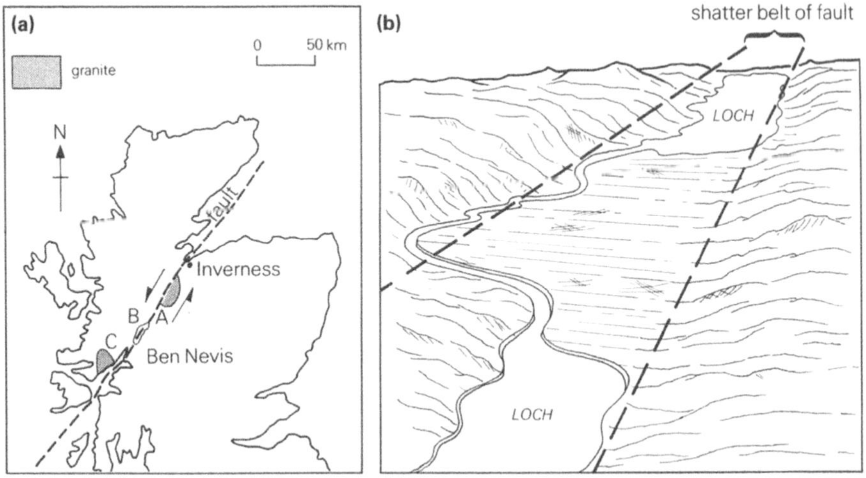

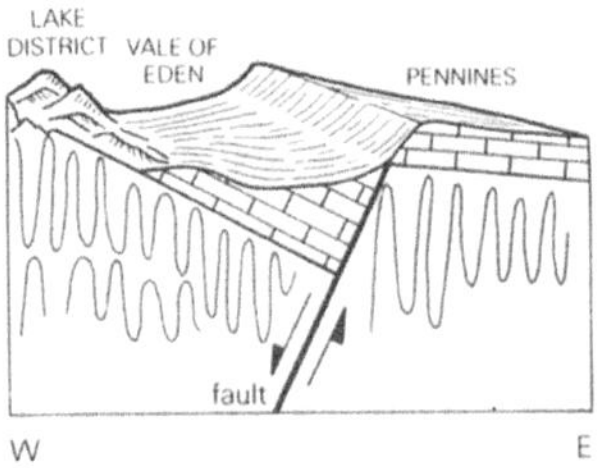

Figure 3.4 Normal faults
Faults have gently lifted the northern Pennines into a block mountain which tilts to the east.

Figure 3.5 The San Andreas Fault
This tear fault still moves about 5 cm/yr on average.

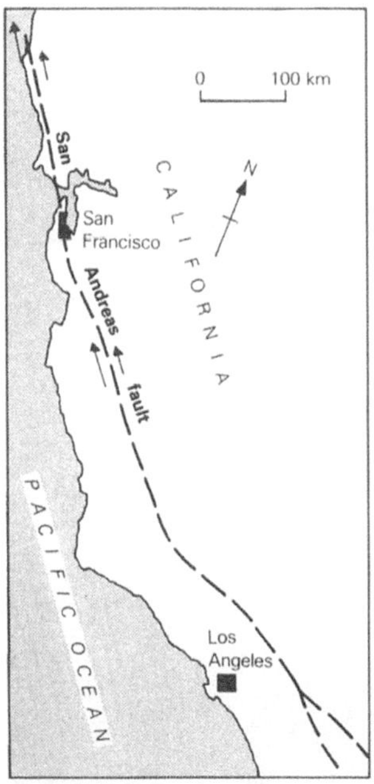

represent the movement of millions of tonnes of rock and give us some idea of the size of the forces that strain and tear at the Earth's crust. Even a single fault at depth sends many shock waves to the surface causing an earthquake to last for several minutes. You can compare this with throwing a pebble into a pond: the pebble strikes the water and is gone, but the ripples it creates wash the shore of the pond for quite a long time afterwards.

Some faults are of a stupendous size. For instance, sub-surface forces in the past slowly tore Scotland in two (Fig. 3.3). As the two halves of Scotland scraped past one another for tens of millions of years, a whole belt of rock was crushed. Erosion has now exploited the weakness of this crushed rock, leaving a gash so big that it takes three lochs just to cover the bottom. In Northern England a series of faults show where the Pennines were pushed up and tilted slightly to the east. The steep sides of these hills show particularly well in the north-west, where the Vale of Eden now separates the Pennines from the Lake District (Fig. 3.4).

Everyone in California, USA, knows about earthquakes and faulting because San Francisco is built across the most famous fault in the world. This, the San Andreas fault (Fig. 3.5), is the same type as the one which tore Scotland in two. The fault line slices through the western coast of the state, showing the place where a small strip of coastal California is moving slowly northwards against the bulk of the American continent in a series of jerks. Although this has been taking place for millions of years, very few months go by without a new movement. Most are very small and only give minor shock waves (tremors). The last time the San Andreas fault had a major movement was in 1906 and then much of central San Francisco was destroyed. Some people predict that another major earthquake will strike San Francisco within the next few years, but as yet there is no accurate way of forecasting this.

As the San Francisco experience shows, faults move in jerks. It may take tens or hundreds of years before forces build up sufficiently within the crust to break slabs of rock, but when the break comes the rock snaps into its new position within minutes.

Earthquakes tell us which pieces of the Earth's crust are most active. It is clearly not a coincidence that, in general, these are the places where volcanoes are most commonly found. This is because a large fault can easily provide a

Exercise 3.1 A major fault in Scotland

The northern part of Scotland once lay further north-east than it does now. A huge tear fault occurred, moving Northern Scotland very slowly in a south-westerly direction over millions of years, shattering all the rocks along the sides of the fault in the process.

1 Copy the part of Scotland north of the fault on Figure 3.3b. Now copy the rest of Scotland, but displace it in such a way that the granite areas match. These granites were once part of the same batholith and they tell us how far the fault has torn. The map you have now drawn shows us what Scotland once looked like.
2 Measure and write down by how much the granite areas have been separated.
3 The tear fault shattered the rocks along its length so that they were weakened and more easily eroded. This zone has subsequently been eroded into a huge trench. Use your atlas to name the three lochs that occupy the deepest parts of the trench.

route for magma to escape to the surface. However, even in places away from major earthquake zones the crust is constantly readjusting itself. If you look in the newspapers you will even find references to minor earthquakes in Britain every few months! Fortunately even 'large' British earthquakes usually do no more than rattle windows or dislodge roof slates.

Although many rocks are very brittle and can yield to pressure only by faulting, many others have simply bent (Fig. 3.6). Bending or **folding** of rocks is a very slow process. It may occur in sediments that have been deposited on a sea floor only recently, or in deeply buried rocks if the pressure is applied over a wide area. Indeed, folding is the dominant process in most major mountain systems. Consequently they are usually called **fold mountains** (Fig. 4.1).

Even rocks that are brittle when cold can fold when they become hot, deep underground. (Compare this with a bar of toffee which breaks when cold but bends when warm.) Folds in hot rocks probably occur deep down in the crust and they are usually associated with the formation of mountains ranges. We only see this type of folding when the mountains are eroded and the 'cores' of the mountains are exposed at the surface.

Figure 3.6 Folded rocks at Stair Hole, Dorset

Folds are permanent deformations of rocks, but a rock can store up energy if bent only slightly in just the same way as a watch spring (Fig. 3.7). This energy can be released again if the bending pressure is removed. The energy is used to return the rock to its original position. For example, only a few thousand years ago huge ice sheets covered the northern continents. Their weight pressed down on the land, forcing it downwards. Now the ice has gone and the land is slowly returning to its original position. Some parts of Scotland are still rising by about 4 cm a century; parts of North America have not recovered as rapidly and the depression remaining is still occupied by the Great Lakes.

Figure 3.7 For caption, see Summary opposite

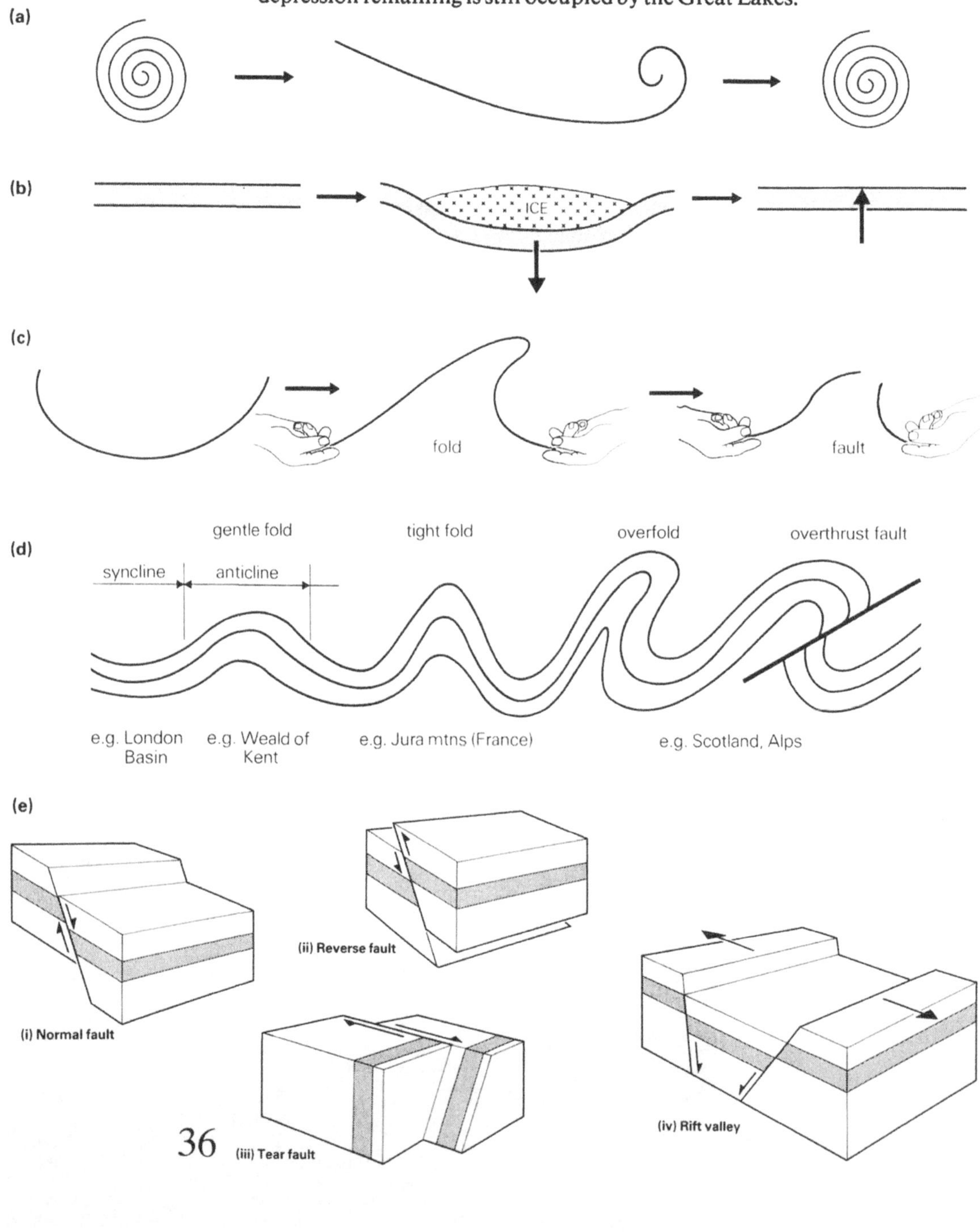

SUMMARY: ROCKS UNDER PRESSURE

If a rock is bent too far it will change its shape permanently This produces a **fold** (Fig. 3.7 c). If bent severely the rock may even break. This produces a **fault** (Fig. 3.7 e).

An upward fold is called an **anticline,** a downward fold is called a **syncline.** The tightness of the fold is an indication of the bending pressure of forces in the Earth's crust (Fig. 3.7d). As the pressure becomes greater the folds become tighter. When pressure is applied from one side more than the other the folds lean over and are then called **overfolds** (Fig. 3.7d). Gentle folds are often very broad. For example the folds that have shaped the London Basin and the Weald of Kent are many tens of kilometres across.

In some cases rocks are too brittle, or the pressure applied is too great, for folds to develop, and then the rocks fault. Most faults move rocks in a more or less vertical direction (these are **normal** or **reversed faults**) (Fig. 3.7e, i & ii). It is rare for faults to occur singly and they are usually found in groups. Large faults can produce distinctive landforms such as **rift valleys** (Fig. 3.7e, iv) (e.g. the Rhine rift valley) or uplifted block mountains known as **horsts** (e.g. the Pennines). **Tear faults** occur when rocks scrape past one another with little vertical movement (e.g. the San Andreas fault) and can usually be seen only when erosion removes the shattered rocks alongside the fault line (Fig. 3.7 e, iii).

Figure 3.7 Folds and faults (*opposite*)
(a) A coiled watch spring can be pulled out, but it will return to its original shape when released. (b) The loading of ice on continents bends the rock downwards temporarily, but it recovers after the ice melts. (c) If a spring is bent too far, it changes shape permanently and may finally break; the same applies to deformed rocks. (d) Types of folding. (e) Types of faulting.

Exercise 3.2 Rift valleys

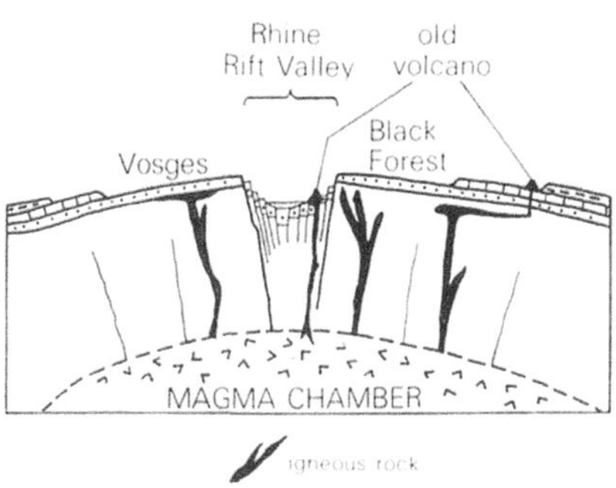

Figure 3.8 The Rhine rift valley

1 Study the diagram of the Rhine rift valley and adjacent areas (Fig. 3.8). What information tells you this is a rift valley?
2 List the other main features of this area that indicate lines of weakness in the crust.
3 Find a map of East Africa in your atlas. Locate Lake Victoria at 2°S 33°E. To either side of this lake are long narrow lakes that occupy the two arms of the East African rift valley. In fact the rift system starts in the Middle East near the Dead Sea on the Israel/Jordan border (32°N 36°E). Follow a line from the Dead Sea to the Red Sea and then south across Ethiopia to join the two regions of lakes.
4 Make a sketch of East Africa and the Middle East, and mark on the rift system from the Dead Sea to Malawi (14°S 35°E). The East African rift system is still active and many small earthquakes occur frequently.

Figure 3.9 El Capitan in the Yosemite valley (California)
Notice the unbroken surface of this rock. The absence of joints is the main reason for its resistance to erosion.

ROGER JONES

Exercise 3.3 The Yosemite Valley – a landscape guided by joints and faults

El Capitan means 'the chief', and it is the name given to the tallest unbroken cliff in the world. El Capitan rises vertically 1000 m above the floor of the Yosemite Valley in California, USA (Fig. 3.9). The most experienced climbers in the world come to pit their skills against this famous slab of granite and it still takes them three days to reach the top. They need a mass of special climbing gear, as natural footholds are so few. This is because the El Capitan granite has not developed many joints since its formation.

And just as a man relies on joints for gaining a climbing foot or hand hold, so nature uses joints for increasing the rate of erosion (see p. 112). Where joints are few, so the effects of rain, river and ice are slow. The Yosemite valley is at its narrowest where the El Capitan granite is found. But just a little further along the valley the rocks change, the number of joints increases and the valley widens out. A closer inspection of the valley shows that whole cliffs are guided by 'master' fractures in the rock. For example the 'Three Brothers' cliff shows how fractures in three directions have been exploited to erode the rock into this unusual shape (Fig. 3.10).

1 Draw a sketch map of the Three Brothers (Fig. 3.10) and mark on the other two directions of master joints that have determined the shape of this cliff.
2 Valley shape and the course of tributaries are also controlled by master fractures. Look at the map (Fig. 3.11a). There are two main lines of fracture at right angles. The Nevada falls flows across one trend line, the Vernal falls flows across the other. Draw the valley and mark in sections that run parallel to the master fracture directions. This shows you the extent to which faults and fractures can influence the shape of a landscape.

ROGER JONES

Figure 3.10 The Three Brothers, Yosemite valley Joint control has determined their shape; the black line shows one direction of 'master' joints.

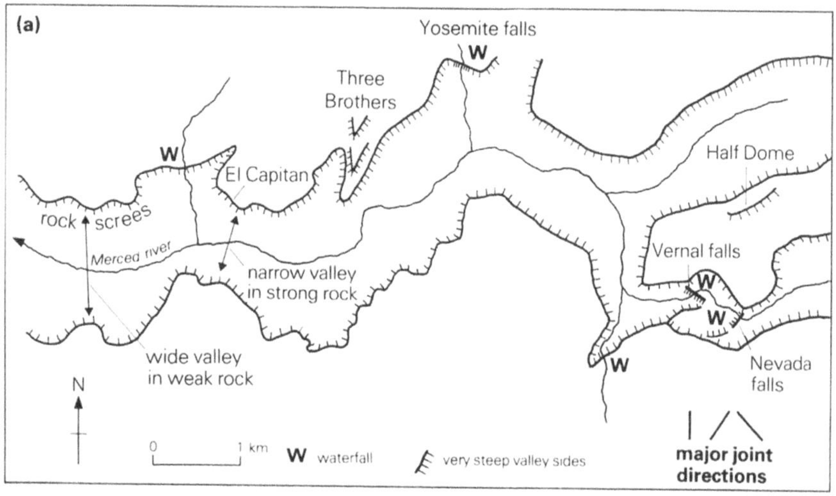

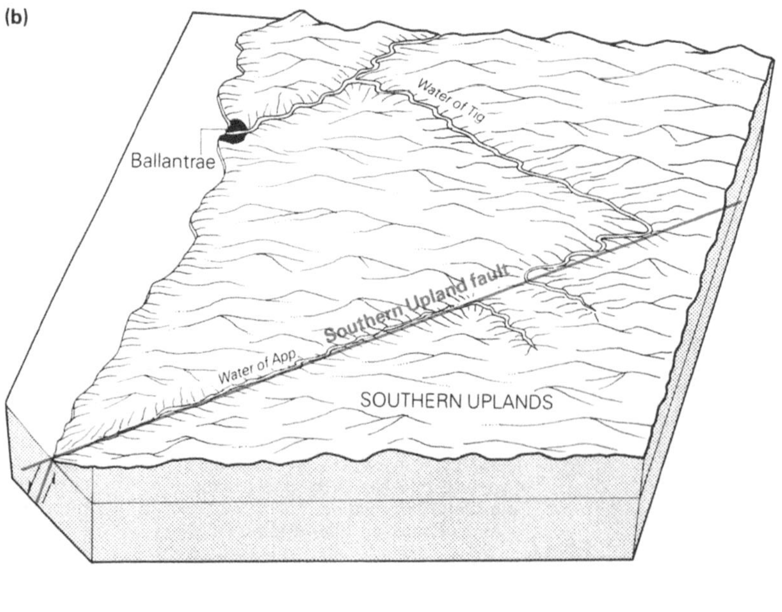

Figure 3.11 Landscapes influenced by joints and faults (a) The Yosemite valley has been eroded by water and ice, but the shape is still largely due to its pattern of joints and faults. (b) Describe how the water of App in southeastern Scotland has been controlled by the major fault line. Why is the valley shape less complex than Yosemite?

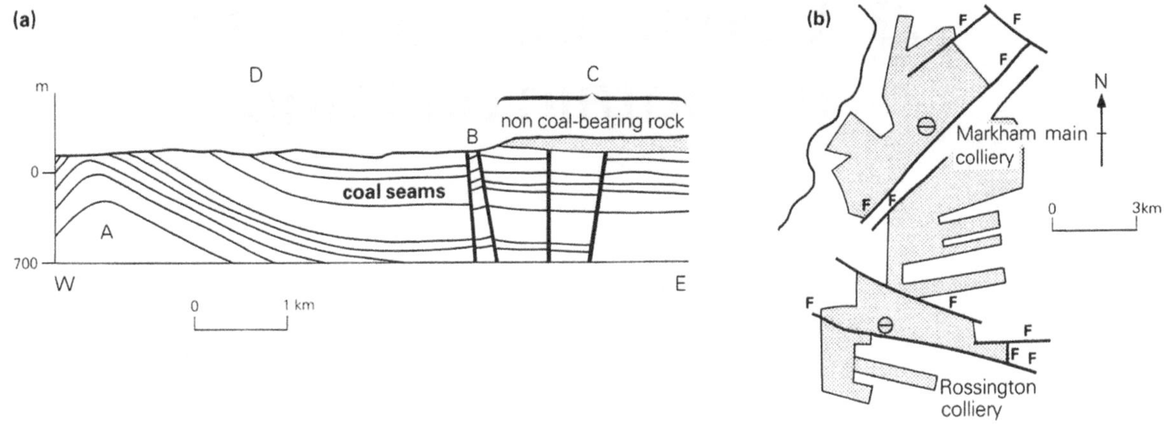

Figure 3.12 Coalfields (a) Part of the Derbyshire coalfield in section. (b) A plan of the Markham and Rossington coalmines in Derbyshire. Faults are marked F; the area of mining is shaded.

Exercise 3.4 Folds, faults and coalmining

The working life of a coal mine may depend not only on the thickness of its coal seam but also on the number of faults and on the dip of the rocks containing coal. In Figure 3.12 the main coal seams and faults in the important and productive coalfield near Chesterfield (Derbyshire) are shown. Notice that part of the coalfield is exposed to the surface and part of it still concealed beneath cover rocks.

1 Write down the names for the features A, B, C, D.
2 What difficulties might result from trying to mine in folded areas?
3 What difficulties might result from trying to mine in faulted areas?
4 Study the maps of the Markham Main and Rossington collieries and explain what has determined the extent of each mine.
5 Briefly write down the best conditions you would expect for a productive coalfield with a long working life.
6 Why were the coal mines in the west of the section the first to be opened and why are new mines now being opened up to the east?
7 Notice that part of the coalfield is described as **exposed** and part **concealed**. What do these terms mean when soil and farmland cover both areas?

4 Continental drift and mountain building

Continental drift

The moving shell

In the previous chapters we have seen that volcanoes and earthquakes are the signs of major lines of weakness in the Earth's crust. We have also seen that these weaknesses make up clear patterns, one of the most striking of which is the Pacific Ring of Fire (Fig. 1.5). These lines of weakness really split up the crust into large slabs, usually called **plates** (Fig. 4.1). Once again, think of the crust as being rather like the shell of a hard-boiled egg that has been cracked with a spoon before being eaten. Jagged edges appear everywhere over this thin and brittle shell, yet individual pieces remain intact; the intact pieces of rigid shell are plates and the jagged edges are called **plate boundaries.** It is along the plate boundaries that we find volcanoes and earthquakes.

On page 7 we noticed that the temperature contrasts between the hot core and the cold crust probably causes convection currents to churn over the rocks of the Earth's interior. These currents would cause a massive drag on the Earth's crust causing whole pieces of it (the plates) to move. However, every part of the Earth's surface is occupied by crust. This means that each time a plate moves it must collide with, scrape past and split away from its neighbours.

Figure 4.1 The shattered shell fragments of the Earth's crust
Arrows show the main movements taking place at present. Only six major plates are needed to account for all of the Earth's major landforms.

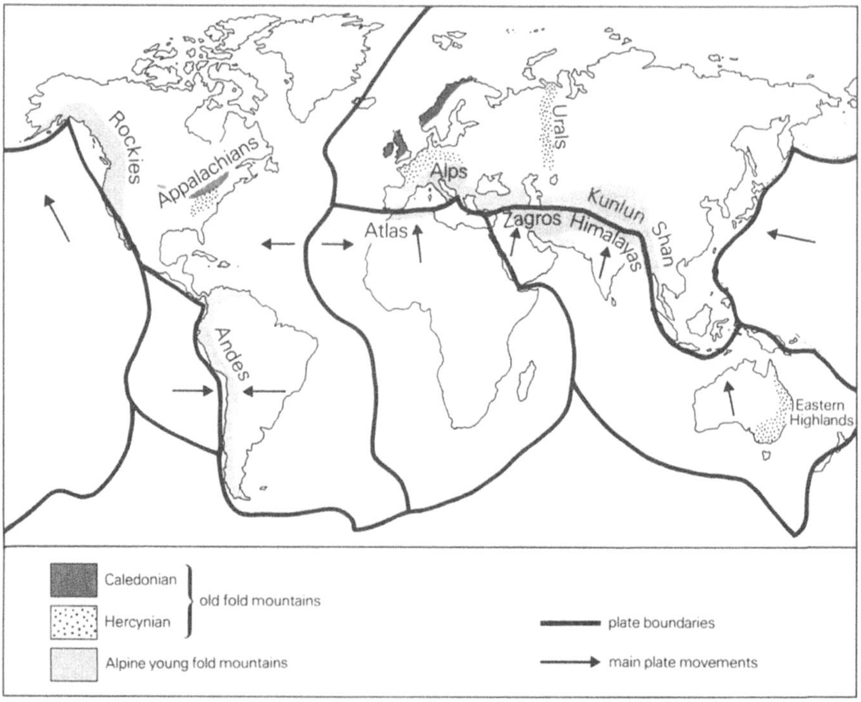

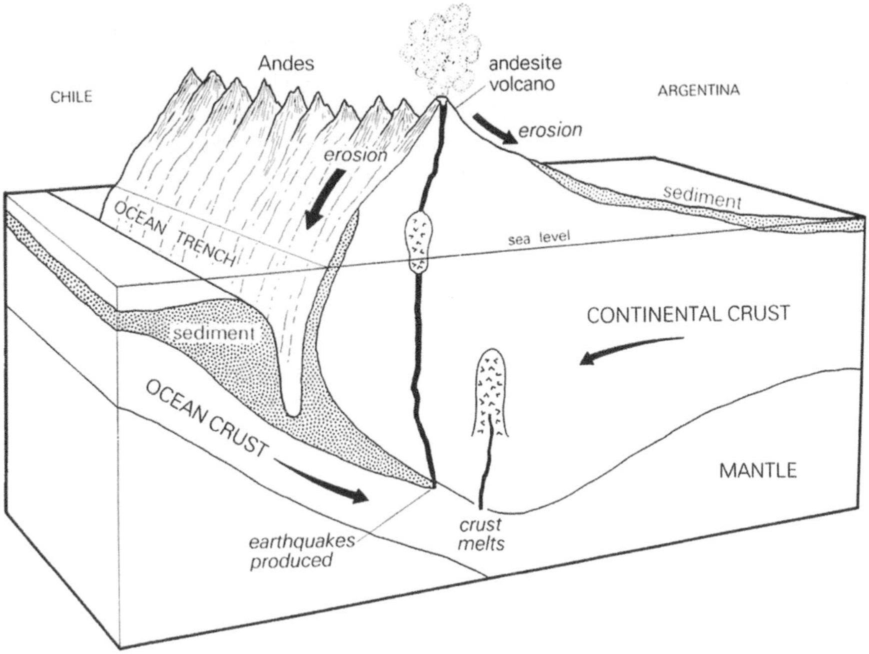

Figure 4.2 The Andes mountains
The Andes form where a slab of ocean crust dives under a continent. Sediment deposited on the ocean floor is dragged along and scuffed up against the leading edge of the continent. The ocean crust slowly scrapes against the continent as it sinks, and earthquakes result. Eventually melting occurs and some material rises up as 'bubbles' of magma to produce volcanoes and batholiths. The western mountains of the USA and Canada have formed in the same way as the Andes.

Oceanic trenches

Way down below

One of the most important results of plate movement occurs when one plate collides with one of its neighbours. When this happens there will be either a massive crumbling of the edges of the plate or one plate must ride over the other. There seems to be evidence for both of these types of collision. Indeed the results of overriding were first discovered by a British ship.

Towards the end of the 19th century, *HMS Egeria* was completing the task of surveying the ocean floor around the Island of Tonga in the Pacific Ocean. The captain was amazed as he watched over 8000 m of wire being payed out before the weighted line reached the ocean floor. He had discovered a part of the ocean that extends farther below sea level than Mount Everest rises above it! In fact further exploration revealed that the Tonga Islands hide a yawning chasm nearly 10 km deep in places.

We know that this gash in the ocean floor – the Tonga–Kermadec trench – is so deep that seven Grand Canyons could be piled on top of each other in it; on land it would extend from the North Cape of Norway to London (a massive 3000 km) and yet it is sometimes only 8 km wide.

The Tonga–Kermadec trench is just one of a whole series of trenches that are to be found along the margins of many oceans (Fig. 4.3). They are all near to lines of volcanoes and earthquakes and so are related to active movements of the earth's crust. In fact trenches are thought to be the surface evidence of the places where a rigid ocean plate is overridden by a rigid continental plate (Fig. 4.2). As the ocean sinks down beneath the continent it is remelted along the line of the trenches and part of the molten material rises back to the surface to give volcanoes. Earthquakes occur as the top of the sinking ocean plate grinds against the bottom of the continental plate.

Here we have the means of explaining the origin of at least some mountains. For example, not only has the collision of part of the Pacific Ocean crust caused the deep Chile trench to form, but in the process the edge of the South American continent and the coastal sediments have been crumpled and lifted up into mountains. The sinking ocean plate has partly melted and the magma produced has slowly risen through the continental rocks over millions of years as giant bubbles of liquid rock. Today, these rocks have solidified and have been eroded into the jagged peaks of the Andes Mountains.

Exercise 4.1 Ocean trenches and ocean islands

1 Copy Figure 4.3, which shows the deep trenches surrounding the Pacific Ocean. Most of the trenches are named after the islands which they parallel. Using your atlas, name the trenches 1–5.

2 Find, name and mark onto your copy several large mountains situated within these island chains. You can expect all the mountains you name to be active volcanoes.

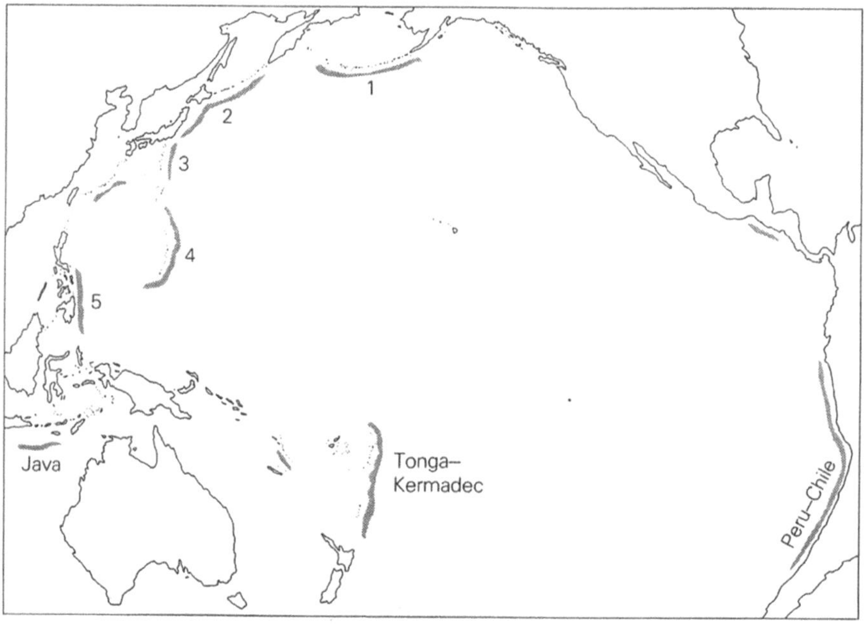

Figure 4.3 Ocean trenches of the Pacific

Fold mountains

Crashing plates

Not all mountains form at ocean/continent margins. The Alps and Himalayas are well away from oceans and so must have been formed in some other way. In these cases scientists believe that two continents have collided. Because both continents are very thick it is not possible for one to be pushed below the other. As a result, when continental plates collide, they mostly crumple up along their leading edges (Fig. 4.4) welding two continents together into a larger unit. India is now known to have collided with the southern edge of Asia, driving up sediments into the great ranges of the Himalayas and the Kunlun Shan. Similarly, Africa appears to have collided with Europe and so crumpled the Alps and the Atlas mountains.

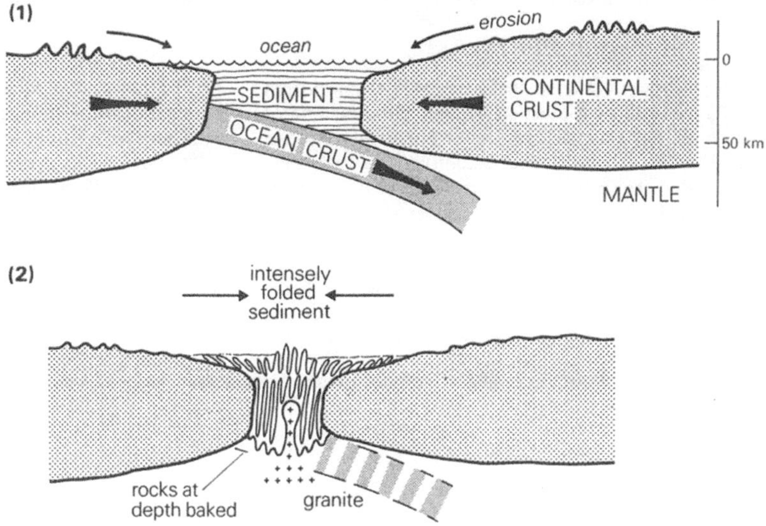

Figure 4.4 How mountains like the Alps and Himalayas were formed (1) Erosion of continents produces much sediment which is trapped as the continents collide. (2) Collision crushes sediments and some become metamorphosed. Granites also form deep in the mountain 'roots'.

It is important to realise that these are not just events of the past. They are happening now. India is still thrusting into Asia at the same rate that it has done in the past and there is no reason to suppose that the Himalayas have finished building. The great volcanoes of Italy tell us that the Alps are still active and that the Mediterranean Sea may be only a temporary feature. It may yet be lifted up to the height of the Tibetan Plateau which itself was a sea only a few million years ago.

Exercise 4.2 Comparing past and present volcanic landscapes

Scientists can measure the way Iceland is growing as the two halves pull apart and lava wells up to fill the gap in the crust.

1 List the features of volcanicity on Iceland (Fig. 4.5).
2 Now list the ancient volcanic features of north-western Britain and Ireland.
3 Write a couple of sentences to say what might have happened to this part of the British Isles to cause the volcanic activity.

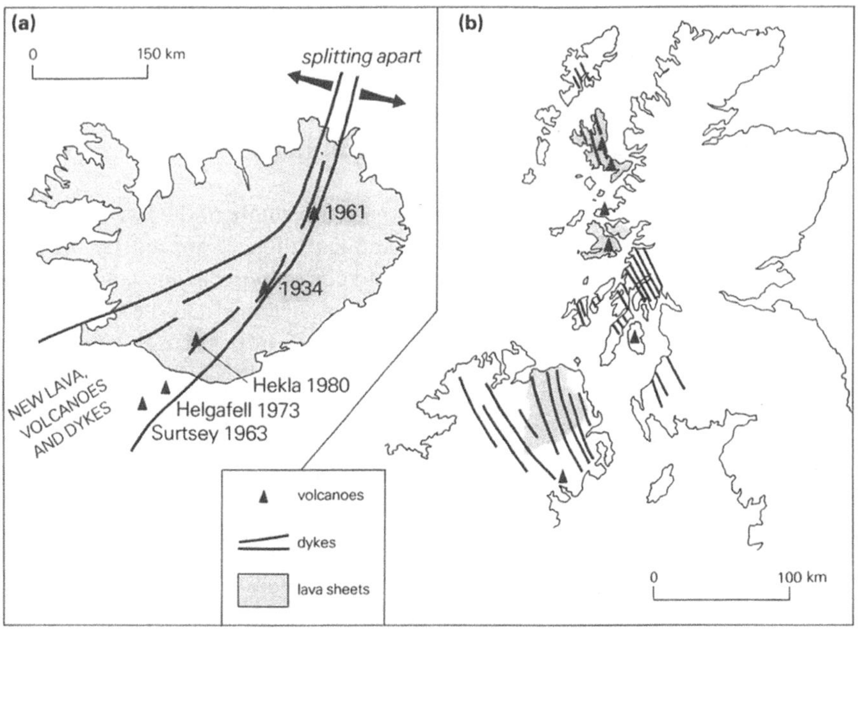

Figure 4.5 Some similarities between Iceland and Scotland

Topic Metamorphic rocks

Metamorphic rocks are those that have been changed from their original state by the action of high pressures and/or temperatures. For example, the rise of hot molten magma associated with volcanoes and batholiths can bake the surrounding rocks into new rocks with new minerals. However, these only create small baked zones a few kilometres wide at most.

The major regions of metamorphic rocks are found where there have been plate collisions. When continents collide the ocean sediments between them are crushed. As a result the sediments first begin to fold and then to melt over regions that may be thousands of square kilometres in area. The main fold mountains are mostly made from metamorphic rocks that were once sediments in ocean basins.

The type of metamorphic rock depends on the degree of heat and pressure. Sometimes the weight of overlying rock alone can cause rocks to change. However, some of our most attractive banded rocks (called schists and gneisses) and some of our most valuable and attractive minerals (e.g. garnet, ruby) form under conditions of intense heat and pressure. Metamorphic rocks are nearly always very hard and resistant to erosion.

Mid-ocean ridges

Plastering the cracks

Because some plates are colliding into others, the trailing edge of a moving plate must leave a huge 'wound' in the crust as it moves away from its original position. This wound must be healed continually by the rise of new magma along fissures in the crust. The volcanic ridge down the centre of the Atlantic Ocean is one place where healing is taking place today. Because North America is slowly drifting away from Europe, there is a continual need for new lava to well up and fill in the gap in the ocean floor. We see this as outflows of lava along fissures such as the one that caused the Helgafell eruption (p. 9). Thus, just as when continents move towards one another oceans shrink, at the same time new oceans are opening in the places from which the continents came. And because magma heals the fissures, all the ocean floors are made of igneous rock. They are covered by thick layers of sediment only near the edges.

Continents of the past

Slow plate to China

We can now see the whole of the Earth's complex development in one picture. Volcanoes and earthquakes are mostly the signs of broken plate margins, and mountains and ocean trenches show where plates have collided. Looking at the distribution of young and old fold mountains (Fig. 4.1) enables us to build up a picture of the shapes of past continents. It seems that sometimes plates were thrust into one another only to be pulled apart again and sent drifting off in new directions when convection currents changed. As a result new mountains have formed constantly through time. And as new mountains were formed so old ones were eroded and the sediment taken out to sea to form new layers of sedimentary rock. It is a long story and it is still continuing. The sediment now building up in the North Sea will probably be crushed up into mountains of the future, and there are already signs that Africa is splitting away from Asia – the Red Sea is a new ocean being born.

SUMMARY: HOW MOUNTAINS AND OCEAN BASINS ARE FORMED

Most of the Earth's surface is covered by six large slabs of crustal rock called **plates.** Each plate may contain a piece of ocean floor as well as a piece of continent. All plates are moved by the drag of convection currents in the Earth's interior. As each plate moves, so it must collide with at least one of its neighbours, causing mountains to form. Ocean basins form where two continents split apart.

(a) If two plates carrying continental crust meet, they crush up any rock between them into new fold mountains (e.g. Himalayas, Alps; Fig. 4.4). Any ocean crust is forced deep back into the Earth to be remelted. Some of this melted rock rises to the surface again to create volcanoes (e.g. Etna and Vesuvius in Italy).

(b) If a plate carrying a piece of continent collides with a plate carrying only ocean crust, the thin ocean crust is usually forced below the continent leaving an ocean trench and a range of largely volcanic mountains (e.g. the Andes) (Fig. 4.2).

(c) Where two plates split apart they leave a wound that is filled with igneous rock giving a thin crust (Fig. 4.6). Water flows in over this new thin crust to from a new ocean. The Red Sea is an example of an early split; the Atlantic Ocean is a split 140 million years old. Very often the centre of this ocean will be the site for a string of new volcanoes (e.g. Iceland, Fig. 4.5).

Figure 4.6 How the Atlantic Ocean formed
(i) Convection currents develop below a large continent consisting of America, Europe and Asia. (ii) The continent is split and magma flows to the surface creating a mid-ocean ridge of volcanoes like Iceland. The continents then drift apart.

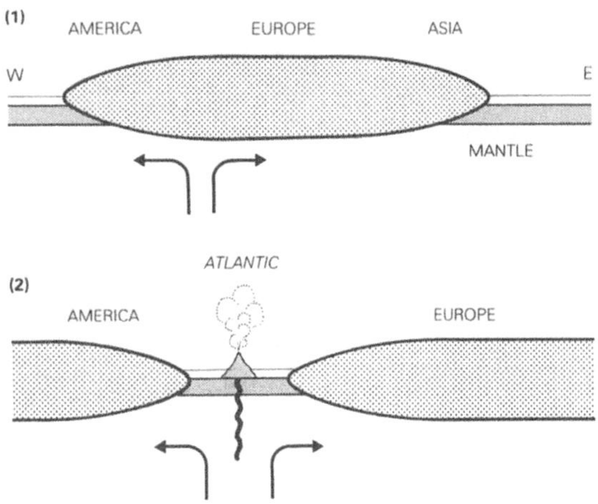

The movement of Britain

Footloose and fancy free?

We are left with one last problem. Can we use the idea of plate movements to explain the very strange collection of rocks that make up countries such as the British Isles? For example, salt deposits and limestones were all formed under tropical or near-tropical conditions. Scientists estimate that, for all these deposits to have formed over the past 500 million years Britain must have drifted as part of a crustal plate from about 30° south of the Equator to its present position at 50–60° north and it is still moving very slowly.

Scientists have found recently that most crustal plates move as fast as a few centimetres a year. This may seem slow but, for example, it would allow a plate to move a distance equivalent to that from London to Edinburgh in 15 million years or from Liverpool to New York in 120 million years. In fact there are many similar rocks in Europe and North America which suggest that these two continents split apart some 120–140 million years ago. It is even possible to fit the plates back to their positions before the movement occurred. When this is done we see a startling picture of all the continents grouped into one supercontinent centred near the South Pole. Some people have referred to the movement of the continental plates as the 'flight from the Pole', but the movement has really been very, very slow (Fig. 4.7).

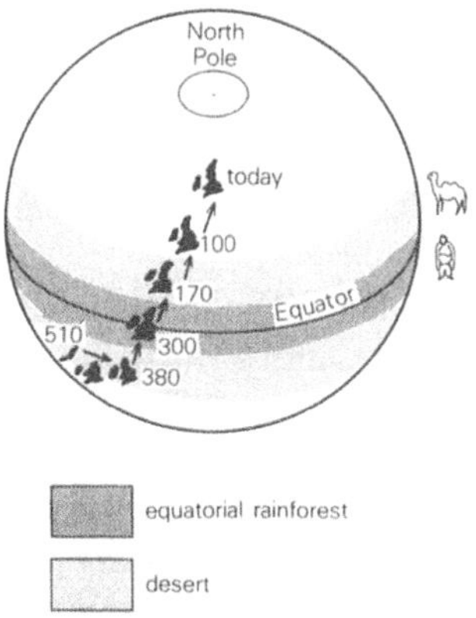

Figure 4.7 Continental drift
During the past 500 million years the British Isles has drifted 9000 km. As it did so it passed through different climatic belts. The numbers are dates in millions of years.

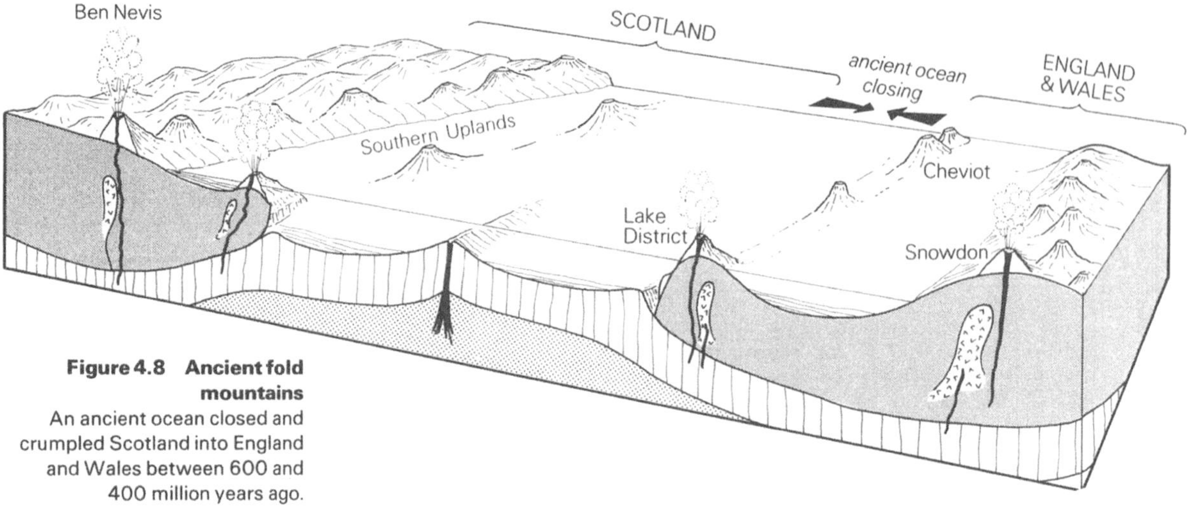

Figure 4.8 Ancient fold mountains
An ancient ocean closed and crumpled Scotland into England and Wales between 600 and 400 million years ago.

Geology and scenery in Britain

Britain's battle scars

We cannot even begin to understand the present landscape of the world without some knowledge of these past plate movements. Without them the arrangement of different rocks, mountains, plains and oceans appears random and meaningless. Today Britain is far from an active crustal plate margin and so we are spared the constant threat of volcanoes and major earthquakes. And yet in the past this has not been so. Britain still shows many scars between crustal plates. For example, about 500 million years ago Scotland (then part of a separate continent) began a slow collision into England and Wales, which crumpled up all the sediments between them into intensely folded mountains (Fig. 4.8). This period of Earth history has been called the **Caledonian** mountain-building period. It lasted for over 200 million years and for much of this time Scotland, northern England, Wales and the northern half of Ireland must have looked like the Himalayas do today. When the collision stopped there were mountains whose now-eroded stumps still make up some of the grandest scenery in the British Isles. Without this event there would be no Grampian mountains, no Lake District, Snowdonia or other mountains for people to admire and walk over.

Another 130 million years went by and the 'Caledonian' mountains were mostly eroded away, their debris being deposited in the sea and formed into new sedimentary rocks. Then another collision began which eventually raised the mountains where we now find South Wales, southern Ireland and the South-West peninsula. This was the **Hercynian** mountain-building period which not only formed new mountains but also pushed the Pennines up into a huge arch. When, after more than 100 million years, this period was over, erosion gradually lowered the mountains and deposited new sediments in nearby seas. No new collisions have occurred in Britain since then, but the 'ripples' of a slow collision between Europe and Africa during the past 70 million years have produced waves of folds over southern England creating domes such as the Weald of Kent and the basins of Hampshire and London. It also jolted the old northern mountains back up from low eroded plains to their present heights.

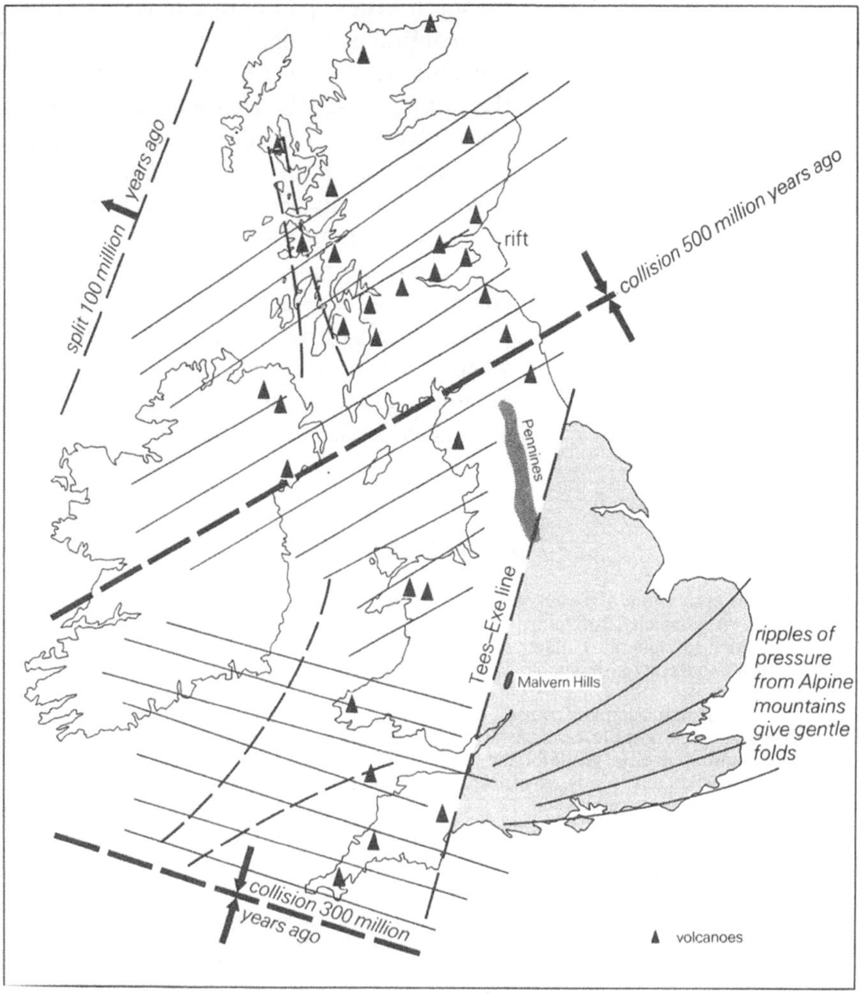

Figure 4.9 The shape of Britain
This is the result of hundreds of millions of years of change. The Tees–Exe line separates the lowland south-east of Britain (where old crushed rocks have subsided and been covered by layers of softer rocks) from the north and west where the old hard rocks still stand proudly as mountains.

To add a further complication, over the past 120 million years at least, America has been tearing itself away from Europe, creating the Atlantic ocean in the process. As the ocean opened a final burst of lava was ejected from many volcanoes in Ireland and western Scotland. All these events provide the foundation from which erosion has carved the varied scenery in the British Isles (Fig. 4.9).

The rate of geological activity

A bit of peace and quiet?

Today when we look around, it appears that all the turmoil of events in the past is now over. However, this is far from the truth. India is still colliding with Asia, and Africa with Europe, at just the same speed as they have done now for tens of millions of years. Mountains are still forming even though we can rarely see the changes in our short lifetimes. For example, the whole of Malaysia has been formed in the past hundred million years. In less dramatic areas at this moment the south-eastern part of Britain is sinking by about 1 mm each year, while the north of Britain is rising by up to 4 mm a year. But 1 mm of sinking a year would cause the North Downs to go below sea level in less than a million years! Indeed, this sinking caused the Victorians to build a wall all along the Thames in London (the Embankment) to reduce the risk of flooding. Today this is no longer enough and a barrier has been built (the Thames barrage).

We have mostly been looking at Britain's history because the detail has been worked out carefully, but it is not exceptional. Indeed, it merely illustrates the effects of plate movement all over the world. You will find that nearly all continents are covered with sedimentary rocks hundreds or thousands of metres thick. These in turn all lie on ancient metamorphic and igneous rocks. So, except when an area becomes part of a mountain system, it is rarely thrust clear of the sea for long. Most non-mountain areas have often been bobbing up and down around sea level, building up rocks that add diversity to the Earth's landscapes.

EXAMINATION-STYLE QUESTIONS

1. Write an essay on one of the following topics: (a) volcanicity, (b) folding, (c) faulting, (d) mountain building. Illustrate your essay with suitable diagrams.
2. Using maps and diagrams of named examples where possible, describe the causes and effects of: (a) gentle folding, (b) earthquakes, (c) explosive volcanoes.
3. Name an area where geysers and mudpools occur. Explain the operation of a geyser, and explain how hot water is used to generate electric power.
4. Name three fold mountain ranges, two volcanoes and one rift valley. Draw diagrams to show (a) a rift valley and (b) a region of folds showing synclines and anticlines.
5. With the aid of diagrams, describe and explain the formation of fold mountains.
6. Explain why some volcanoes erupt violently and have steeply sloping cones, whilst others have gently sloping cones and produce mostly lava flows. Name examples of each type of volcano.
7. Name two areas in the world where there are many volcanoes. Explain what is meant by the terms dyke, sill and batholith. Give examples of each type and describe their influence on the landscape.
8. Draw a labelled diagram to show the main features of a volcano. Give an example from the British Isles of an old volcano, a dyke and a sill. Describe the differences between dykes and sills.
9. Name a coal field and describe the geological conditions that have affected the mining of the coal. Draw a diagram to show how coal and natural gas are related in the rocks, and explain why both gas and oil are found only in small isolated areas. Describe how this has affected the exploitation of the world's oil and gas reserves.
10. Explain how the following materials are used: limestone and chalk, salt, iron ore and clay. For any two of these state a major area of extraction and describe the way the materials influence the location of industry.

Part two **Atmosphere and ocean**

The circulation of air and water is a never-ending cycle driven ever onwards by the Sun.

Figure 5.0 The swirling pattern of cloud as seen from space

NASA

5 The weather machine

The origins of meteorology

The Gunpowder Plot

Figure 5.1 Lightning fashion!
(Ann Ronan Picture Library)

In the Middle Ages bell ringing was a dangerous job. Bells were rung in church towers and steeples to disperse the thunder – then regarded as the work of the devil. Unfortunately bell ropes kept getting wet during thunderstorms and wet ropes hanging from bells in tall towers provided an easy path for lightning to reach the ground. Bell ringers kept being killed – from electrocution!

Stormy weather was a major hazard to armies and navies as well. There was a constant danger that a flash of lightning might strike a gunpowder store and cause an explosion. During thunderstorms navies very often had to disengage at the height of the battle because of the risks of their ships blowing up from 'natural causes'.

One day a lightning flash hit a gunpowder store in Northern Italy and it was blown to bits. It was this event that led to the widespread use of lightning conductors on buildings. At the end of the 18th century it was even possible to buy ladies' hats and gentlemen's umbrellas complete with their own lightning conductors (Fig. 5.1).

But lightning was not the only natural hazard that a ship at sea might have to face. In 1854 the French Navy was struck by a hurricane-force wind while anchored off the Crimea and a ship was lost with all hands. This was too much. A means of predicting the weather and especially the possibility of storms was urgently needed.

Early weather measurements

All hot air

In the meantime many other people became interested in the atmosphere. Hot air balloons had been flying since 1783. In 1803 two people in Germany rose up to the staggering height of over 7000 m in a balloon, discovering that temperatures decrease with height almost to the point of freezing them. In 1823 the British Meteorological Society was formed to try to organise all the enthusiasm of the new amateur meteorologists and in the 1850s the first weather maps were produced. Now people were able to see how the weather changed over the Earth's surface and, to a limited extent, up in the air. The science of **meteorology** had arrived.

It seems a far cry from bell ringers and hot air balloons to the computer-made weather maps we see today. Satellites orbiting above us now provide weather pictures on our television screens within a couple of hours of their being taken. Ideas are changing fast. For example, we now know that ground surface measurements of the atmosphere are not enough. What is happening near the ground is really controlled by events high up in the sky where unseen winds weave about and drag our weather with them. The weather man might still make some errors but his forecasts are becoming more and more accurate all the time.

The Earth's heat balance

A balancing act

Over the past few years the Mariner and Voyager spacecraft, sent out on exploration missions from Earth, have greatly increased our knowledge of the planets. Nevertheless, even without the pictures that are now being transmitted back through millions of kilometres of space, we knew that sunbathing on the Sun-facing surface of Mercury would be, to say the least, a somewhat uncomfortable affair. Mercury is the planet nearest to the Sun and, rather in the same way as when we sit too close to a fire, it gets uncomfortably hot. The surface temperature at lunch time on Mercury is a little over 400°C so there would be no trouble roasting a turkey even without an oven! By contrast, a few astronomical steps will take us to one of the outer planets, Saturn. Here the sun appears as just a bright star in the sky and its heating effect is so small that the 'surface' temperature is as low as –150°C.

So distance from the Sun determines the energy received by the planets. Fortunately the Earth is at an intermediate distance and the solar energy we receive (called **insolation**) neither roasts nor freezes us.

Storing solar energy

Greenhouse Earth

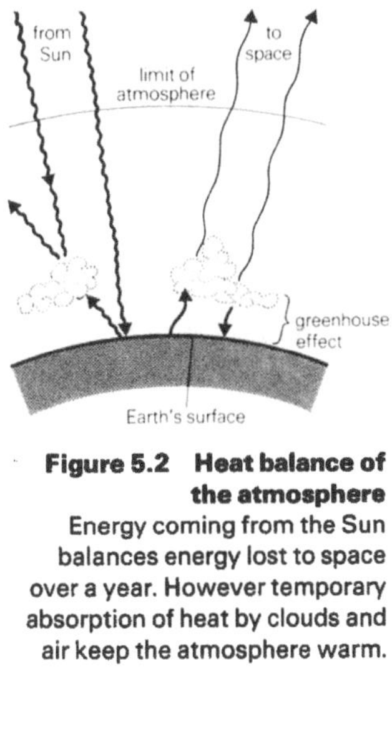

Figure 5.2 Heat balance of the atmosphere
Energy coming from the Sun balances energy lost to space over a year. However temporary absorption of heat by clouds and air keep the atmosphere warm.

In the long term, the Earth's surface and atmosphere seem to be in balance, becoming neither hotter nor colder. The surface heats up because it receives energy from the Sun. The Earth in its turn radiates the received energy back out to space until there is a balance between income and output. But whereas the atmosphere is fairly transparent to incoming solar radiation, a lot of the outgoing radiation is temporarily trapped by water vapour and carbon dioxide gas in the air. As a result the atmosphere receives an extra source of heating (Fig. 5.2).

The role of water vapour is seen very well by comparing clear and cloudy nights. On a clear night the cloud 'blanket' absorbs heat from the Earth and radiates it back again; as a result the weather is much less cold. Even when clouds are not present, water vapour and carbon dioxide in the air still prevent some of the heat loss. Indeed, the way the atmosphere allows solar radiation in, but partly stops it from being lost again to space, is superficially so like the way glass helps heat a greenhouse that it is called the **greenhouse effect.** In fact the greenhouse effect is responsible for keeping air temperatures 25°C higher than they would be otherwise.

Changes in the Sun's heating power

Rays of warmth

There is a lot of energy available in the rays from the Sun. Sometimes this is dramatically illustrated, no more clearly so than in the sleepy hills of south west France, near the town of Odeillo. Here, high up in the Pyrenees, a hillside has been covered with stainless steel sheets adjusted in such a way as to make a vast concave mirror. These mirrors are slowly turned and the Sun's rays, caught by each shimmering plate and focussed on a single point, instantly provide enough heat to melt steel (Fig. 5.3a).

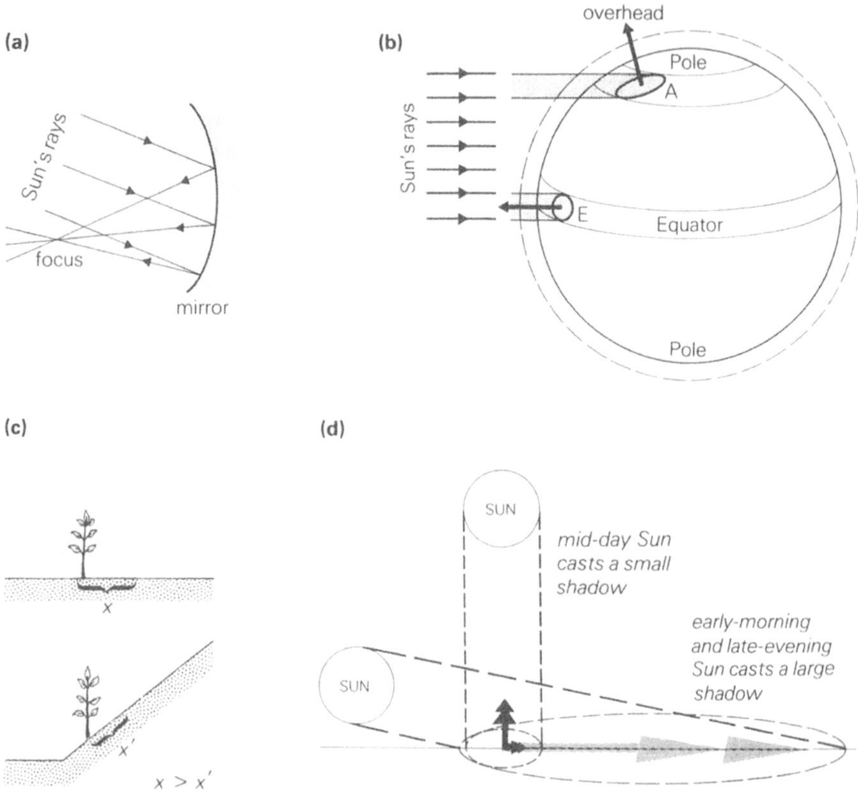

Figure 5.3 Examples of energy concentration
(a) Concave mirrors concentrate the Sun's rays. (b) At E, near the Equator, the Sun is overhead and energy is not spread out; at A, nearer the Pole, the Sun is not overhead and the same energy that heats E now has to heat a greater surface area. As a result A is warmed less than E. (c) When landsurfaces are tilted, the same Sun's energy is used to warm a smaller area of ground and so crops grow more readily. (d) Long shadows in the morning and evening show how the Sun's energy is being shared over large areas. As a result the Sun's rays appear to be less warm than at mid-day. The same reasoning applies to a comparison between summer and winter.

On the Earth's surface the most concentrated rays occur when the Sun is overhead and they become progressively more spread out as the Sun's angle to the vertical increases (Fig. 5.3b). It is this that makes the Earth much hotter at the Equator than at the Pole, and midday hotter than early morning. On a local scale the amount of heat received depends on the direction and tilt of a hill slope. Farmers in upland areas all know the importance of slope and aspect to their activities (see Ex. 5.1).

Exercise 5.1 The importance of aspect

Aspect is the compass direction that a slope *faces*. The vines that grow in the Rhine gorge of Germany are amongst the most northerly in the world. As a result they need all of the Sun's heat they can get. Figure 5.4 shows part of the gorge near its southern end at Bingen. Here, besides aspect, the tilt of the slope toward the Sun is also important.

1 Copy the cross section and mark on the regions covered by vines and forest.
2 Draw a circle with a diameter of 5 cm. Assume the '12 o'clock' position is north. Now use a protractor to find the most northerly-facing slopes still used for vine growing. Draw these aspects as radial lines onto your circle. Shade in the sector of the circle between these extremes. This indicates the limiting conditions for vine growing.
3 Find the direction of the slope where vineyards are planted *highest* up the gorge sides. Mark this as a radial line onto your circle.
4 Using the results of your drawing, describe how aspect controls the growth of vines in the gorge.
5 From the cross section and Figure 5.3 suggest another physical factor that might be important in controlling the growth of vines.
6 Can you explain why villages and farms in a steep-sided Alpine valley often occupy one side of the valley, while only forests occupy the other side? What aspect would the forested slope have?

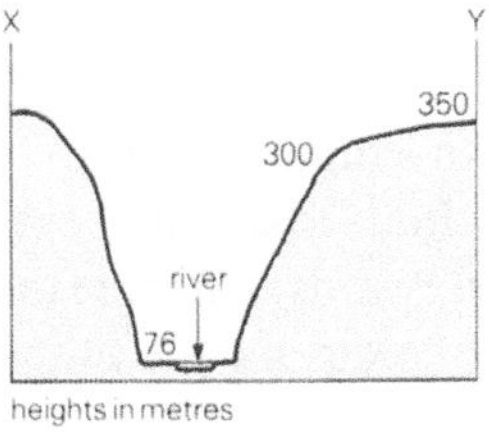

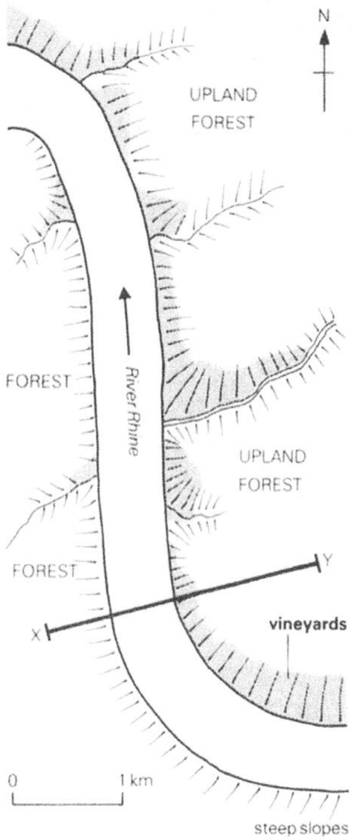

Figure 5.4 The Rhine gorge (Germany)
Right. In plan and, *above*, in section. *Far right.* Vines growing on a terrace. Notice the north-facing slope in the background is used only for scrub and forest.

SUMMARY: HEATING THE ATMOSPHERE

The atmosphere is primarily heated from below. Solar **radiation** warms the ground. Air in contact with the warm ground is warmed by **conduction.** The warmed air is less dense than the air above and it rises through it by **convection.** Winds also bring air across the ground surface, mixing warm and cold air together by **turbulence** (Fig. 5.5). When clouds form yet more heat is provided to the air by **condensation.**

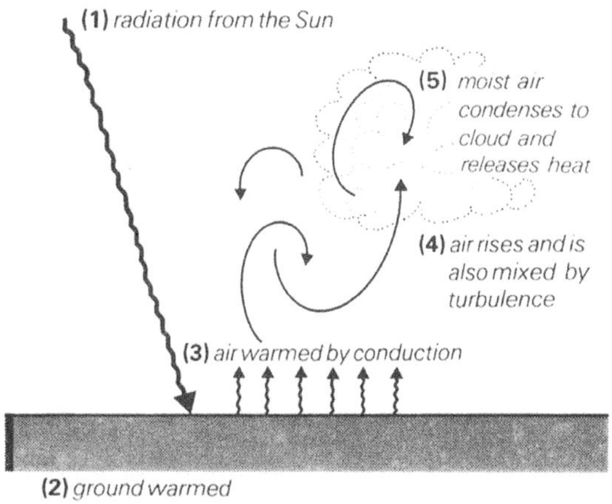

Figure 5.5 The ways the atmosphere is heated

SUMMARY

Air circulation patterns

The restless atmosphere

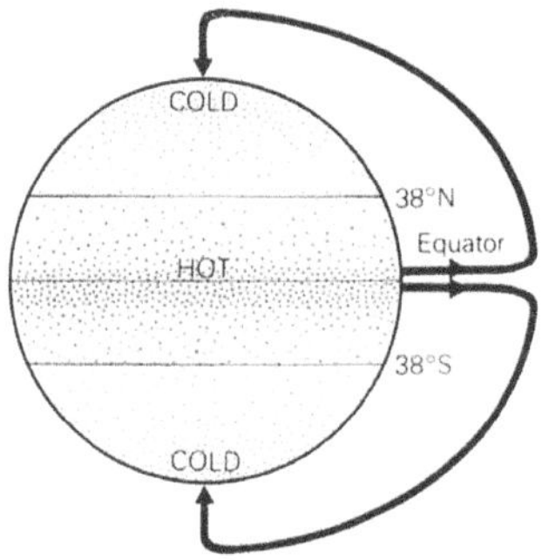

Figure 5.6 Convection in the atmosphere
The Earth's surface receives more heat than it loses equatorwards of 38°, and less polewards. If the Earth were stationary, a simple convection system of one cell for each hemisphere might develop as shown.

Although a long-term balance is achieved between Earth's input and output of energy as a whole, all areas polewards of about 38° of latitude lose more heat to space than they receive from the Sun, and all areas equatorwards gain more than they lose (Fig. 5.6). So why don't the Poles get colder and colder and the Tropics hotter and hotter?

The answer is provided by the atmosphere and oceans, which share out the surplus solar energy of the Tropics and moderate the extremes of temperature that we might otherwise expect.

We can get a good idea of how the oceans and atmosphere work because the same processes operate in many homes as over the Earth as a whole. In the living rooms of most mid-latitude homes there is an equivalent to the Tropics – a heat source such as an open, gas or electric fire, or perhaps some sort of hot water central heating (Fig. 5.7). In fact, although they all **radiate** energy out into the room, a large part of the energy is used in warming up the air in contact with them by a process called **conduction.** The warmed 'jacket' of air surrounding the heat source then expands to become less dense than the surrounding air and it rises, causing fresh cool air to take its place and be warmed in turn. This is just the same way that air circulates in the atmosphere.

In such living rooms the coldest surface is usually the window and here, the 'polar region' of the room, air is colder and therefore more dense than its surroundings and sinks. The heat source and the window together set up a circulating or **convectional** flow of air around the room. The Tropics provide the heating source for the Earth's atmosphere, and oceans leaving the cold polar regions to provide the temperature contrast which sets up the convection. As a result there is a tendency for two basic convectional flows to be established around the Earth to share out the heat: one from the Equator to the North Pole and one from the Equator to the South Pole (Fig. 5.6). They are more or less mirror images of one another.

The effects of the Earth's rotation

Turn, turn, turn

So far we have not considered how the rotation of the Earth might affect convection. You will already realise how difficult it is to cause liquids and gases (fluids) to move in straight lines – even water gurgling out of a bath will always swirl round and round. Fast-flowing rivers are another example: not only do they flow in a turbulent way, but they insist upon following a winding path. Whenever wind whisks up leaves from a pavement they are lifted in curved paths. So we should expect the atmosphere and oceans to behave in the same way.

Sometimes this is easy to see and tornadoes, typhoons and hurricanes are well known for their spiralling winds (Fig. 5.25). Less well known but much more important are meandering (winding) tunnels of air high up in the sky (called **jet streams**, p. 56) and swiftly moving tunnels (streams) of water set within the oceans called **ocean currents** (such as the famous Gulf Stream, p. 82). Wherever you look, fluids are on the move; we saw this to be true even with the convection currents within the Earth (p. 7).

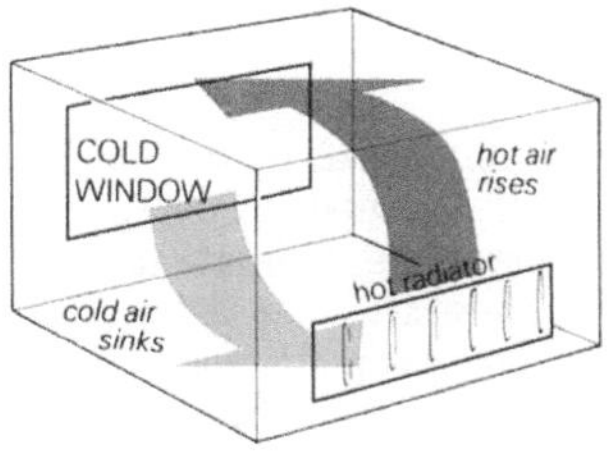

Figure 5.7 A room convection cell
When heating source and window are on opposite walls, cold draughts are generated around the feet. Can you think of better places for the heating source to be to prevent draughts.

Exercise 5.2 Local-scale convection

1 Look at Figure 5.7. Explain why many people say they get hot heads but cold feet while sitting in a heated room such as this.

2 Builders usually place radiators underneath windows. What effect would this have on convection currents? Would it be a good or a bad thing?

Exercise 5.3 Coastal breezes

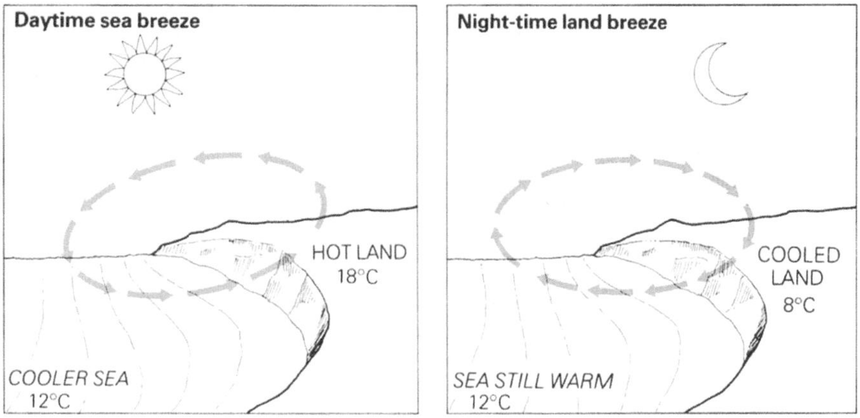

Figure 5.8 Local-scale convection along a coast

Convection operates at all scales from the size of a room to the size of the Earth. Figure 5.8 shows the convection that often occurs along sea coasts.

1 Why in the summer is the daytime breeze called a sea breeze and that at night called a land breeze?

2 Sea water absorbs a lot of heat in its upper layers and keeps a fairly constant temperature day and night. By contrast land can store very little heat, so it becomes very hot by day but loses heat rapidly at night. Using the temperature figures on the diagram, explain why convection changes direction between day and night.

3 Would these diagrams still be correct in winter?

These swirling patterns are enormously important in that they control our daily weather. Their effect is easily seen by comparing the simple pattern outlined in Figure 5.6 which was drawn for a 'stationary' Earth with the picture of the real spinning Earth shown in Figure 5.0. The basic convection currents are still the main driving force of the circulation. That has not changed. But on a spinning Earth the most effective way of transferring air from one place to another is to use the apparent chaos of giant swirling movements.

High and low pressure

What goes up must come down

If you look at a cross section of the atmosphere (Fig. 5.9) you will find that over some parts of the Earth the air is rising, while in others it is sinking. On a global scale, however, all the ups are balanced by the downs, despite these regional differences.

Sometimes the air sinks faster than it can push the air below out of the way. It is rather like a lot of people trying to leave after a football match: they all crowd together as they try to push through the gates. In the atmosphere the build up of air is called a **high pressure region.** Conversely air can be drawn away from a region faster than it can be replaced. This gives a **low pressure region.**

There are some parts of the world where air is always building up faster from

Topic Global winds

Eventually all the air that rises at the Equator and flows at high levels towards the Poles has to return near the surface to complete its global circulation.

High in the sky we might expect the air to be streaming polewards. This is shown on the cross section, Figure 5.9. However, if you were to go up to these high altitudes in a plane you would find that the air appears to be moving largely to the east and not polewards at all!

Still air at the Equator is actually moving very quickly. It has to move at nearly 2000 km an hour just to keep up with the spin of the Earth! You can see this on Figure 5.10a. Any wind we experience is due to small changes from this need to keep up with the spinning Earth.

Now suppose the upper air moves from the Equator, where it is going fast, towards the Poles, where it only needs to go slowly. As soon as the air leaves the Equator it is going too fast for the latitude it arrives in, so it gains on the Earth. So all the time air moves to the Pole to share out the Sun's heat, it keeps going too fast, turning a simple direct motion into a very long curved path (Fig. 5.10b). In practice the curve is a very gentle one with air moving faster in an easterly direction than it does polewards. We feel this excess speed as a **wind.** Because we describe winds by the direction from which they come, this is called a *westerly* wind. Conversely, air returning to the Equator is always moving more slowly than the latitude it reaches. We feel this as an *easterly* wind. The surface wind patterns of the Earth are shown on Figure 5.9.

It turns out that a nice even westerly or easterly wind cannot be maintained on a spinning Earth without developing places of slack and places of fast movement. This is just the same as the uneven flow in a river. There are some fast-moving 'streams' embedded in the main flow. In the atmosphere some of these have such fast-moving air currents in them that they have been called **'jet streams'** (Fig. 5.10b). There are two jet streams in each hemisphere. One steers high and low pressure systems over the mid-latitudes. The other is above the regions that receive the monsoon. Today weather forecasters keep as close an eye on high-level jet streams as on ground-level observations.

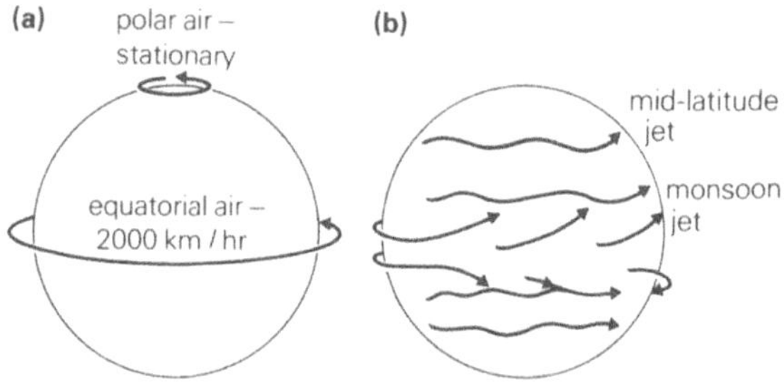

Figure 5.10 Global winds
Air really takes a long curved path on its way to the Poles at high altitude.

Figure 5.9 Circulation in the atmosphere
(a) The major features of the global circulation;
(b) a simplified view of the circulation patterns.

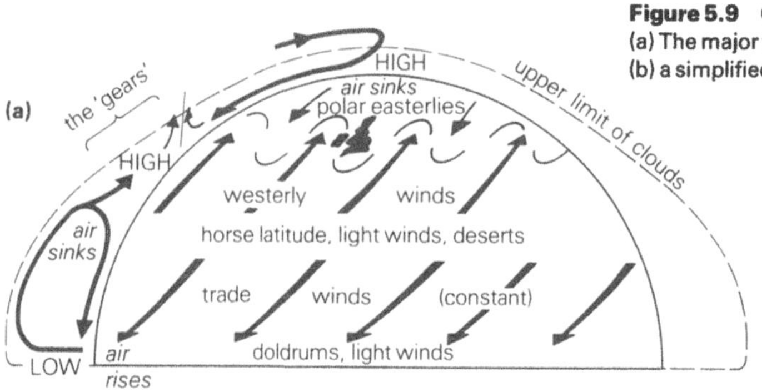

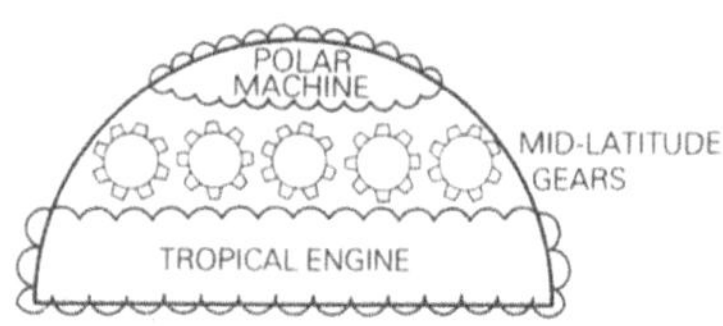

above than it can get away at ground level. These are the *permanent* high pressure regions. A series of such regions are found about 30°N and S of the Equator, encircling the Earth. Underneath them lie many of the world's great hot deserts. They are the cloudless belts on Figure 5.0. Each polar region is also a place of high pressure, and here we find the great ice deserts.

There is only one region that has permanently low pressure. This is a belt of air near the Equator, which moves seasonally with the overhead noon Sun. In this region heavy rain is very frequent. Its position is shown on Figure 5.0 as a narrow band of cloud.

You can see these belts again on Figure 5.9 and they show that the atmosphere contains very definite patterns in its air movements, driven, as it were, by the great heat energy of the Tropics. In fact the whole global circulation can be thought of as an engine powering a machine. Near the Equator there is the engine, providing the power to move the rest of the system. This power is transmitted through a set of 'gears' (the swirling patterns on Fig. 5.0) to drive the machine (the polar regions) which continually need energy pumped into them. People in Britain live in the region of the 'gears' and thus have the most variable weather on Earth.

Changes in air pressure with height

Floating on air

So far we have looked at the way in which convection and a spinning Earth have given rise to places of high and low pressure. These determine whether we shall have rain or drought. Our next task is to examine the properties of air pressure to see how it is related to rainfall.

When Mexico City was chosen as the location for the 1972 Olympic Games all the participants knew they were going to have special problems. Mexico City is over 2000 m above sea level. The athletes in particular knew they would have to arrive at the Olympic village several weeks before the games started just to acclimatise themselves to the thin air. Otherwise they would have been gasping for air and would have given poor performances.

Mountaineers also know this problem well: they stay for days at one high-level camp before climbing further and establishing a new camp. Again it is a matter of acclimatisation. Air becomes 'thinner' (less dense) the higher you go above sea level. It doesn't change much in composition, but there are just fewer molecules of it in every cubic metre as compared with sea level. Go up in an airliner and you will be sitting in a pressurised cabin, because at 10 000 m you cannot breath in enough oxygen from the thin air to keep you alive. And it is colder too, partly because we are out of the greenhouse region of the lower atmosphere. This is why aircraft cabins are also heated.

Forms of water

A reservoir in the sky

Air pressure and temperature are very closely related and in the atmosphere both usually decrease with height. In the first part of this chapter we read that two German balloonists made the first ascent to over 7000 m in 1803. Not only did they get very cold, but one of them found that he could no longer get his hat on – his head had become swollen because there was less pressure on it!

Water vapour is one of the gases in the air that is important in the greenhouse effect. It is also important because it can be turned into liquid water droplets. Water droplets are easily seen whereas water vapour is invisible. A mass of water droplets suspended in the air is called a **cloud.** But what causes water vapour to turn into droplets and what has it to do with balloon ascents?

One way of producing water droplets is to cool the air because *the cooler air becomes the less water vapour it can hold.* So if the air temperature is lowered sufficiently the air becomes **saturated** (filled to capacity) with water vapour. Any further decrease in temperature will then result in water being condensed from the vapour state and converted into liquid droplets.

Air cooled to saturation is quite a common phenomenon. This is what happens to the jacket of air around a glass of iced orangeade when it makes the glass wet on the outside. **Dew** on the grass forms because of the same effect when vegetation cools overnight. But when there is no cold surface for the air to condense on and a whole thickness of air has cooled down to its saturation level,

Figure 5.11 Cloud formation
Hot moist air from a power station readily produces cloud.

the droplets remain suspended in the air as cloud. You produce a 'cloud' when you breathe out on a cold day. The warm moist air from your lungs is quickly cooled until water droplets form. Steam from a boiling kettle and clouds being formed above a power station show the same effect (Fig. 5.11).

Topic How raindrops and snowflakes form

In the atmosphere as air rises it cools until it falls to a temperature at which it can no longer keep all of its water content as vapour, and it has to shed some as water droplets or snowflakes. The myriad of droplets or snowflakes we see as a cloud. But each droplet or crystal is tiny – a matter of only millionths of a metre across – so how do they form into raindrops and snowflakes big enough to fall to the ground?

It may come as a surprise to find that most rain over non tropical regions starts as snow, even on a summer's day.. Whether you receive snow, sleet (semi-melted ice) or rain largely depends on the air temperature below the cloud. If the air is warm the falling snowflakes melt to become rainfall at ground level.

Temperatures in a cloud are very low because clouds are so high (Fig. 5.12). In many cases parts of clouds have temperatures as low as –30° to –40°C. Because of this, much water vapour is shed directly into tiny snowflakes (see Fig. 17.4 for a typical shape), rather than liquid droplets. As a result the top of a cloud is sometimes all snowflakes while the (warmer) bottom is all water droplets. This leaves a mixture of the two in the middle.

In a cloud, air is continually moving upwards. The upward movement brings with it water droplets. However, in these middle regions temperatures are really too low for the water droplets wafted up from below and so they tend to 'condense' on the snowflakes already there, causing the snowflakes to grow. Eventually some snowflakes grow to a size when they can no longer remain suspended by the upward air movement and they fall out of the cloud. So although there are many water droplets at low levels, they often do not grow; it is only the snowflakes in the mixture region that grow big enough to fall out. A deep cloud, with air currents rising to regions high enough to have snowflakes growing as well as a good supply of vapour to produce droplets lower down, looks dark and grey from the ground because its thickness reflects so much light from the Sun. This is why we say that dark clouds threaten rain but that white clouds are a sign of fair (or at least rainless) weather.

In the Tropics, even deep clouds do not cool enough for snowflakes to grow. In these regions, upward air movement is often very fast and water droplets are dragged together. When sufficient droplets have bumped together, very large droplets form. Droplets have to be very large to fall out of a cloud with rapidly rising air currents. This is why many tropical storms have torrential rain falling as large drops.

Figure 5.12 Cloud types

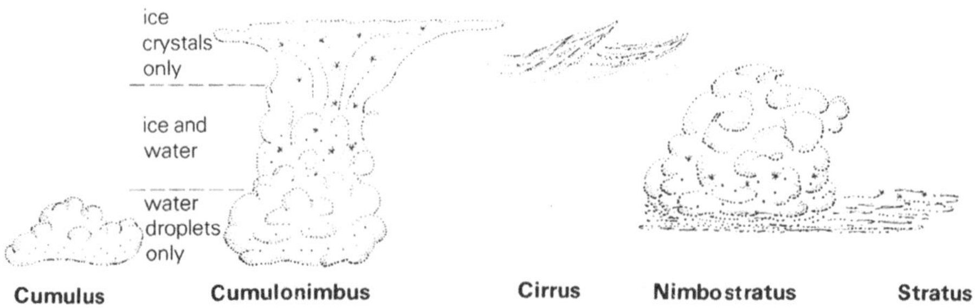

Cumulus   Cumulonimbus   Cirrus   Nimbostratus   Stratus

Top. Small **cumulus** clouds only contain water droplets and will not produce rain. *Centre left.* Deep **cumulonimbus** cloud has a mixing region of ice crystals and water droplets and will produce heavy rainshowers, thunder and lightning. *Centre right.* High-level **cirrus** cloud contains only widely scattered ice crystals and will not produce rain. *Foot.* This **nimbostratus** cloud is thick enough for ice crystals to form; it produces widespread rain, typical of a warm front; usually only the lower part of this cloud can be seen from the ground. **Stratus** cloud is thin and it will only produce drizzle.

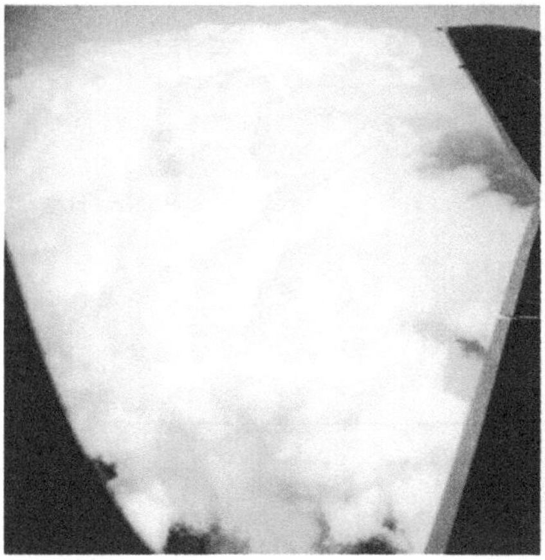

Up, up and away

When air moves towards a mountain range it has no choice but to go over it. This makes it cooler and brings it ever closer to saturation. At some height up the mountainside saturation is reached and then water vapour is turned into droplets and a cloud is born (Fig. 5.13). Cloud formed in this way is called **relief cloud**, and any rain that may fall from the cloud is called **relief rainfall.**

However, when the air has passed over the highest peak it starts to sink down the other side and, therefore, it becomes warmer. The warmer the air, the more moisture it can hold, and so droplets are no longer produced. We see this as the end of a cloud on the leeward (sheltered) side of a mountain. Sometimes the leeward end of a cloud bank is dramatically clear, like a wall. Also with much moisture lost to rain over the mountains there is little left to give any rainfall to lowlands in the lee of the mountains. Low rainfall regions leewards of a mountain barrier are said to be **rainshadow** regions. Where mountains are very high they can even cause the leeward side to become a desert.

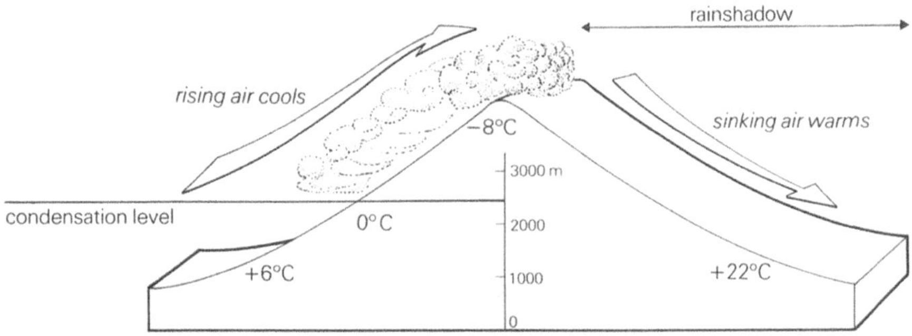

Figure 5.13 Relief cloud Relief cloud is produced when air is forced to rise over a mountain range. A rainshadow zone extends out on the lee side.

Exercise 5.4 Rainshadows

Figure 5.14 Cloud and rain over California

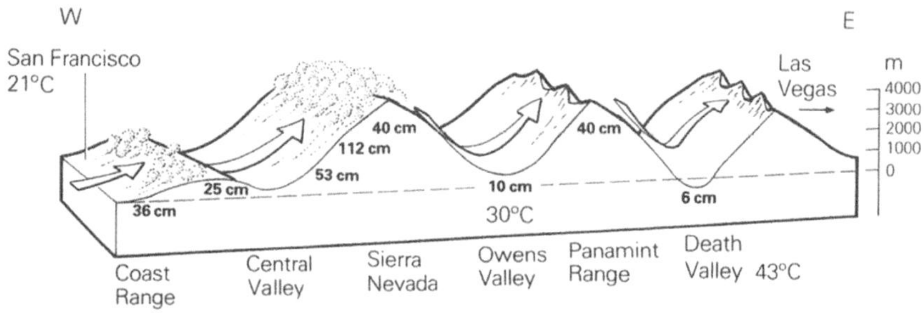

Death Valley lies to the east of several ranges of high mountains in California (USA). Moisture-laden winds blow from the west. Copy out and complete the following description in order to explain the low total rainfall in Death Valley, which is a desert. Use Figure 5.14 throughout.

As moisture-laden air blows in from the Pacific Ocean over California it is forced to rise up the side of the Coast Range. Here the air cools slightly and so some moisture is released as droplets to form clouds. As a result the western side of the Coast Range has 360 mm of rain a year. As the air sinks into the lee of the hills, the air warms and the Great Central Valley has only 250 mm of rain a year. The air is now forced further east and up the side of the 4000 m high Sierra Nevada mountains . . .

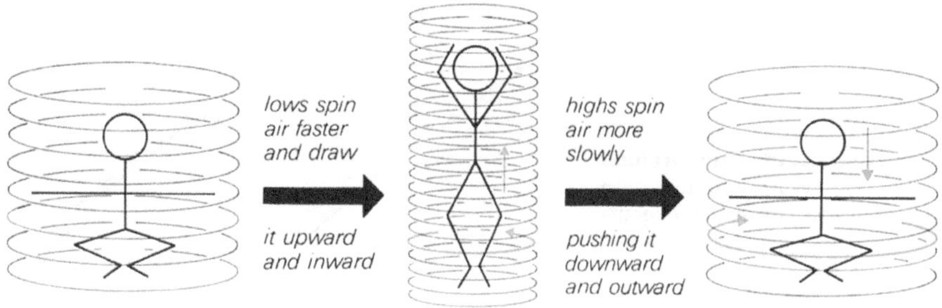

Figure 5.15 Highs and lows The spinning ice-skater illustrates some features of highs and lows.

Frontal clouds

Battle of the masses

Air is forced to rise, and so to produce cloud and rain, in many circumstances other than over mountain barriers. Most commonly, air rises within the spiralling air stream of the mid-latitude 'gears'. These spirals are very large – often well over 1000 km in diameter – and are really columns of revolving air stretching up from the ground surface to great heights (Fig. 5.15). The columns not only revolve in different directions (clockwise or anticlockwise) but can either 'tighten up' to stretch further upwards or 'open out' to become shorter so that the air in them moves more slowly. In many ways this can be likened to ballet dancers and ice skaters standing up and bringing their arms in to increase their spinning, or opening their arms to slow down. Columns of air with downward spirals push air downwards and warm it, which results in a high-pressure region on the ground with clear skies above; we call them **anticyclones**. But those with upward-spiralling air (called **depressions**) pull air up from the ground, creating a low-pressure region and at the same time cooling the upward-moving air to give cloud and rain. Depressions are the major sources of rain in the mid-latitudes.

Because these air columns move over mid-latitudes from west to east, countries such as Britain tend to have rainy spells when an upward-spiralling

Topic Cloud types and rain

A cloud is the visible form of millions of minute particles of water or snow in the free air.

Clouds are classified into three main groups (Fig. 5.12):

Cirrus clouds have a fibrous or feathery appearance and usually form as detached white bands. No rain falls from these clouds.

Cumulus clouds occur as detached clouds, being dense with sharp outlines. They develop vertically in the form of rising mounds, domes or towers, of which the bulging upper part often resembles a cauliflower. The sunlit parts of these clouds are mostly brilliant white; their bases are relatively dark grey and nearly horizontal. The rain that falls from well developed cumulus clouds is heavy and it occurs as showers.

Stratus cloud is generally grey cloud, lying in a widespread level sheet. It often gives prolonged rain or drizzle.

The word nimbus is added to cumulus and stratus clouds when they become thick, grey and capable of producing rain. **Cumulonimbus** clouds produce showers; **Nimbostratus** clouds produce prolonged rain.

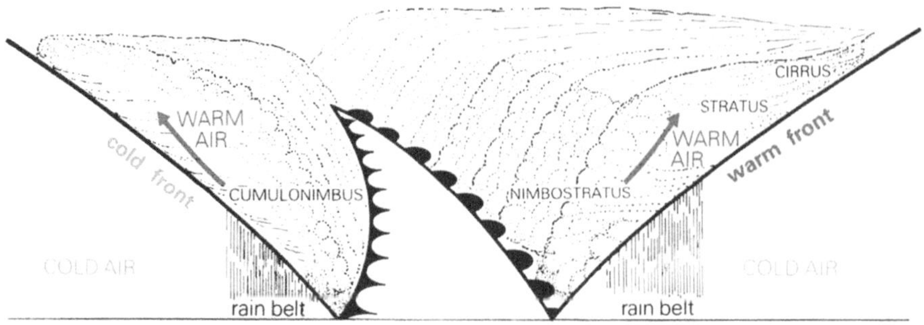

Figure 5.16 Frontal cloud Warm air drawn into a depression rises up over the cold. As it rises it cools and produces widespread cloud banks over the frontal contact region.

depression drifts over, separated by dry spells when a downward-spiralling high-pressure anticyclone brings clear skies. Actually depressions are more complex than simply columns of rain-laden cloud. They draw in air from all directions; they are, after all, the chief means of transferring energy from the Equator to the Poles. Air pulled in from the Tropics is warm and it contrasts with cold air brought down from polar regions. It takes time for these very different types of air to mix and so for a long time there are sharp changes across the boundaries where they meet. These boundaries are called **fronts.**

Because warm air drawn from the Tropics is less dense than the cold polar air that it meets within a depression, the warm air will rise faster than the cold as they are both pulled up into the tightening spiral of the depression (Fig. 5.16). As a result its place on the ground is taken by the cold air. Eventually the warm air ends up on top of the cold air, but until this lift is complete the warm air occupies a wedge-shaped area on the ground. The frontal regions show that the biggest lift occurs either side of the wedge of warm air and they are associated with thick cloud belts and rain. This rain is called **frontal** rain and it is mostly from cloud of a stratus type, occurring in belts perhaps 500 km long. Therefore, frontal rain affects large parts of countries such as Britain at the same time. Notice also that this type of rain is not dependent on relief, so it falls on lowlands and uplands alike.

Cumulus clouds

No man's land

We have been discussing the way in which moving columns of air are associated with special types of weather: highs with clear weather and lows with fronts, cloud and rain. However, between low pressure regions and anticyclones there are quite large areas which are neither one thing nor the other – a sort of 'no-man's land'. There are no fronts in these regions to pull warm air up to make clouds. Similarly the air is not being pushed down, keeping the skies clear. Without any special control from the 'gears', the weather is able to develop in a variety of ways – usually fairly unpredictable – and to give a slice more variety to our weather.

Very often this sort of air region has patchy cloud in it and sometimes this cloud can develop showers of rain, for this is where we most commonly find cumulus cloud, especially if the air is flowing from the north. Unlike stratus cloud, which results from a widespread organised uplift of air over mountains or at a front, cumulus cloud is the result of local ground and air conditions, each bubble of cloud developing spontaneously at some moment when conditions are

Topic Pressure and weather maps

Atmospheric pressure – the weight of air above a particular level – is measured with a **barometer.** Scientists have decided that the reference value by which to judge all pressure values should be called 1 bar. This is divided into a thousand parts, each called a **millibar (mb).** The millibar is the unit used to show pressure on the ground surface charts we call weather maps.

With this system, the lines of equal pressure (called **isobars**) are rather like contour lines on a map of the land surface. Therefore you can think of a high pressure region in the same way as a hill on a contour map – the highest pressure is in the centre and values get smaller outwards. Low pressure regions are the opposite – more like basins – with the lowest value in the centre.

Air flow from regions of high pressure to regions of low pressure, just like water, would run down a hill into a basin. The wind speed depends on the pressure gradient, so the closer the isobars, the stronger the wind. The flow of air is called a wind but the wind direction is not at right angles to the isobars as you might at first suppose. This is because the spinning Earth itself causes the air in both high and low pressure regions to spiral as well, which makes an air stream take a much longer 'corkscrew' route from high to low, rather than the steepest and shortest path. Any wind direction on a weather map is therefore at an angle to the isobars, inclined *inwards* to the centre of a low pressure region and pointing *outwards* from a high pressure region (Fig. 5.17).

Weather maps mainly show isobars so that you can work out the way airstreams are flowing. However, they also show temperatures, the type and amount of cloud, and the present weather. This information is conveniently grouped around the wind speed and direction arrow (Fig. 5.18). Very often curved lines are also drawn on these maps to separate regions of quite different types of air. These lines are called fronts and are marked with the special symbols shown on Figure 5.17.

A depression and part of an anticyclone are shown in Figure 5.19. Here the inward-spiralling air of the depression is picked out by the spiralling cloud pattern. Notice that the warm sector contains much less cloud than at the fronts. Most of the cloud is related to the occluded front. Ahead and behind the fronts are clearer skies with patterns of isolated clouds. These are cumulus shower clouds. Another depression can be seen in the top left corner of the photo. France and Spain are in the influence of a high pressure region. There is very little cloud, except from relief effects over the Pyrenees and the Massif Central.

Figure 5.17 Weather systems
(a) Depressions are characterised by inward moving air; the meeting of warm and cold air and the production of fronts, stratus type cloud and continuous rainfall with quite strong winds. (b) Anticyclones are characterised by outward moving air, clear skies and light winds.

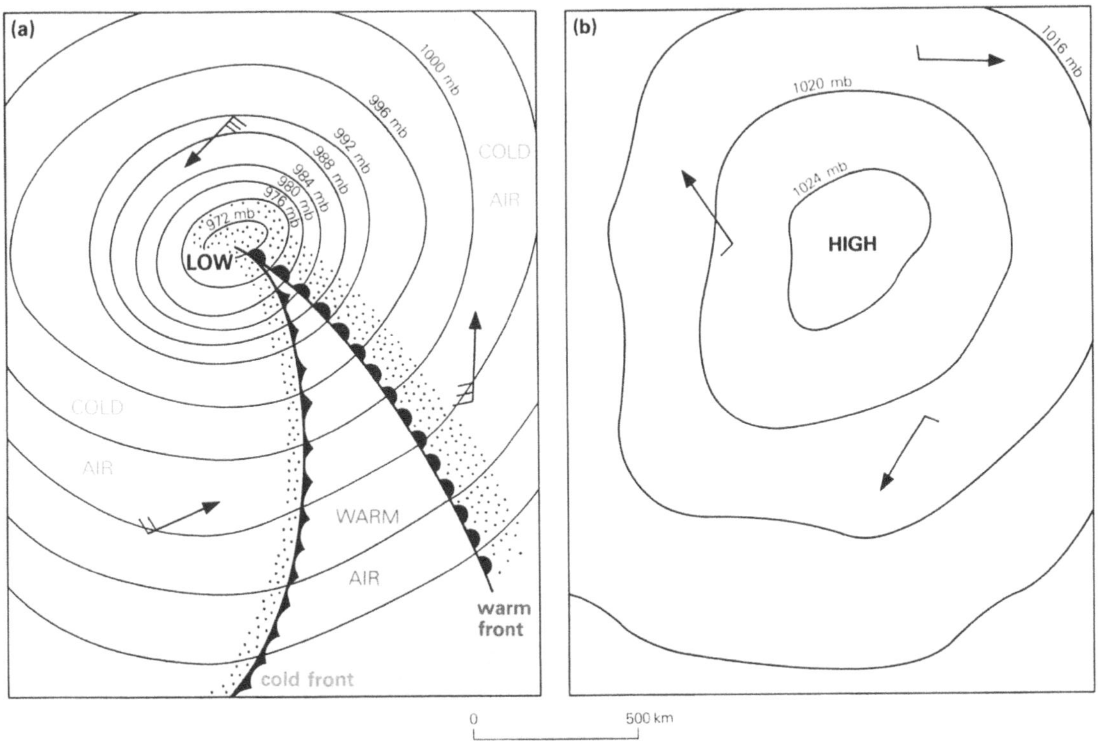

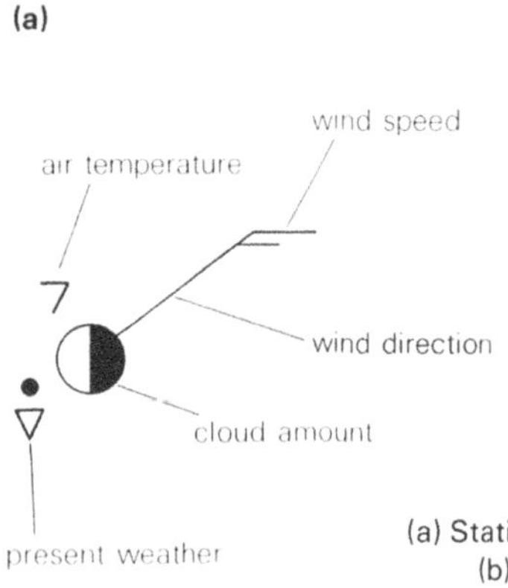

(b)

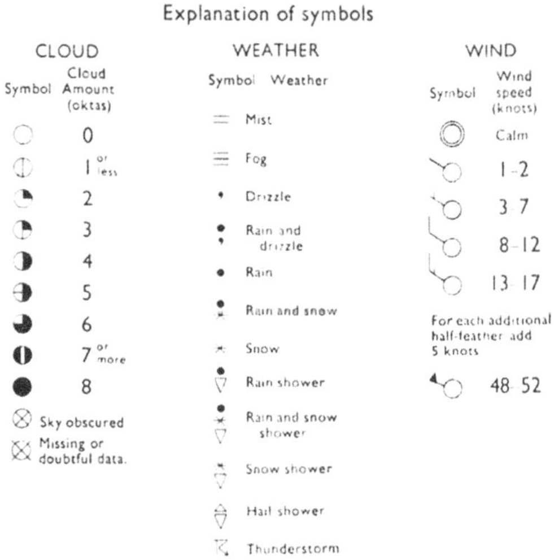

5.18 Weather symbols
(a) Station data for a weather chart.
(b) Definition of symbols used.

Figure 5.19 A depression seen from space

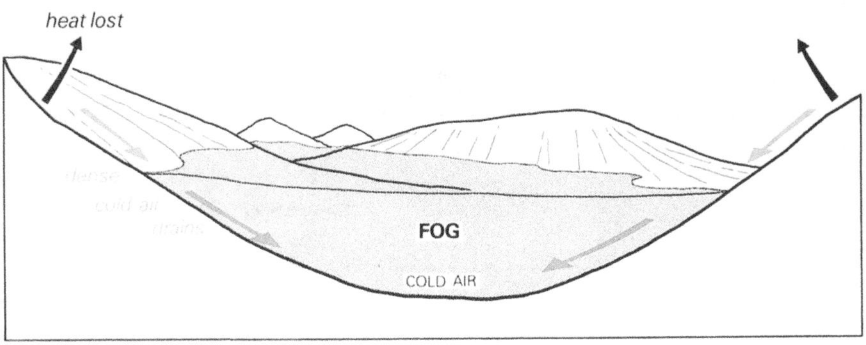

Figure 5.20
Radiation fog
This forms on a clear cloudless night in valleys. Hillsides lose heat by radiation; air in contact cools and sinks into the valley. When temperatures fall far enough for condensation to occur in the air, fog forms. Radiation fog is common under anticyclonic conditions.

Topic Fog

Fog is a cloud on the ground, made from tiny droplets of water. How far you can see in fog depends on the concentration of these droplets: the more droplets, the denser the fog. This in turn depends on the amount of moisture originally held in the air as vapour. A fog occurs if visibility is less than 1 km.

Because air over water holds the most moisture, the denser fogs are usually those at sea (advection fogs, see below). However they can be dense inland (radiation fog) especially if added to by smoke and other man-made pollution.

Radiation fog (Fig. 5.20). December 5th 1952 was a very calm cold day over London and the Thames Valley was full of cold air. As a result a **radiation fog** developed, but because there was no wind and the Sun was too weak to warm the air, there was nothing to disperse the fog. It brushed against pavements and buildings and made them damp while all the time factories, houses and vehicles poured tonnes of smoke and fumes into the air – air which was still. By the next day the pollution accumulating in the cold fog had turned it from grey to brown, and by the next day the visibility was cut to centimetres.

Eventually a wind did build up and disperse the pollution-laden fog (called **smog**) but the damage had already been done and at least 4000 people died from chest and breathing illnesses. Here is an example of a natural phenomenon whose effects had been turned into a disaster by man's activities. It was this event that led directly to the creation of clean air zones in our cities where only smokeless fuels may be burned. In Britain fog still occurs, but its effects are no longer so damaging to health.

Advection fog (Figs 5.21a and b). There is another type of fog called **advection fog** which plagues coastal regions such as California and Newfoundland. It even affects many British coasts. In these cases warm moist air is pulled across cold sea, lowering the temperature of the air until droplets of water are released to produce fog. If the air blows from the sea onto the land, inland the Sun will warm the air again, dispersing the fog. If warm air moves from land to sea, the fog bank forms some miles off the coast and is not so readily dispersed.

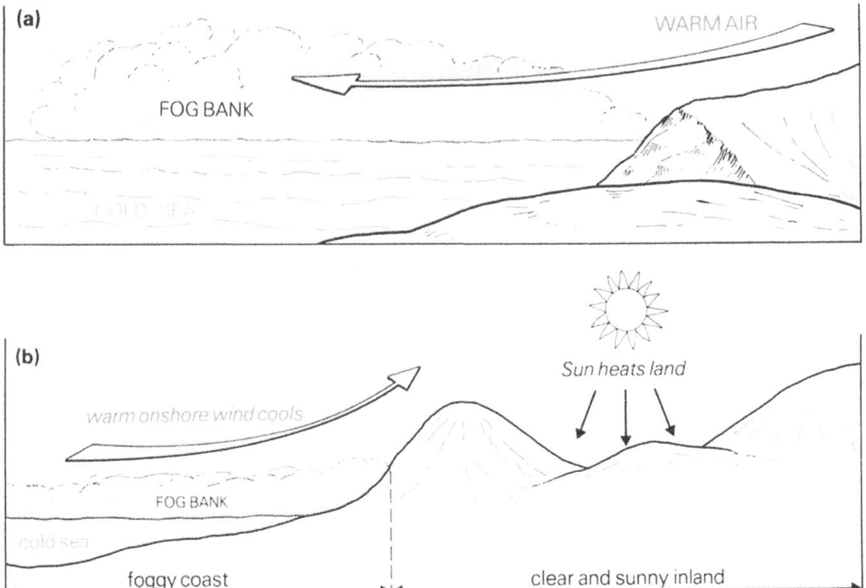

Figure 5.21
Advection fog
This forms (a) over seas or (b) along coasts when warm moist air travels across a cold surface. This type of fog is not easily dispersed by winds because they just bring more warm air to be cooled by the huge store of cold water. The famous fog banks off the Labrador coast of America can last for months.

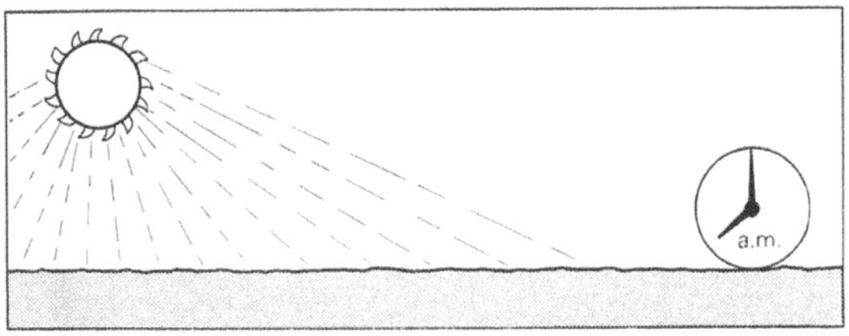

(1) Clear sky; Sun begins to heat ground. Some parts become hotter than others.

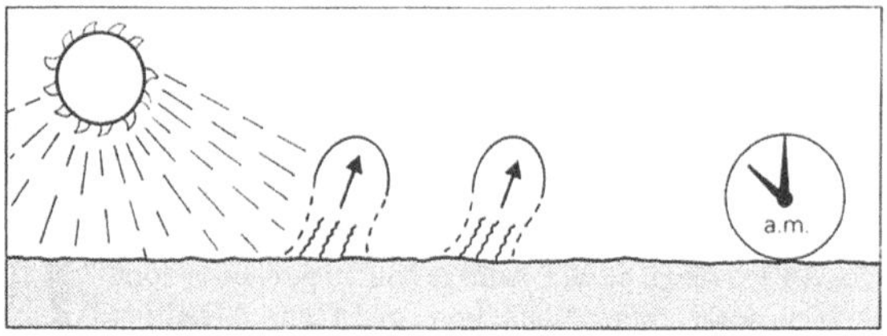

(2) Bubbles of air in contact with hotter ground start to rise.

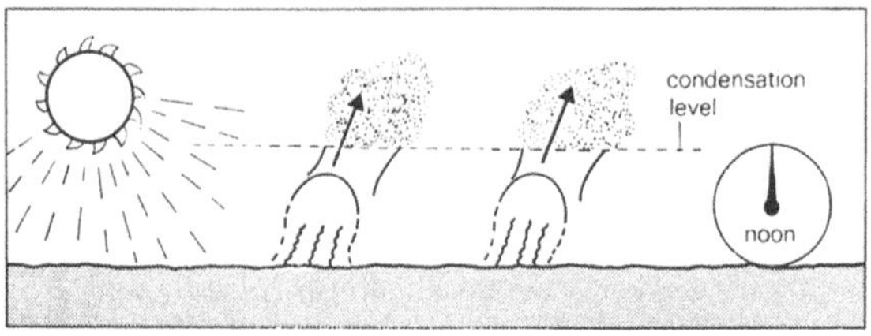

(3) Rising air bubbles reach the condensation level and cloud forms. Bubbles continue to rise, making clouds deeper; new bubbles become heated.

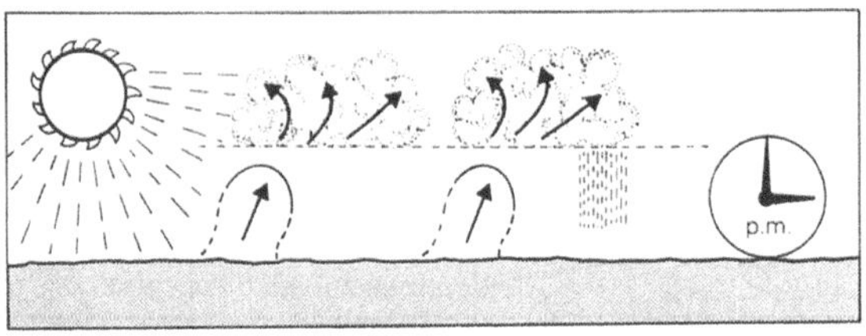

(4) More and more clouds form and begin to obscure the Sun. Some are deep enough to give showers and perhaps thunder.

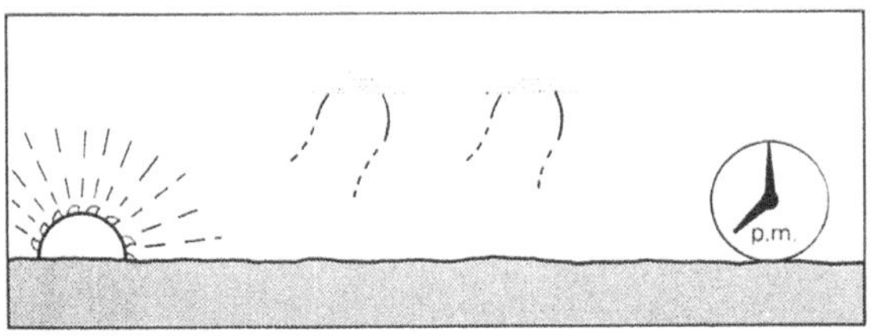

(5) Sun sets, ground cools and the last of the heated bubbles 'melts' away into the surrounding air.

Figure 5.22 The formation of cumulus clouds

right (Fig. 5.22). The 'right' conditions are numerous, but we can consider one of the causes of cumulus cloud that occurs on some fine and sunny summer mornings when the air is crisp and the visibility 'too good to be true'. On such mornings the first cumulus clouds usually appear by about 10 am and gradually enlarge through the morning until by mid-afternoon the sky can be virtually obscured. As late afternoon approaches, the possibility of showers or even thunderstorms looms large, but with the onset of evening the clouds seem to evaporate, get smaller and eventually fade away to give a clear sunset. Many of these clouds are in some way associated with the daily surface heating effects of the sun because they get bigger as the Sun rises in the sky and disappear as the Sun sets and the heating is lost. Here, the convection system is at work again, this time causing only local bubbles of air to rise as they are warmed. Cumulus cloud produces showers. The rainfall is, not surprisingly, called **convectional rainfall.** It is this type of rain, together with the effects of frontal rain and relief, that determines the total rainfall at any place in the mid-latitudes. However, the really important regions for convectional rainfall are nearer the Equator where frontal effects are missing and surface heating more intense.

Topic Thunder and lightning

Motor cars need to generate over 8000 volts just to create a spark a few tenths of a millimetre long and explode petrol in a cylinder. Just imagine how much energy and how many million volts must be involved when a flash of lightning sends a spark thousands of metres through the air and causes great reverberating rumbles of thunder (Fig. 5.23).

When air carries a heavy current (lightning), it gets hot just like an electric wire. The air very near a lightning flash heats to about 30 000°C and, as a result, expands as a shock wave. When the transmitted energy of this shock wave reaches our ears, we hear it as a clap of thunder.

To produce lightning a negative charge of about 300 million volts must build up in the cloud. We think this begins in the lowest levels of deep cumulus clouds where snowflakes form. The Earth becomes positively charged below a thunder cloud, so eventually a spark jumps the gap between cloud and Earth in a series of jerky steps. In fact a lightning flash is many of these jerky flashes separated by only milliseconds of time, so we think we see only one spark, but each jerk and flash generates a shock wave and we hear these as a confused 'roll' of thunder. Also because sound waves travel slowly and the flash might be 5 km long, sound from the heated air near the cloud will take longer to reach us than the shock wave from the same heated air near the ground. This, too, makes the thunder last longer.

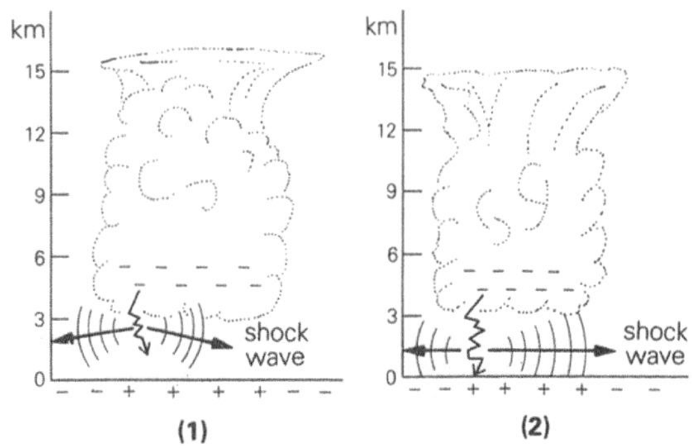

Figure 5.23 How lightning and thunder are produced

(1) A spark jerks down toward the ground, heating air and sending out shock waves as it descends. (2) The spark reaches the ground and the flash is completed. Heated air near the ground sends out further shock waves. These are heard as thunder.

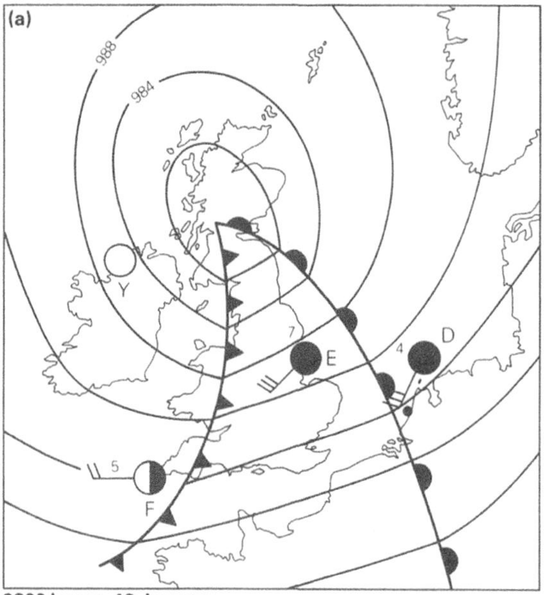

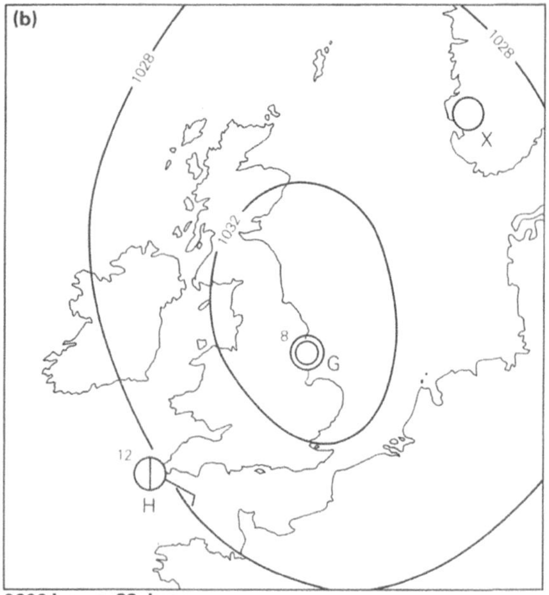

Figure 5.24 Weather charts for January and June

Exercise 5.5 Weather charts

Study the two weather charts A and B, Figure 5.24. Make a careful note of the time of year when each chart was made and think about the consequences of this before you answer the questions below.

1. Name the type of weather system shown on charts (a) and (b).
2. Explain what the line with the half circles along it is meant to represent.
3. What is the difference between this line and the line with triangles?
4. Write down the wind direction you would expect at stations X and Y and state whether you would expect light or strong winds in each case.
5. Explain why a symbol for rain is shown against the station D.
6. Give reasons to explain the weather reported at station G.
7. Explain why there is a difference in temperature between stations D, E and F.
8. Describe the weather you would expect to experience at station X.

SUMMARY: TYPES OF WEATHER IN MID-LATITUDES

Depressions (low-pressure regions) have upward-moving air which spirals anticlockwise and inwards in the northern hemisphere. They drag air in from all directions, but the different types of air take a long time to mix. Because air from the Equator is warmer and less dense, it rises over cold air from the Pole along inclined planes called **fronts.** Warm air at fronts rises and cools, producing great sheets of stratus or nimbostratus cloud and rain.

The leading edge of the warm air wedge is called a **warm front**, the trailing edge is called a **cold front**. Between them lies the **warm sector**. When the warm sector rises completely off the ground an **occluded front** forms, accompanied by a wide band of cloud.

The sequence of weather associated with a depression moving past an observer on the ground (Fig. 5.16) is:

(1) Cold air with thin high level cloud approaching.
(2) Cloud becomes lower and thicker and is accompanied by a belt of rain.
(3) The warm front passes, wind changes direction and comes from a more tropical direction and the temperature rises. Cloud becomes more broken and sunny periods occur.
(4) The cold front arrives abruptly, the air turns colder and the wind alters to a more poleward direction once more. This is accompanied by a further belt of rain.
(5) Rain ceases and cloud becomes higher and thinner with only scattered showers.

Anticyclones are quiet interludes between the hectic and often stormy events of depressions. They have downward-moving air which, in the northern hemisphere, spirals clockwise and outwards from the centre. As the air moves downwards it warms and all its moisture can be held as vapour. There is no chance of rain. Skies are usually clear, and winds are light because the subsiding air spreads out slowly. With clear skies summer anticyclone weather is hot and sunny, but in winter, when the angle of the Sun is low and daylight much shorter, less heat reaches the ground than is radiated to space, so anticyclones bring cold, frosty weather. Indeed the lowest air might cool so much overnight that **radiation fog** forms.

Topic Recipe for a hurricane

Tropical storms – hurricanes, typhoons and cyclones – are features of the climate of the eastern shores of America, Asia, and Australia, respectively. They form as small low pressure 'cells' that bud off the equatorial low pressure belt.

To make a hurricane, mix one large ocean having a surface temperature above 27°C with one fairly small tropical low pressure system. Allow to simmer gently over the ocean for about four days gathering energy and moisture and sending towering clouds into the sky (Fig. 5.25). Stir vigorously using the rotation of the Earth.

Now comes the tricky bit. Examine the developing tropical storm closely every 24 hours to see if the clouds are still forming in strength. Many tropical storms just fade out at this stage. When enough energy has been accumulated for winds to reach more than 110 km per hour, move it gently across the ocean and allow it to be directed by the upper air flow towards an inhabited coast (Fig. 5.26). Make sure there is a good newspaper reporter on hand:

> 'The storm has now reached the island of St Lucia in the Caribbean Sea. Called Hurricane Allen its winds have gathered strength and at over 240 km per hour they are tearing up houses from their foundations and ripping the banana harvest from the trees. The fifth of August 1980 is going to be remembered throughout the long months of restoration that the island faces. Now the people of Jamaica are hoping that it will veer away from them and strike out into the open ocean again . . .'

To destroy a hurricane, first remove the source of energy that drives it by taking it away from the ocean and over the land. Allow to cool. With luck it will be a mere shadow of its former self within a couple of days.

NASA

Figure 5.25 Hurricane Gladys over the Caribbean

Figure 5.26 The development of a hurricane

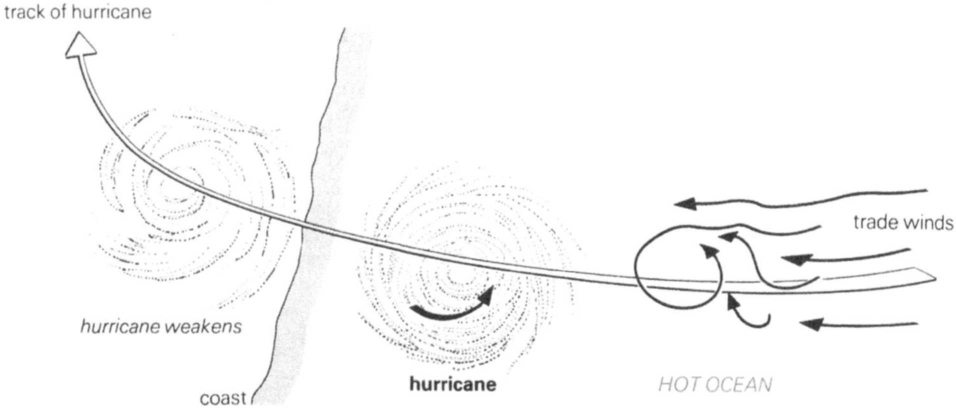

SUMMARY

Exercise 5.6 Routes over the Atlantic

Figure 5.27 Aeroplane routes over the Atlantic

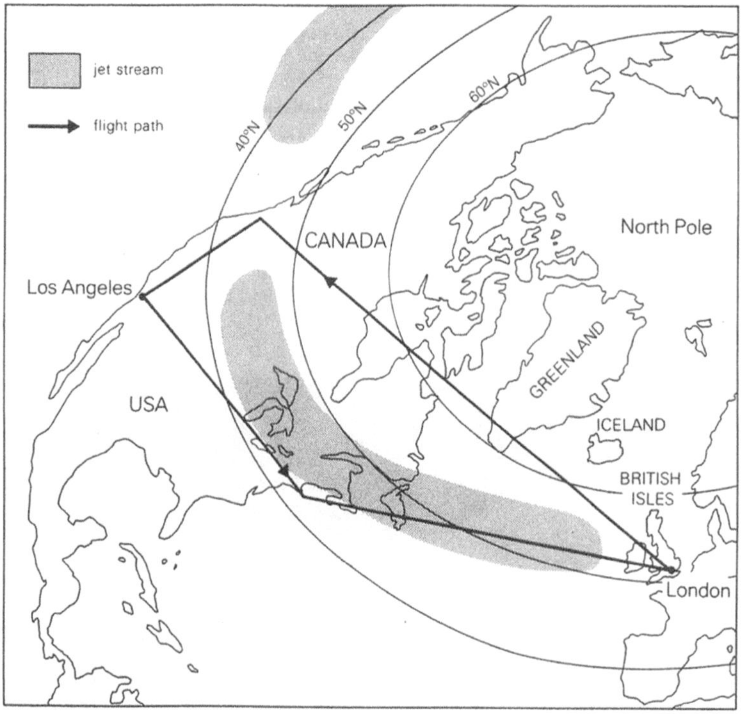

One of today's major forms of travel is by aeroplane. Millions of people fly from country to country for trade, a holiday or to visit friends. One of the busiest routes is over the Atlantic Ocean between Europe and North America. With 10 000 litres of fuel used per hour by the larger planes the most economical flight path must be chosen. This may not be the shortest distance however because of a variety of weather conditions, as we shall see. For example it is important to fly through 'thin' air where frictional resistance to movement is less, but at the same time it is vital to find a route either in the same direction as the global winds or at least a route with low head winds.

1 To obtain 'thin' air would you fly at a high or low altitude (p. 57)?
2 In which direction do the main upper air winds blow over the North Atlantic (p. 56)?
3 Are there any special regions of high winds which could be a help (p. 56)?
4 If you were choosing a west-to-east route across the Atlantic from Los Angeles to take advantage of winds, where would you aim to fly?
5 If you were travelling from east to west, what wind would you aim to avoid?
6 Figure 5.27 shows the main type of air routes followed over the Atlantic to Los Angeles. Explain why different paths are chosen for the east and west directions.
7 Why are these routes changed daily after the air flight controllers have received the daily weather forecasts?
8 In what way might these differences in wind be responsible for a flight from Los Angeles on the west coast of the United States to London taking 10 hours, but a flight in the opposite direction taking 11 hours?

Exercise 5.7
Geography behind the news

We can often tell a lot more from a newspaper item than is written down. The clip here is from the *Daily Telegraph*.

1 What kind of weather pattern must have prevailed in the North Sea – a depression with low pressure or an anticyclone with high pressure (p. 63)?
2 Given such strong winds, were the isobars close together or far apart?
3 What is the relationship between the wind speed and the height of the waves?
4 What time of year was the storm? Explain the reasons for your answer.
5 From the clip we can also tell whether Denmark has flat open country or hilly land. What clues are given?

2 HURT ON STORM-HIT FERRY

By ALAN COPPS

TWO British women were recovering in a Danish hospital last night after being taken by helicopter from a car ferry riding out a storm in the North Sea.

The storm driven by force 10-11 gales also lashed the coast of south and east England causing thousands of pounds worth of damage, mostly at Margate, Ramsgate, and Southwold.

Both women were said to have fallen heavily when a giant wave violently shook the Dana Regina, 12,000 tons, just before it hove to about 50 miles west of the Danish port of Esbjerg.

The ship on a regular sailing from Harwich to Esbjerg was carrying 785 passengers, about 500 of them British.

A sister ship the Winston Churchill, 8,700 tons, on its way from Newcastle to Esbjerg with about 400 passengers was also forced to ride out the storm.

50ft waves

Last night the Dana Regina was starting cautiously under way again and was expected to arrive at the Danish port early today, nearly two days late. The Winston Churchill was expected to arrive shortly afterwards.

Mr Jens Mogensen, captain of the Dana Regina, said waves were still 45-50ft high. Some of the cars on board had been damaged, but the ship herself was unharmed.

In Denmark ,the hurricane force winds whipped up thousands of tons of newly-sown soil, costing farmers millions of pounds. The dust cloud, at points reaching 7,000ft, turned the sky brown and yellow and was said by meterologists to be the worst for 50 years.

6 The world's weather

Variations in weather

A global view

We have looked at key parts of the global weather machine and how it gives particularly complex weather patterns at latitudes similar to that of Britain. Now we shall consider how the global pattern affects other latitudes (Fig. 6.1).

In Chapter 5 we thought in terms of a simple pattern of air rising over the equator. However, heating is not always greatest at the Equator: the zone of maximum heating moves as the overhead noon Sun changes in latitude during the year. So the basic global circulation must include seasonal variations.

Seasonal changes in mid-latitude countries such as Britain involve slow temperature changes between summer and winter. There are no significant wet or dry seasons because depressions track across the area throughout the year. This type of 'subtle' change is quite different from areas nearer the Equator. For example, the reason people go on holiday to the Mediterranean coast resorts is because they know that summers are invariably hot, sunny and dry. Sandy beaches can be found around Britain, but a 'guaranteed' dry season is what sends people on package tours to the Spanish coast. In these regions we find a fundamental difference betweem summer and winter, not only in temperature but also in rainfall.

However, it is one thing to have a dry season that brings in tourists and quite another to have to cope with a dry season if you rely on agriculture for survival.

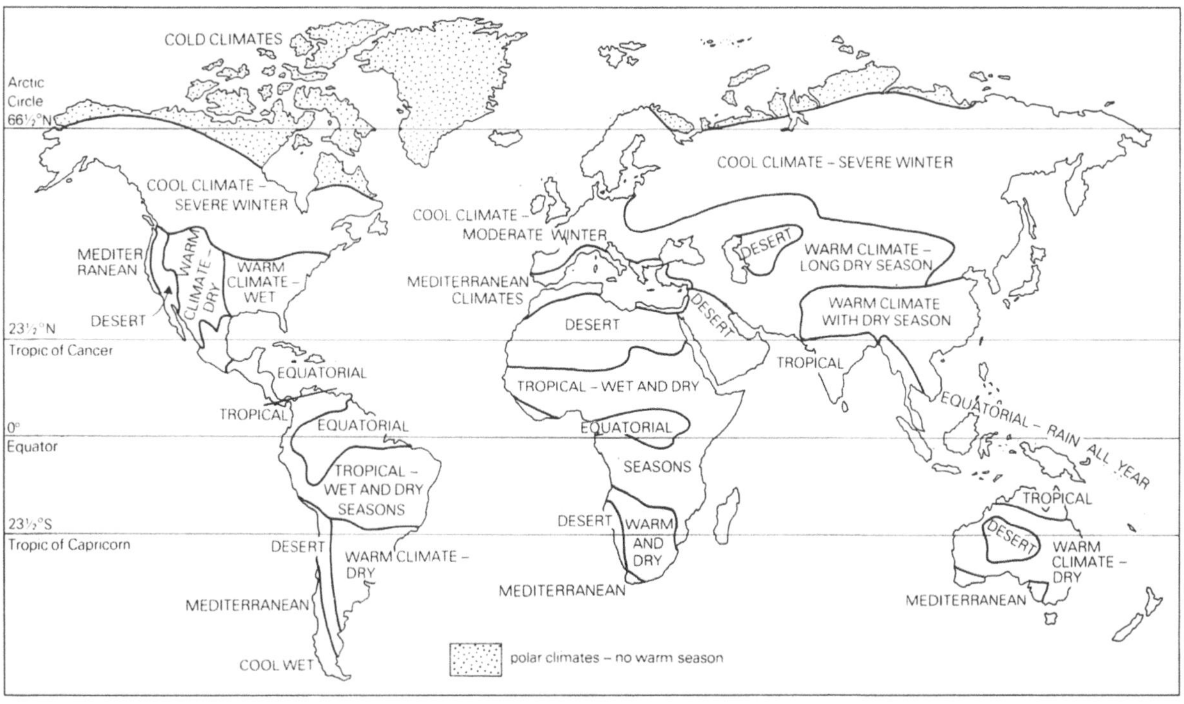

Figure 6.1 The world's major climatic regions

Topic The seasons

The Earth spins on an axis which is tilted at 23½° to the plane of its orbit around the Sun. The Sun's radiation reaches the Earth in parallel rays, making the lighted half of the earth *day* and the shaded half *night*. Because of the tilt, different places have different lengths of day and night as the Earth completes an orbit around the Sun each year. At the polar regions (beyond 66½°N or S), they receive 24 hours of daylight in summer and 24 hours of darkness in winter (Fig. 6.2). The time during each 24 hour period in daylight, and the angle of the Sun's rays to the surface, determine the amount of solar radiation received to heat up the surface at any place.

Seasons are caused by the differences in day length and the angle to the overhead Sun, and not by the different distances between Earth and Sun throughout the year. Notice that the tilt of the Earth causes the northern hemisphere to receive more overhead Sun in June (summer) whilst the reverse is true for December (winter).

We call the longest and shortest days of the year the **solstices;** they occur on about 22 June and 22 December. Days of equal periods of light and darkness are the **equinoxes;** they occur on 22 March and 22 September. All areas polewards of latitude 66½° receive 24 hours daylight or 24 hours darkness during the solstice; everywhere on the Earth receives 12 hours day and 12 hours night during the equinoxes.

Figure 6.2 The seasons

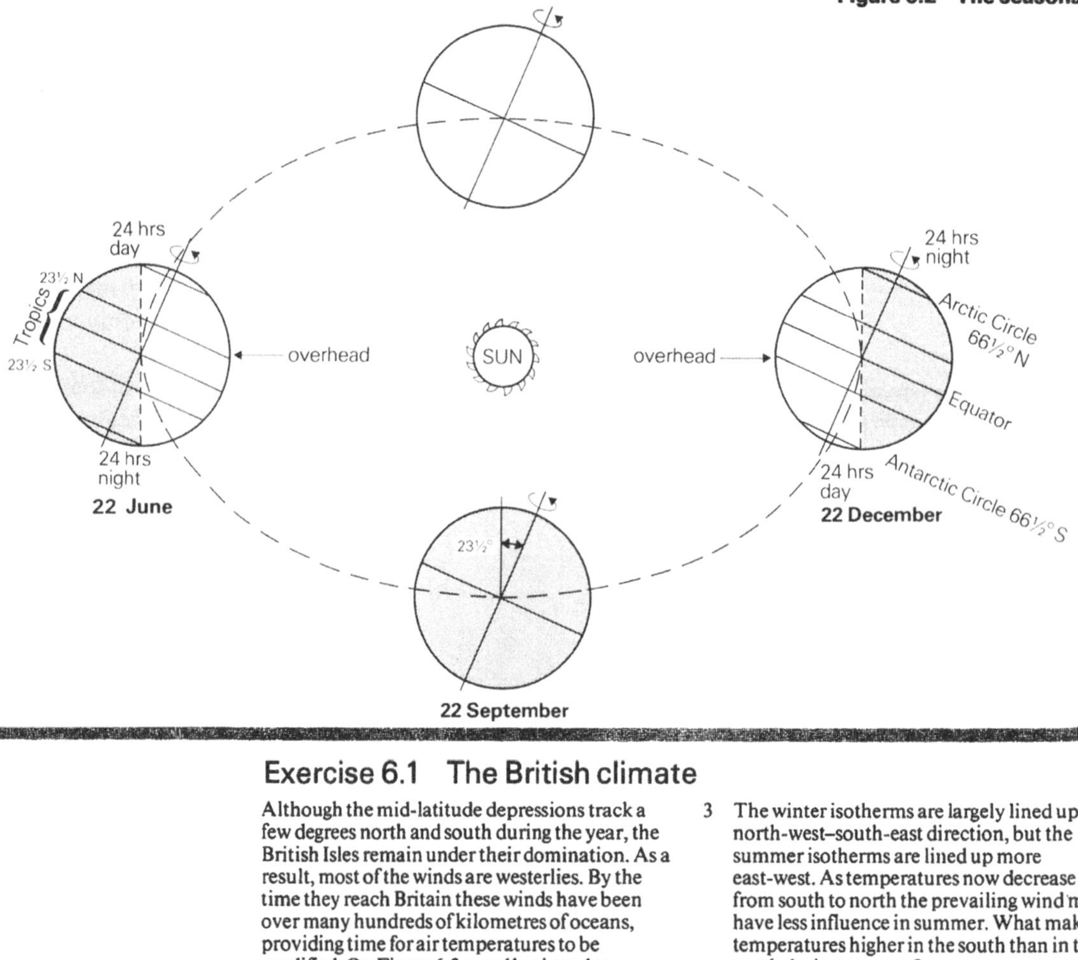

Exercise 6.1 The British climate

Although the mid-latitude depressions track a few degrees north and south during the year, the British Isles remain under their domination. As a result, most of the winds are westerlies. By the time they reach Britain these winds have been over many hundreds of kilometres of oceans, providing time for air temperatures to be modified. On Figure 6.3 a and b, air and sea temperatures are shown for winter and summer conditions.

1 Study the pattern of isotherms (lines of equal temperature). In winter is the sea warmer or colder than the land? Is this still true in summer?
2 In winter, what is the relationship between sea temperature, prevailing wind direction and land temperature?
3 The winter isotherms are largely lined up in a north-west–south-east direction, but the summer isotherms are lined up more east-west. As temperatures now decrease from south to north the prevailing wind must have less influence in summer. What makes temperatures higher in the south than in the north during summer?
4 What causes the high rainfall values shown on Figure 6.2c?
5 What term would you use to explain the low rainfall of the English Midlands?
6 Explain how the differences in climate may affect farming in the British Isles.

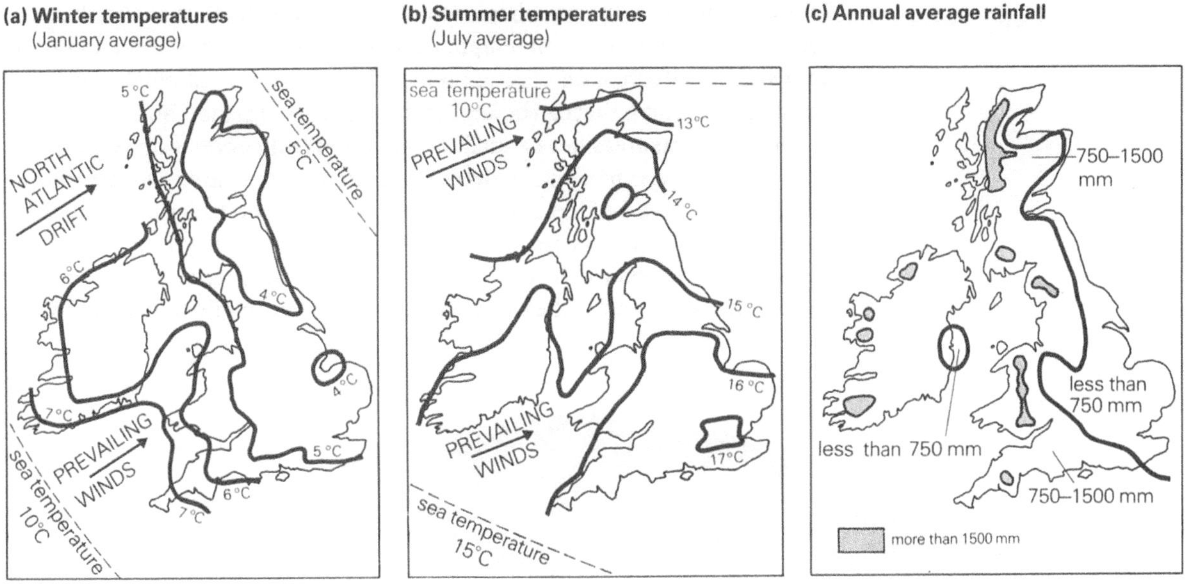

Figure 6.3 **The climate of Britain**
(a) Average temperatures for January. (b) Average temperatures for July. (c) Average annual rainfall in Britain.

Exercise 6.2 Temperate continental and maritime climates

We normally describe regions often influenced by onshore ocean winds as **maritime.** This is because the ocean has moderated the air temperature, making it either slightly cooler in summer or warmer in winter. As a result, **maritime climates** do not experience as large a range of annual temperature as do continental interiors, which are far from oceanic influences. Similarly rainfall is usually plentiful and evenly distributed throughout the year. **Continental climates** occur when the moderating oceanic effects are lost.

1 Study the climatic data for Valentia in Ireland, Cambridge in England, and Hanover in Germany (Fig. 6.4).
2 Using your atlas make a rough sketch map of Western Europe to show the three places. Mark them on your map with their initial letters.
3 Now write the total rainfall value beside each station. Draw arrows between the stations from the largest to the smallest. Can you suggest an explanation for the observed variation?
4 Measure the annual range of temperature (maximum *minus* minimum) for each station. Record these values on the map and draw arrows as before. Why does the range increase towards the interior of the continent?
5 State the main differences in rainfall distribution between the three stations.
6 Briefly summarise what you think are the main characteristics of (a) a maritime climate and (b) a continental climate.

Figure 6.4 Climatic data Valentia, Cambridge and Hanover

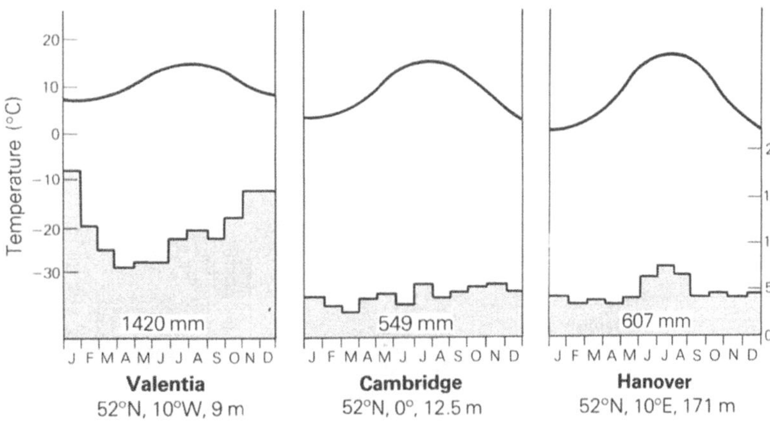

In some tropical countries (those within about 20° of the Equator) a dry season can be entirely rainless with temperatures continuously above 30°C. This has a very significant effect on vegetation and the growth rate of plants, and on the numbers of animals that can be kept or the types of crops that can be grown.

Some tropical areas have a gradual transition from dry season to wet season; others suddenly experience a change overnight. These areas have a **monsoon** climate. At the end of the dry season the ground is parched and dry, with soils cracked open, and grass brown and dormant. Then one day rain falls so fast it cannot be absorbed by the soil and it runs in brown streams across fields and roads. This is the start of the wet monsoon season, when rain falls virtually every day.

Winter

So in Britain the continual variability introduced by the swirling 'gears' reduces seasonal rainfall variation to a minimum, in contrast with areas nearer the Equator where rainfall seasons are much more marked. It is obviously important to understand why this happens.

Seasonal changes in climate

Follow the Sun

Look at Figure 6.5 which shows two satellite photographs of the global circulation. Where air rises, cloud and rain are found; where air sinks, clear skies and drought prevail.

For a place X at about 10°N the change is dramatic throughout the year. In December X is under a sinking air flow and drought prevails, but in June it is

Figure 6.5 The apparent movement of the overhead sun (Photographs: European Space Agency)

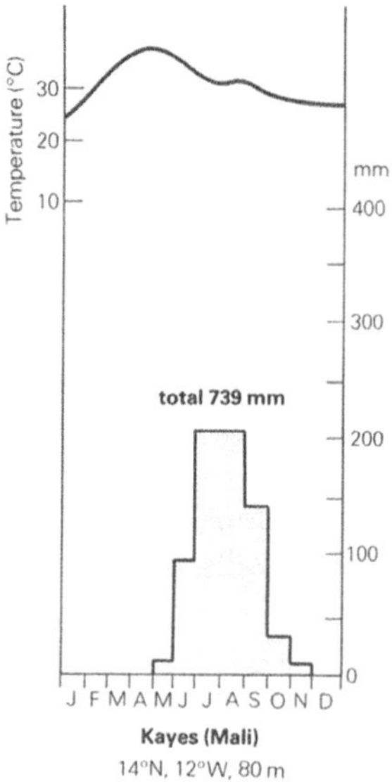

Figure 6.6 Climatic data for Kayes (Mali)

under a rising air flow and it receives rain. As a result station X has a wet season and a dry season. It is typical of much of the tropical world (Fig. 6.6).

Station Z, at about 25°N, never emerges from within a region of sinking air. This gives it a **desert climate** (Fig. 6.10). Station Y, at the Equator, is under rising air throughout the year, so rain falls in every month. This is characteristic of an **equatorial climate** (Fig. 6.7) and it provides the moisture for the rainforests of South-East Asia, South America and Central Africa.

The variation in pressure systems from December to June is clearly connected with the movement of the overhead noon Sun. On 22 June (summer solstice) the Sun is overhead at midday at 23½°N of the Equator, bringing the zone of greatest heating with it. The coldest regions are polewards of about 70°N. As a result, all the patterns of air flow in the northern hemisphere are shifted north and crushed up between 20°N and 70°N. It is this movement that provides the rainy season to the northern hemisphere Tropics.

Because the cold polar region is of a fairly constant size, the mid-latitude depression belt is not able to move more than a few degrees northwards. This is the reason we do not see very much change in the British weather pattern, which remains dominated by weather systems from the Atlantic

On 22 December (winter solstice) the Sun is overhead at midday 23½°S of the Equator, taking the zone of greatest heating with it. As a result the pressure belts in the northern hemisphere expand southwards, with the systems near the overhead Sun moving the greatest amount. Thus, in the Tropics, what were rainy areas in June come under the influence of sinking air, and the dry season spreads to the northern Tropics. As a result the wet season begins in the southern Tropics.

Of course it is not surprising that all of the world's climates are linked together in this way. Basically climates exist in latitudinal belts or zones around the Earth. This is clearer in the southern hemisphere (which is mostly water) than in the northern hemisphere where large landmasses make the pattern more confused.

The influence of land masses

Hot stuff!

Figure 6.7 Climatic data for Singapore

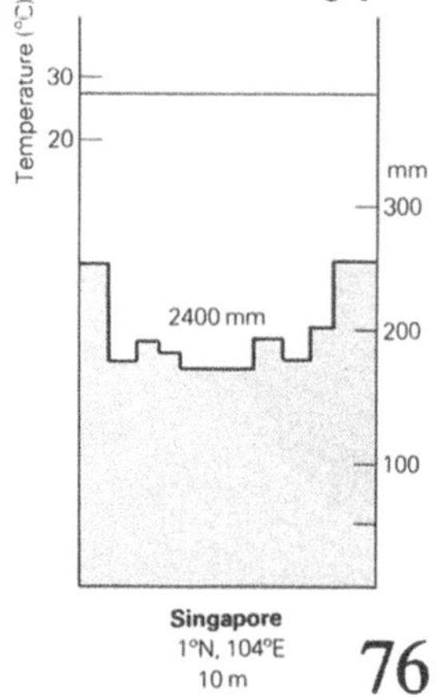

The main difference between oceanic and continental heating has already been discussed. Water absorbs and stores heat in its surface layers. As a result water temperatures do not vary much all year in the oceans. By contrast, the land conducts heat very poorly; its surface heats up considerably in summer and cools down dramatically in winter. Heated surfaces well away from the sea cause air in contact with them to warm and rise, thus creating low pressure regions; the reverse is true for cooled surfaces, which create high pressure regions. Indeed the major landmasses are so big that the effects of their heating and cooling are very influential and often outweigh the effects of the main global circulation patterns. Only the subtropical high pressure belt is strong enough to overcome the effects of continental heating. As a result deserts are found even though surface temperatures are very high.

Asia, the largest landmass, is the most important land area distorting the global picture in this way, producing an intense low pressure system in summer and a high pressure system in winter. The effect of the summer low is to draw air in over the continent from nearby oceans. Some of this air has enough moisture to bring considerable quantities of rain with it, often in the form of heavy and prolonged downpours. By contrast, in winter pressure becomes high with subsiding and outflowing air as the continent cools down quickly. As a result, most of the moisture-laden air is blocked out, and a dry season prevails.

Upper-air winds

It's an ill wind . . .

It is not possible to separate the patterns of weather on the ground from the movements of air high in the sky. Nowhere is this more true than in monsoon lands.

Monsoons are characteristic of some areas with distinct wet and dry seasons (Fig. 6.8). In such areas the wet season arrives very suddenly. For example, the leading edge of the summer monsoon in India usually reaches the southern tip of the country on about May 15th, after a couple of months without any rain at all. It is so regular and it starts so suddenly that people refer to the monsoon as being early or late if it is only a few days different from normal. This suddenness is surely not explained just by gradual seasonal shifts in the global circulations; some other force must be involved.

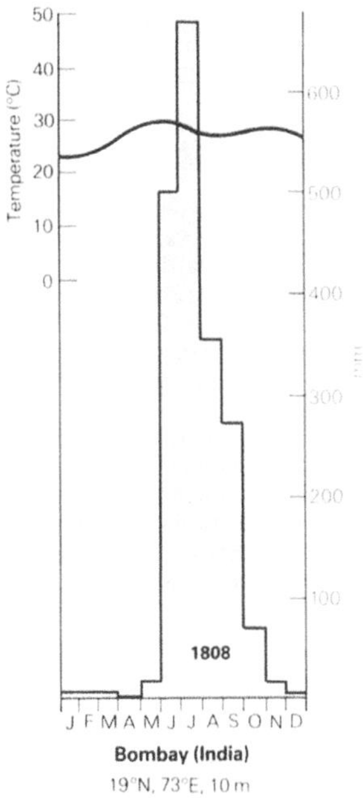

Figure 6.8 Climatic data for Bombay (India)

In India, for example, the effect of a jet stream on the monsoon is very important. In December the northern hemisphere monsoon jet stream flows at its nearest to the Equator (Fig. 6.9). This brings it over northern India, south of the Himalayas. At this time of year a high pressure develops and the dry season prevails. In the spring the air patterns over the Earth all start to move polewards, as described earlier. But over India this movement is hampered by the Himalayas and the jet stream appears to get stuck to the south of the mountains. As the summer approaches, this jet stream is farther and farther out of place until it finally 'flips' and suddenly moves up to the north of the mountain range. Although we are not sure precisely how this happens, the result is certainly dramatic. One day there is a jet stream containing lots of air; the next day it is

Figure 6.9 The causes of the monsoon
The rainy season is due to summer heating over India causing a low pressure region and pulling moist air in from the ocean. The abrupt start to the monsoon is caused by an upper air movement of a jet stream.

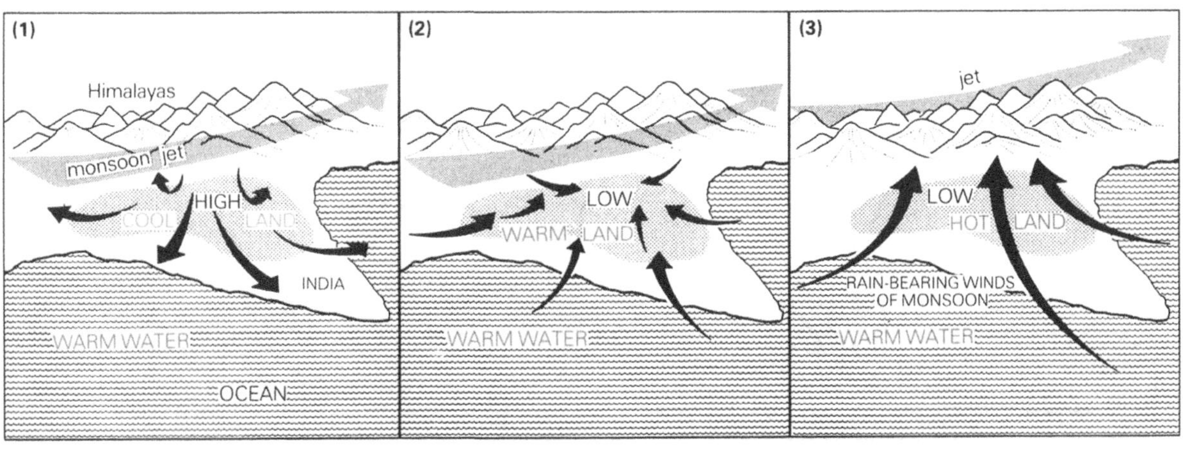

gone and there is a great lack of air over Northern India. Moist (rain-producing) air now rushes in from the Indian Ocean to fill up the space left by the jet and in doing so starts the monsoon. Once started the normal summer heat keeps air rising over the land and drags more air and rain in to prolong the rainy season.

Polar climates

Real cool!

The complex tropical and sub-tropical lands lie to one side of the mid-latitude gears and occupy more than half the Earth's surface. The cold lands lie to the other side, but their extent is very small, being less than 10% of the Earth's surface. Here there are fewer than four months with temperatures above 10°C

Figure 6.10 Climatic data for Ain Salah (Algeria)

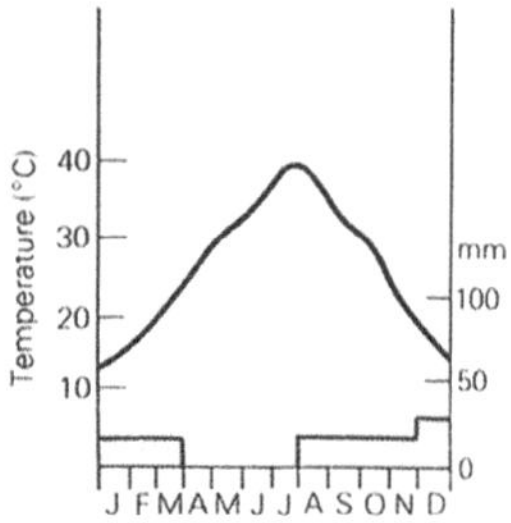

Exercise 6.3 Tropical climates

Figures 6.6, 6.7 and 6.8 show climatic information for three places in the Tropics.

1 Describe the pattern of temperature at the three stations.
2 Explain why Bombay and Kayes (Mali) have a much larger range of temperature than Singapore (look carefully at the latitude of each place).
3 What is the name given to the abrupt change from dry season to wet season shown in Figure 6.8?
4 Explain why many tropical places have a wet and dry season.

Exercise 6.4 Desert climates

1 Study Figure 6.10. Hot deserts are often defined as having less than 250 mm of rain a year. Add up the rainfall bars. Does Ain Salah come within this classification?
2 With a very low rainfall how much cloud might this area experience?
3 What pressure belt influences this area and keeps rainfall low?
4 Temperatures vary widely between summer and winter. Why should this be so?
5 Deserts can be sustained by high pressure zones, rainshadows or simply distance from the sea. Make a rough copy of Figure 6.11 and indicate the probable cause of each of the deserts marked.

Figure 6.11 The world's deserts and areas liable to desertification

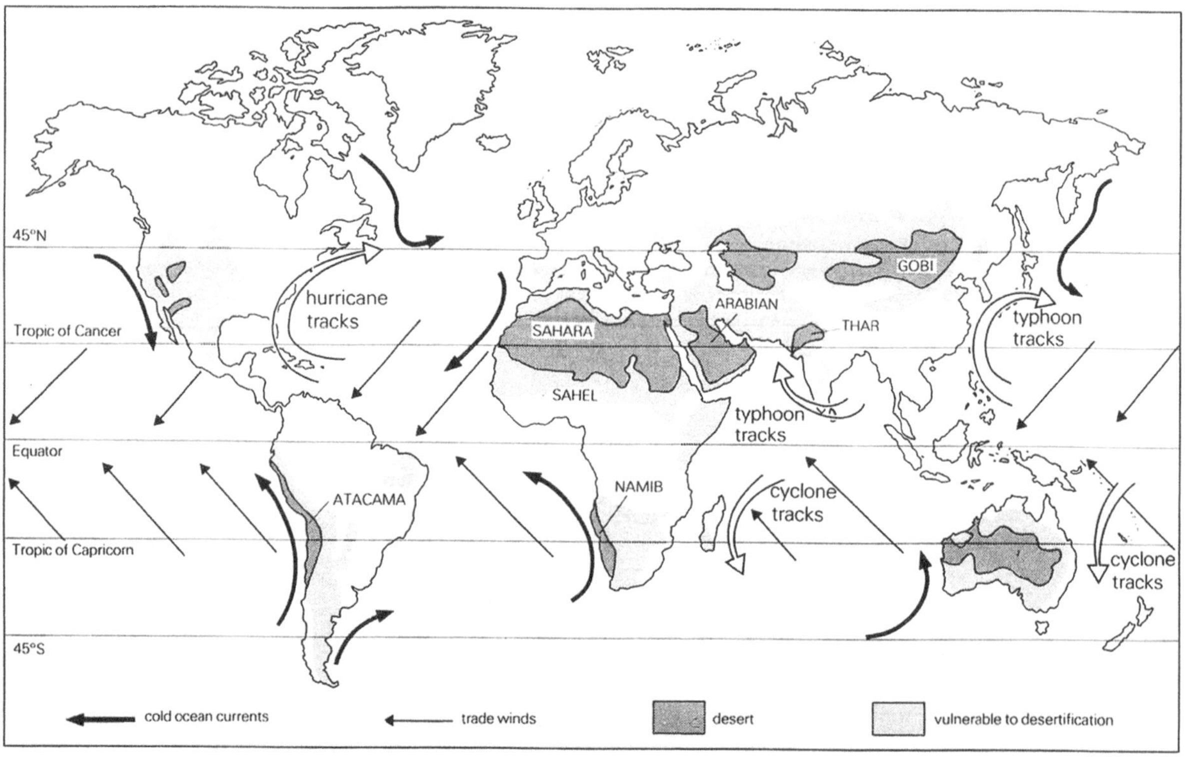

and the weather is dominated by cold winds from the Poles. Nevertheless in these regions, where long periods of clear skies prevail because of the domination of sinking polar air, there is a great range of temperature. In winter −30°C would not be an unusual temperature when darkness occupies much of the 24-hour period and the Sun only nudges above the horizon. In contrast, temperatures on some summer days can rise to 20°C because of the long hours of sunshine. Tomsk (Fig. 6.12) lies at the southern limit of this area, to the north conditions become even more severe. Real polar areas have no month with an average temperature above 10°C, so their temperatures range from this value downwards. The lowest temperature recorded at the South Pole is −150°C!

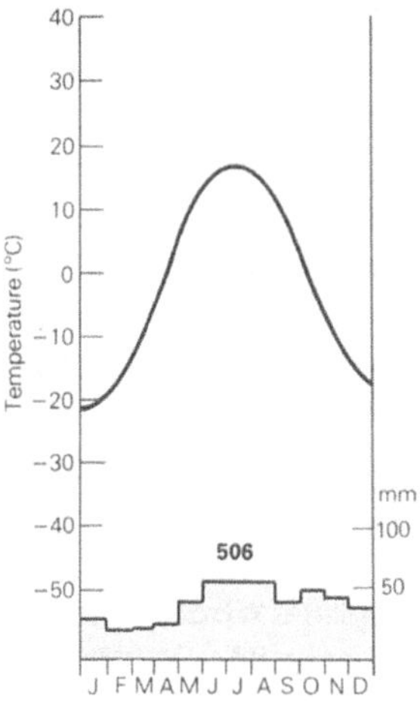

Figure 6.12 Climatic data for Tomsk (USSR)

Exercise 6.5 Cold climates

Cold climates have very few months above freezing and therefore a short growing season. Tomsk, in Siberia, (Fig. 6.12) is in the coniferous forest belt of Asia, but only a little way north conditions become too severe for trees to grow, and tundra conditions begin (see p. 132).

1 If plants require a sustained temperature of over 6°C for growth, how long is the growing season in Tomsk?
2 For how many months is it likely that precipitation will fall as snow?
3 For how many months is the temperature above freezing?
4 Find Tomsk in an atlas. From its position in the Asian continent can you suggest why it has such a low total precipitation?
5 Climatic conditions such as these must play a large part in determining the type of houses and the lifestyle of the people. Suggest what some of the main problems must be and the way they could be overcome.

SUMMARY: THE WORLD'S CLIMATES

Hot climates. Hot climates have a coldest month above 18°C and they occupy a broad belt well beyond the Tropics, thereby including over half the Earth's surface. Seasonal and daily changes in temperature are controlled by the movement of the overhead Sun. This gives two maxima of temperature (spring/autumn) and rainfall in equatorial areas and single summer temperature/rainfall maxima in other tropical areas.

(a) *Equatorial:* With the sun overhead in spring and autumn there is no summer maximum and fairly uniform temperatures prevail all the year. Rainfall occurs in all months but is heaviest in spring and autumn just after the Sun has passed overhead.

(b) *Tropical:* These regions lie at the edges of the equatorial rain belt. They experience dry trade winds for most of the year, with rain bearing winds at other times. Many eastern coastal margins experience **hurricanes** and **monsoon** rainfall. There is considerable temperature range through the year, with 27°C being a common maximum for the hottest month.

Warm temperate climates. Warm temperate climates lie between the trade winds and the westerlies and they are transition climates, having a summer 'tropical' season and winter similar to temperate areas.

(a) *Mediterranean* (western margin): These are coastal climates, with hot dry summers (over 20°C in warmest month) and wet mild winters (above 6°C in coolest month).

(b) *Eastern margin:* These areas have mild winters and hot summers but the rainfall patterns are reversed. Low winter rainfall and plentiful summer rains are the rule, but there is no long dry season as with Mediterranean-type areas.

Cool temperate climates. Cool temperate climates have a cold season when plants do not grow. Seasonal changes in temperature are more important than changes in rainfall. These regions are dominated by weather from depressions which track along in the belt of Westerly winds. Continental areas have a large seasonal temperature range. Rainfall occurs in all seasons but decreases away from the coasts; interior regions have a summer maximum because convection rain is most important.

Cold climates. Cold climates are dominated by depressions and anticyclones in the same way as cool temperate regions, but as longer and more severe winter increases most outdoor activity is restricted.

Desert climates. These climates occur in diverse regions where temperature and rainfall vary widely. Often annual rainfall below 250 mm coincides with these areas, but a scant vegetation is the best indicator of desert climates.

(a) *Hot deserts* do not have cold seasons below 6°C. They occur in subtropical high-pressure regions and are reinforced at coasts by cold offshore ocean currents.

(b) *Cold deserts* have at least one month below 6°C and are mostly in rainshadow areas or at the hearts of continents.

. . . and it will be normal for the time of year'

The years 1975 and 1976 were a period of absolute delight for the journalists of Europe. There was no shortage of news or stories of human interest. First, it was the driest period for 50 years, then 100 years, then 200 years . . . or was it really 500 years? And while the scientists enjoyed themselves by bandying various predictions about, the levels of water in Europe's reservoirs went down and down. Eventually the government stepped in with restrictions on the amount of water that could be used, as parts of Europe headed for the greatest drought that had ever been known. Standpipes were erected in the streets and some areas had their mains water turned off. People queued for water with buckets; older people talked of 'nothing like it since the war', while others had baths in a couple of inches of water and then laboriously scooped out their used bath water into buckets for their garden flowers. Eventually the British Government appointed a Minister for Drought, and within two days the sky opened up, the rain fell, and the new Minister was left without a job.

Of course, we would describe this as a period of abnormal weather because, on average, about 25–50 mm of rain a month are expected in Britain. We listen to the weather forecast on the television or radio and are told 'the temperature will be below normal for the time of year' or (seemingly much more rarely) 'it will be much sunnier than usual'. Now the terms *normal* and *average* have hidden meanings. They subtly reinforce our beliefs that the climate is constant and unvarying except in the form of small fluctuations about some long-term average . . . the value that gives us our *normal weather*. However, what we call 'normal' readings are, by international agreement, only 30-year averages. So the 'normal weather' for Christmas Day 1982 is the average for the period 1952–82, but in 1972 the average was for the period 1942–72. This false impression of 'normal' is given more strength by the shortness and biased nature of our memories. When we take a close look at the weather records for the past 100 years we see that the weather of the past 10 years is not even typical of this relatively short period. If we go back somewhat further, we find that only 20 000 years ago northern Britain was covered by ice! In fact our present weather becomes less and less typical, or 'normal', the greater time span we look at.

Exercise 6.6 Mediterranean climates

In the table below, the temperature and rainfall data are given for Marseilles in Southern France (43°N, 5°E).

1 Draw these figures up as a bar graph similar to all the others in this chapter.
2 A Mediterranean climate has (a) most of its rainfall in the winter with very little in summer; (b) hot summers with the warmest month above 21°C and mild winters with the coldest month above 6°C; and (c) much sunshine, especially in summer.
Show how Marseilles fits into a Mediterranean type of climate.
3 State how you can tell it is sunny from the information in the table.
4 The minimum temperature for plant growth is 6°C. The Mediterranean coast of France, especially the area to the west of the River Rhône, is well known for its vegetables. At what season of the year do you think growers try to sell their crops in order to get maximum profits from the markets of Northern Europe?

	J	F	M	A	M	J	J	A	S	O	N	D	
Temperature (°C)	7	8	10	13	16	20	22	21	19	15	11	8	
Rainfall (mm)	43	36	48	56	43	28	18	20	61	96	71	53	*Total* 574

7 The waters of the oceans

Oceanic circulation

Swirling waters

When large storms develop in the open ocean they can hurl water about in tumultuous waves up to 20 m and 30 m high; they can even break the back of a supertanker. But observed on a global scale, even these giant waves are nothing more than surface ripples driven by the friction of winds on the ocean surface. The bulk of the ocean water remains completely unaffected by the surface waves – it is involved not in these local disturbances but in a worldwide transfer of energy. Only in shallow seas does the friction of wind on the water reach down to the sea bed.

Figure 7.1 Heating the ocean
The ocean absorbs heat when the Sun is overhead, but very little when it strikes at a shallow angle. Arctic oceans receive very little heating even though the Sun shines 24 hours a day in summer.

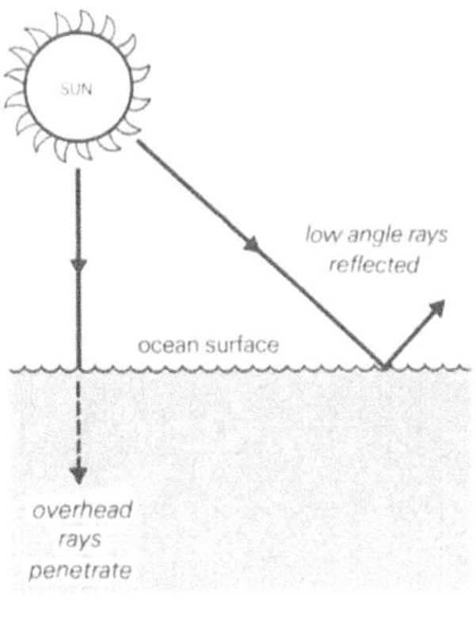

The movement of large volumes of water is just as important to the world's heat balance as the movements that take place in the atmosphere. Two thirds of the Earth's surface is water and this receives two thirds of the sun's energy reaching our planet. However, water absorbs this energy quite differently from land. At some time, such as when the Sun's rays strike the water surface obliquely, virtually all the energy is reflected back to space (Fig. 7.1). However, when the Sun's rays come from overhead, 90% of the energy is absorbed by the surface layers of water, providing light and warmth for millions of sea creatures and also a vast store of heat for the Earth.

Notice that it is the upper layers of the ocean that absorb the Sun's radiation and warm up. Warm water is less dense than cold, so the heated surface waters stay on top of the oceans. This is very different from atmospheric heating, in which air is warmed by contact with land or sea and so is heated from below. With air warmer near the ground and colder at height, the warm air rises and produces all sorts of turbulent effects in the atmosphere. With warm water on top, oceans are rarely affected by this sort of turbulence.

Although oceans are heated from above and the atmosphere from below, both receive much more heat near the Equator than near the Poles. As a result oceans are warmest in equatorial regions and coldest near the Poles. As with the air,

Topic Why the sea is salty

When a volcano erupts it throws out great volumes of steaming water along with gases, ashes and lava. All the water in the oceans, on land and in the atmosphere originally came from volcanoes in this way. New water is still being added to the Earth's surface by the same means.

Every new addition of water contains a great variety of minerals dissolved in it, but somehow the oceans are dominated by just common salt (sodium chloride, NaCl). Similarly rivers contain a great variety of minerals in solution eroded from the surface of rocks. Yet the flow of river water in the oceans does not upset the salt concentration at all.

Most of the minerals carried into the oceans, from either volcanoes or rivers, are used to make the shells of sea creatures or else are absorbed by freshly settling sediments on the ocean floor. Some minerals such as manganese even manage to concentrate themselves into nodules which are left to litter the sea bed. But sodium and chlorine are not completely used up in this way and so are left as the main components of sea water. When water evaporates to form moisture in the air, the salt is mostly left behind. Any salt that gets into the air as spray from breaking waves is washed out by rain and returned to the sea by rivers. It is this long-term balance that keeps the sea salty.

these temperature contrasts cause worldwide convection currents to form in an attempt to spread out the heat energy more evenly (Fig. 7.2).

We see the ocean convection as huge volumes of water constantly on the move; these are called **ocean currents** and many of the largest have been given names by the sailors who make constant use of them. Names such as the **Gulf Stream** are familiar to most people. This is a large volume of warm water that travels from the Caribbean Sea along the east coast of the United States and finally crosses the Atlantic to provide warmth to the shores of Europe.

However, warm surface waters such as the Gulf Stream are only part of the worldwide circulation. They must be balanced by returning currents of cold water. Warm currents are usually fast-moving, relatively narrow bands of water; cold currents move much more slowly, but spread out more.

Figure 7.2 The main ocean currents of the world

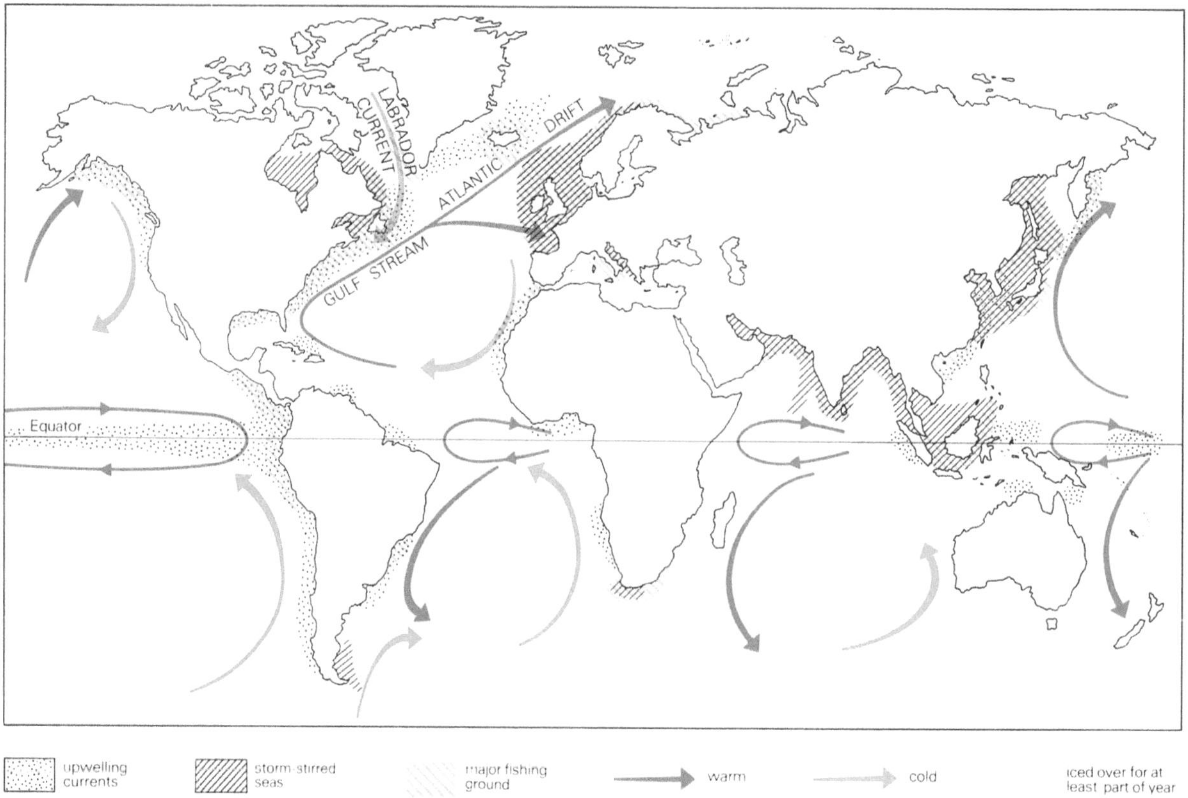

Exercise 7.1 Ocean currents

1 Look at Figure 7.2 and concentrate on the southern hemisphere first. Describe any current pattern you see in the South Atlantic Ocean. Can you find the same type of pattern in the other southern oceans?
2 Now compare the northern and southern Atlantic ocean currents. What is the basic difference?
3 What happens to the ocean flow near the Equator?
4 Complete Figure 7.3, which is a simple drawing of the ocean currents. Notice each ocean has a *figure of eight* pattern of currents. This may help you remember the ocean currents more easily.
5 In the past many voyages of exploration have used ocean currents to advantage. Describe a route you would use to go from Plymouth in England to Sydney in Australia, making full use of currents.

N.B. In the past, sailing voyages were closely dependent on finding favourable winds and ocean currents, but even today steaming against a strong current can cost very much more than a slightly longer course that avoids it. The shipping industry is still influenced strongly by the distribution of the main ocean currents of the world.

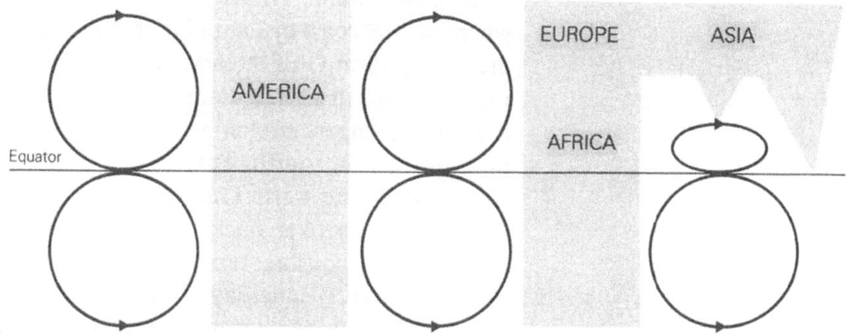

Figure 7.3 A simple model of ocean circulation

Wind and convection

Current affairs

Because the oceans are warmed by the Sun's energy more near the Equator than near the Poles, large-scale convection currents develop. The currents that even out the oceans' heat are often well defined 'threads' of water. Those bringing warm water to the polar regions are called **warm ocean currents,** while those bringing cold water to the Equator are **cold ocean currents.** Convection currents cause the main north–south flows, but many of the east–west flows are partly dragged along by winds such as the trades.

Convection and wind drag effects are clearly seen, for example, in the North Atlantic Ocean. Here, off the coast of Venezuela, the trade winds blow so constantly from east to west that they drag the surface water from Africa westwards until it piles up in the Gulf of Mexico (Fig. 7.2). All the while the water is heated by the Sun and it gradually becomes warmer and warmer. But water cannot collect in the Gulf of Mexico indefinitely, so some of it pours continuously out to the north as the trade winds drag more in from the east. You can form some idea of the enormous volume of water involved when you know that the Gulf water is 20 cm higher than the Atlantic. The sea level actually slopes down to the north-east allowing gravity to help water to pour out. About 26 million cubic metres of sea water flow from the Gulf to the Atlantic every second. This is the water that starts the famous Gulf Stream.

Convection currents help transfer water towards the Poles and create a return current of cold water down the coast of Africa. As the warm water moves north it helps warm up the air in contact with it. This warmed air then flows over North-West Europe to give an exceptionally mild climate for this latitude.

Fronts in the ocean

Icy fingers

Most of the ocean currents are quite uniform over large distances, but problems come when water that is cold at the surface meets warm surface water. The Arctic Ocean is always cold, from the surface down. Indeed, because water is most dense at 4°C, the deepest part of the Arctic is actually warmer than the surface. Because it is so cold, the surface of the Arctic Ocean is often a mass of pack ice and icebergs (Fig. 7.4).

The cold Arctic water is constantly drawn southwards as part of the worldwide convection in the oceans. With it comes a liberal surface coating of broken **pack ice** together with **icebergs** that have broken away from the glaciers around the coast of Greenland. This cold water comes farther and farther south until, somewhere off the coast of Labrador or Newfoundland (called the Grand Banks), it meets the warm Gulf Stream. With cold and warm water both on the surface, mixing occurs in an irregular way like the interlocking fingers of two hands – one a set of icy fingers stretching out southwards, the other a set of warm fingers pushing their way northwards.

Eventually the warm Gulf Stream water ends up on top of the cold Arctic water, and the broken pack ice and the icebergs melt. It is this region of mixing that has been a constant terror to generations of North Atlantic seafarers (see also fog, p. 65). On Wednesday 10 April 1912 the newly built ocean liner Titanic sailed on her maiden voyage from Southampton to New York. At 60 000 tons she was the largest ship then afloat, rising 25 m out of the sea and extending 300 m from bow to stern. At just after twelve o'clock on a calm, starlit night the ship's engines were switched off and the huge vessel came quietly to rest, not 48 hours steaming time from New York.

Full of curiosity, many of the 2208 passengers and crew came up on deck to find out why the giant ship had stopped. Most were unaware that, a quarter of an hour previously, the ship had struck the submerged part of an iceberg and torn out much of her bottom. She was such a large ship that there had been no crash, no judder, nothing. A passenger later reported 'Suddenly . . . a hissing roar . . . made us all turn (to see) a rocket leap upwards . . . and then an explosion . . . and a shower of stars sank slowly down and went out one by one. And with a gasping sigh one word escaped the lips of the crowd: 'Rockets'! Anybody knows what rockets at sea mean!'

With water pouring into the ship from its holed keel, a full-scale evacuation took place and by 2.15 am the ship had almost sunk.

'And then, as we gazed awe-struck, she tilted slowly up, revolving apparently about a centre of gravity just astern of midships, until she attained a vertically upright position . . . then she slid slowly forward . . . and the sea closed over her.'

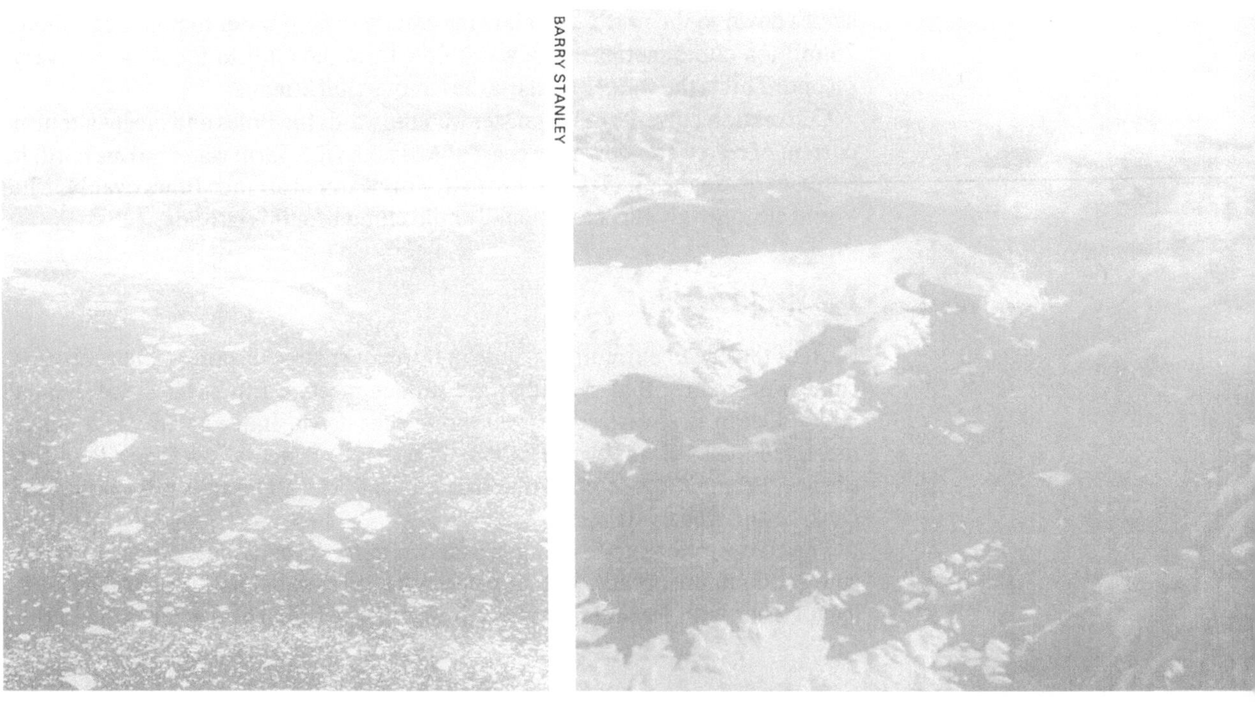

BARRY STANLEY

Figure 7.4 Pack ice and icebergs
Left. Pack ice breaks up into fragments. *Right.* These icebergs calving from glaciers in Greenland present a danger to shipping as they float southward between Greenland and America. These views are from an aircraft 10 km above the ice.

Topic The reasons for tides

Movements of ocean currents and atmosphere are mainly driven by energy from the Sun, but tides are most strongly influenced by the gravitational pull of the Moon.

Everything in the universe has a gravitational field and attracts and is attracted to everything else. The strength of the gravitational pull depends on the size of the body and its distance from those it is attracting. So the larger the body, the greater the pull – at least as far as our everyday experience is concerened. Because the gravitational pull by the Moon is less than that of the Earth, for example, astronauts have been able to make very high jumps on the Moon. We, however, are used to everything being firmly anchored by the Earth's gravitational pull. Indeed, if it weren't for gravity the spin of the Earth would throw everything off into space. Yet although the main force on the Earth is our own planet's gravity, the pull of the Moon is still great enough to attract the ocean waters (Fig. 7.5).

When the moon is overhead the moon's gravitational pull drags water towards it. We call the piled-up water a **high-tide.** As the Moon goes round the Earth so the high tide goes with it; and as the Moon encircles the Earth every 28 days but the Earth spins on its axis every 24 hours, the Earth and Moon get out of step by 52 minutes a day. As a result high tide is 52 minutes later each day (Fig. 7.6).

There are two high tides every lunar day. The other high tide occurs exactly half way between those caused by the Moon's attraction, but is explained by the very peculiar way the Earth and Moon revolve together. It does not necessarily produce the same height of tide as that due to the Moon's attraction.

The pull of the Sun is much less than the Moon because the Sun is so much farther away. But when the Sun and Moon pull in the same direction a very high tide occurs – a **spring tide;** conversely, when the pull of the Sun is at right angles to the pull of the Moon, water is pulled from two directions at once and a low tide occurs – a **neap** tide. Neap and spring tides occur alternatively about every 14 days.

If the Moon drags a high tidal flow of water across the Earth's surface and into a funnel-shaped piece of coast, the tides will become higher and higher towards the narrow end of the funnel. This is why the coasts of the Bristol Channel and the English Channel have such high tidal ranges. These high tides can push water back up river channels for many miles. One of the most famous examples is the Severn **bore,** which is a wave of water pushed up the river beyond Gloucester.

Tides can be an advantage or a disadvantage to man. A high tidal range is used to drive turbines and produce electricity in the Dinard estuary in Brittany, France. Here the estuary has been dammed and the water flows in and out through turbines (Fig. 7.7).

However, the same high tidal range presents great difficulties for shipping. London, Rotterdam and Le Havre are examples of ports which large ships can enter only at high tide. But Marseilles is a Mediterranean port where tides are very small and ships can enter at any time. This is a major economic advantage because ships waiting out at sea for a high tide are not earning money.

The growth of Southampton has also been affected by tides. Because of the Isle of Wight water reaches Southampton by two different routes, Southampton has extra-long high tides every day. Ocean-going liners and cargo vessels can make good use of these bonus tides for entering and leaving the port.

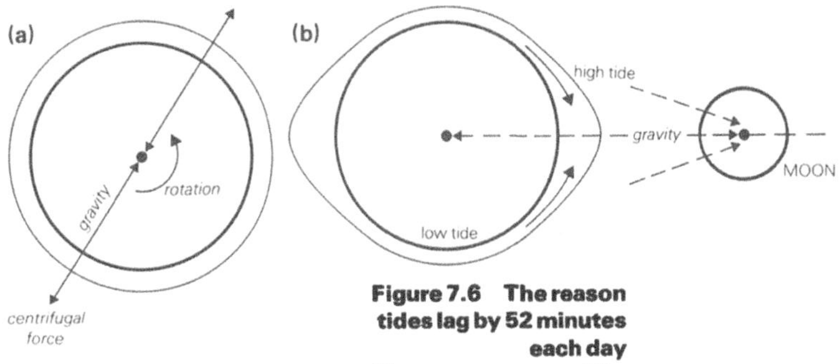

Fig. 7.5 The tides
(a) The Earth spins on an axis. In doing so it tends to throw everything outwards (by *centrifugal* force). This is more than balanced by the gravitational pull towards the centre of the Earth. (b) As the Moon moves round the Earth its gravitational field upsets the balance of forces shown in (a). It opposes the Earth's gravity and allows water to pile up facing the Moon as a high tide. The Sun's effect is the same, but smaller.

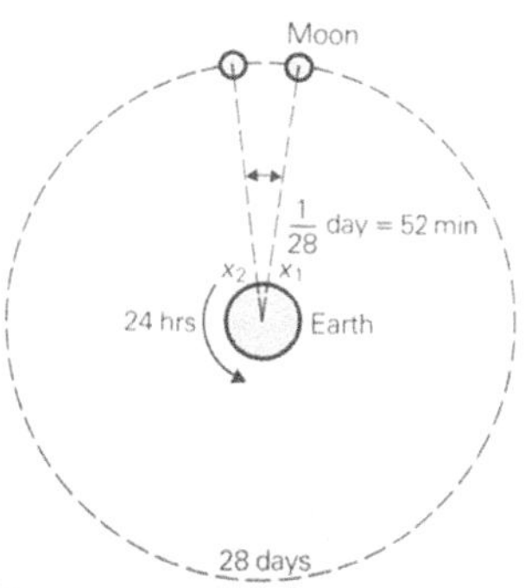

Figure 7.6 The reason tides lag by 52 minutes each day
The moon goes round the Earth every 28 days; the Earth revolves every 24 hours, hence point X beneath the Moon will be below it again in $(24\text{ hrs} + \frac{24}{28}\text{hr}) = 24$ hr 52 min later.

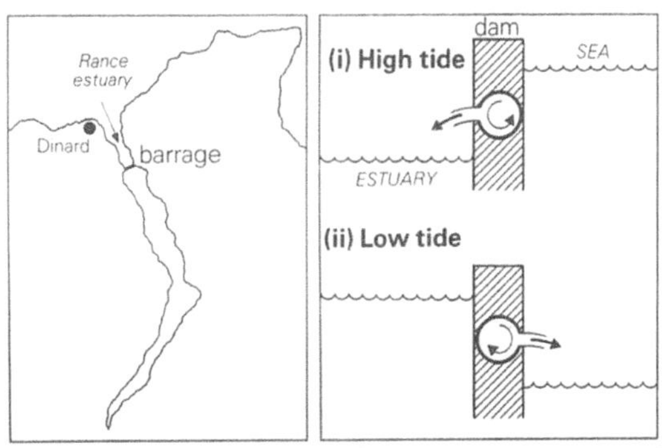

Figure 7.7 The tidal barrage in France
(1) High tide and sea water drives turbine as the estuary is filled. (2) Estuary full and a turbine is driven as the water flows out at low tide.

At the time of the Titanic disaster (in which 1500 people drowned) the only way to detect an iceberg was to mount a lookout (with a telescope) at the bows of the ship. Radar was not invented until 30 years later. It was not even possible to detect the presence of icebergs accurately by observing sea temperatures. Because this is a region of mixing, temperatures can vary from 1°C to 14°C and back within a kilometre or two and icebergs can easily float out of the cold water and into the warm. Indeed, they can sometimes float south for hundreds of kilometres beyond the mixing zone before being melted away completely. The Titanic disaster happened at latitude 42°N, about the same as northern Portugal (Fig. 7.10).

Global fishing grounds

Where shall we fish tomorrow?

In the spring of AD 29 Simon Peter said 'I am going out fishing'. 'We will go with you' said the other disciples. So they started and got into the boat. But they caught nothing. Morning came and there stood Jesus on the beach, but the disciples did not know it was Jesus. He called out to them 'Shoot the net to starboard and you will make a catch'. They did so and found that they could not haul the net aboard, there were so many fish in it. (From the Gospel according to St John.)

Fishing is a risky business and it is quite common for fishing boats to go on an expensive voyage and return with very little. Yet the distribution of fish is well known. It is just that the fisherman has to guess where, within a specified area, the fish will be at any one time as they swim through the water grazing on the countless floating sea creatures which are their food.

In fishing, knowledge of the movements of ocean waters and of the biological life cycle overlap. We can see this by following the fate of a fish that dies. Burial at sea occurs in a relentless way as soon as the dead creature sinks to the sea bed. Here minute particles of sediment are constantly settling out of the water and building up on the sea floor (Fig. 7.8). Eventually they become sedimentary rock. If the rate of sediment build-up is fast enough, the dead body will be encased in the sedimentary ooze and then be subject to a slow 'embalming' process which will eventually turn it into a fossil. However, the sea bed is usually a very active place, for it is the home of millions of scavenging creatures whose entire existence depends on finding nice juicy dead bodies to eat. Much of this work is achieved by bacteria, which, as part of their 'eating' process, release large quantities of nutrients into the sea water. But these are the very nutrients that are needed urgently back in the

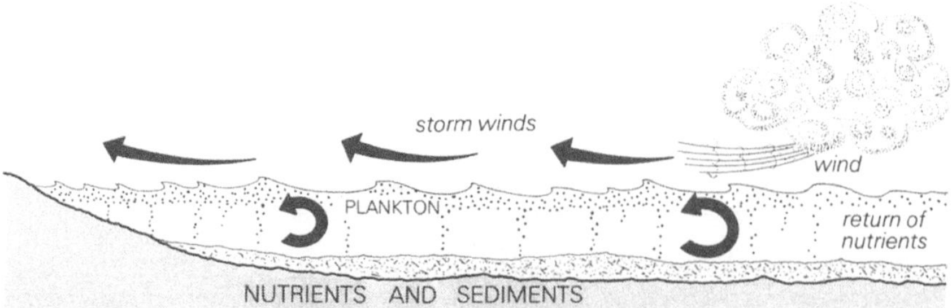

Figure 7.8 Storm turbulence Winter storm winds stir up shallow seas so that nutrients are brought off the sea bed and returned to the surface.

warmer waters near the surface to allow the growth of plankton. **Plankton** are microscopic sea creatures that float in the upper waters of the oceans. They are the main food of most fish. As a result, continued survival of plankton, fish and scavengers depends on getting the nutrients from the sea bed back to the surface. With warm water on the surface anyway, there is no convection mechanism to overturn the water, stir up the bottom and bring the nutrients to the surface, so an external force is needed to stir up the water.

Fishing in shallow seas

When the north wind blows

With no internal mechanism, the stirring can occur only through the help of some external force. In fact two main forces operate: the wind and mixing ocean currents, the first only in shallow seas and the second in either seas or oceans. So the distribution of plankton concentrations, fish and the world's fishing industry depend on finding places where the effects of winds can reach the sea floor or where ocean currents mix.

The North Sea provides an example where strong winds reach to the sea bed. The strong winds that blow over the North Sea in winter are part of the deep depressions that send whirlpools of air across Britain and Europe. As they blow they transfer some of their energy to the ocean by friction, stirring it up and creating waves (Fig. 7.8). But stirring the surface of the shallow North Sea can produce a 'knock-on' effect right down through the water to the sea bed. As a result the nutrients of the sea bed are brought to the surface by turbulence. In winter this gives the North Sea its greeny-brown and cloudy look, because sediment and nutrients are all stirred up together. But in the spring, when the winter gales die down and the Sun warms the surface again, the nutrients are still there in suspension and can provide the food for plankton. As the plankton multiply (called 'blooming') so their presence attracts the hungry shoals of fish. Of course some fish short-circuit this food chain and feed directly from the bottom, but they can do this only in shallow, nutrient-rich waters. Thus sea-bed turbulence produced by strong winds provides the basic cause of the North Sea's prolific fishing grounds, alas now suffering from over-exploitation.

North Sea winds blow from many different directions and stir up the sea bed by force, but even a gentle wind can drag sea water along and stir up the bottom of a shallow sea if it blows constantly for long periods of the year. Thus the coastal regions of Arabia and the Bay of Bengal are also important places for fishing, even though they do not receive many storms (Fig. 7.9).

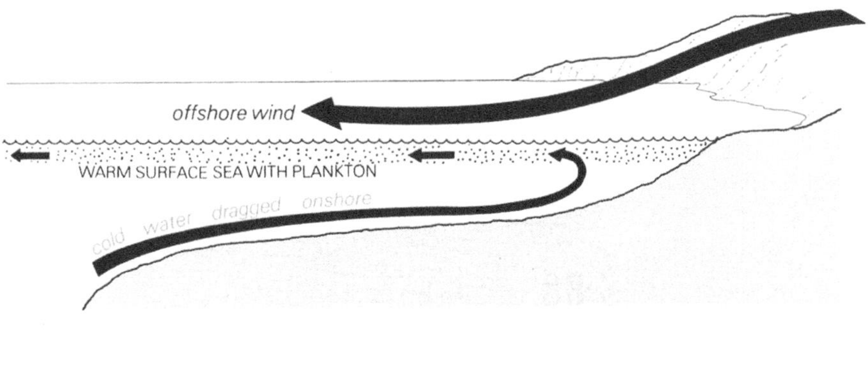

Figure 7.9 Offshore winds
When offshore winds blow constantly they push the warm surface water away from coasts. Cold, dense water from the depths rises to replace the warm water and in doing so drags up nutrients from the sea bed. In this way the basis for plankton growth is provided.

Deep-sea fishing

Fruits de mer

In just the same way that there are frontal zones which separate regions of warm air from regions of cold air, so there are frontal zones in oceans separating regions of warm water from regions of cold. These are places where two masses of water collide and struggle past each other, usually in an irregular way as we saw in the Titanic disaster. Where two strong currents meet there can even be regions of rough sea. Where two currents collide all sorts of floating particles are concentrated, including oil slicks. But, of special interest to fishermen, they are also places where plankton become concentrated and which, therefore, attract fish. As a result ocean fronts are some of the world's most productive fishing grounds and their study is of vital importance.

The fronts can occur in the middle of oceans, as Thor Heyerdahl found on his Kon-Tiki expedition across the Pacific. His observations, although indirect, show the effects well. When still 1 000 km off the coast of South America he encountered a region of very rough water. This was the turbulent zone where the fronts met.

"The crest of the sea reached two metres above the level of the roof of the bamboo cabin, and if two vigorous seas rushed together, they rose still higher in combat and flung up a hissing watery tower which might burst down in unexpected directions. Next day we noticed more life than usual in the sea . . . during the afternoon we saw a big fish approaching the raft close to the surface . . . When we were having a rather wet and salty midday meal . . . a large turtle was lifted up by a hissing sea right in front of our noses . . . (then) . . . it was gone as suddenly as it had appeared (and) . . . we saw the gleaming whitish-green of dolphins' bellies tumbling about in the water below. The area was unusually rich in tiny fish an inch long which sailed along in big shoals . . . At 6.20 pm (Erik) reported our position as latitude 6°42′S and longitude 99°42′W . . ."

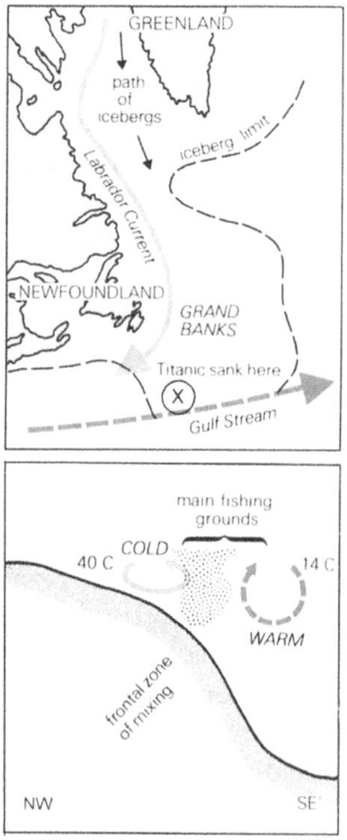

Figure 7.10 The Grand Banks fishing zone, Labrador This major fishing zone results from a meeting of cold and warm water. Inshore the sea bed is also stirred up by storms.

One of the most prolific fishing regions occurs just off the coast of Newfoundland in about the same place as icebergs are found. Here the cold Labrador current mixes with the warm Gulf Stream. Together with storm winds, turbulent mixing drags nutrients from the floor of the continental shelf and sends them high up into the warm surface waters where they are eaten by plankton (Fig. 7.10).

Exercise 7.2 Fishing grounds

1 Copy the fishing grounds from Figure 7.2.
2 Using Figure 5.9 separate the shallow sea areas into those with winter storms and those where the wind blows constantly.
3 Why are some potential fishing grounds not exploited. (You will need to consider each location separately.)

8 The water cycle

Rainfall patterns

Enough is enough

November 1974. Darkness was falling quickly as a Jumbo jet prepared to land on the main runway of Gatwick airport. The plane flew on through dark grey cloud, guided by radar and navigation directions from the airport control. 'You are cleared to land on runway 26; wind 15 knots from the west; heavy rain with much surface water on the runway.' 'Roger, Gatwick control; making my final approach now.' Slowly the giant plane appeared through the thick low cloud and drifted down into the gathering gloom of the November evening (Fig. 8.1).

For those people working on the airport services, it had been another difficult day. Heavy rain always made outdoor activities slower and it had been raining every day for the past ten days. However, today was the worst of all; there had been so much water on the runway that the drains had hardly been able to cope.

In the shopping centre of nearby Crawley, wet shoppers trudged home as water cascaded down from the roofs and disappeared underground. On a farm next to the airport, the farmer was becoming worried. The River Mole flowed through his land and he had been nervously watching the river level rise and fall with each storm over the past few days. Each time, the peak flow had crept a little nearer the top of the banks. Now it had rained even more and his experience told him that all the signs pointed to a flood. He went out into the rain to move his cattle from the pastures by the river . . .

Figure 8.1 Approaching Gatwick airport
This pilot's-eye view shows the green grass beside the runway, typical of a season with plenty of rainfall.

BARRY STANLEY

Figure 8.2 Drought
It is a surprise to see the bed of a river. There is too little water to flow over this weir.

August 1976. As the farmer stepped over the trickle of water that was usually the River Mole, he could not remember a time when the level had been so low (Fig. 8.2). True, it would always decline steadily when there was a period without rain, but usually a fairly low stable value was maintained. Now, as he turned his head to watch the Jumbo come in to land at the airport, he was standing beside fields of dry and parched land whose grass had hardly grown all summer.

The talk in the shops in Crawley centre was on only one theme – the difficulty of keeping the garden flowers alive. Waiting at the supermarket checkout, the shoppers were faced by a large poster which cautioned 'Don't waste water' and showed many ways of saving water, such as sharing a bath.

In the cloudless evening sunset, the pilot in the aircraft saw the approach lights of the airfield shine out from the sides of the tarmac. To each side of the runway were areas of yellow-brown grass, similar to those he had been travelling over ever since he crossed the British coast. If he hadn't a map, he could have believed he was over the dry plains of Africa . . .

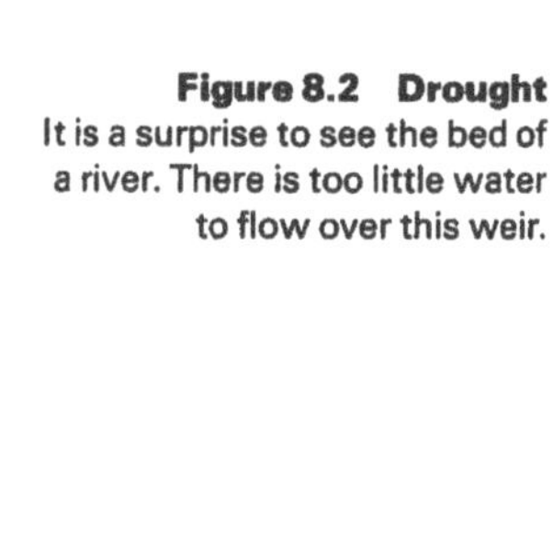

Figure 8.3 River channel area
The area of a river channel is only about 1% of the total area of the drainage basin.

These two contrasting scenes of Britain are from recent years. As it happens the end of 1974 was particularly wet, but the following eighteen months were exceptionally dry. Both events made news because they were uncommon. In general Britain would expect periods of rain, separated by a few dry days throughout the year. And because of these fairly regular alternations of dry and wet weather, we expect the river levels to change a little, but not to rise sufficiently to cause floods, nor fall enough to cause droughts. So what we really expect is that the landscape, and rivers in particular, will smooth out these rather extreme alternations of rain and no rain. In doing this we expect rather a lot, as we shall see.

The water cycle

The obstacle course

In general, observation shows that a river smooths out variations in rainfall. But first let us be quite clear about the relationship of a river, stream or brook to the area it drains. The surface area of the water-filled channel is really quite small. In Figure 8.3 a region is shown with only its rivers drawn in. A thick line has been drawn to show the region of land from which water drains to river. We call this area the **drainage basin** of the river. As you can see, the area of water surface is only about 1% of the drainage basin, so although rain falls directly into the river channel, by far the largest amount falls onto the land around. Most of this land is covered by soil and vegetation, so the rain falling has to overcome a sort of obstacle course to get to the river banks.

Infiltration

The first hurdle

Stand out in the rain and the first hurdle of the course becomes clear – you get wet. If you were to stand out in the rain for long enough the rain would soak all of your clothes and then drip off onto the ground, but you would have absorbed a lot of water that will never get to the river. But whereas you can move out of the rain, plants can't, so they get very wet every time rain falls. And each time, they trap some water before the remainder trickles down leaves and stems and enters the ground (Fig. 8.4).

Soil moisture

The giant sponge

The wetting of vegetation is an important delay mechanism and a source of water loss. Nevertheless it is only a minor hurdle in comparison with what is to come. As the water trickles into the soil it enters a world full of narrow tortuous passages, constrictions and dead ends.

A soil is really a giant inflexible sponge. It has all of the properties of a washing-up sponge, but you cannot squeeze it to force the water out. Firstly, it soaks up water – pour a little water on at the top and nothing comes out of the bottom. Even when you pour enough water on to cause an outflow there is an inevitable delay as all of the tiny passages are negotiated. Only then does the water come out – the rest is held in the sponge against gravity. Of course it is very

Figure 8.4 Vegetation surfaces
Water is held as droplets on vegetation. The droplets on these rose leaves will be evaporated when the storm is over.

important that this should be so, for where else would plants obtain their water?

If you fill a sponge with water and then hang it up it will drain for a long time – a kitchen sponge (10 cm × 10 cm × 0.5 cm) drained for 20 minutes in a classroom experiment. The soil that makes up the drainage basin in Figure 8.3 is 10 km by 10 km by about 2 m thick. A rough calculation shows the soil is about four billion times larger than the sponge. Now perhaps it will be less of a surprise to find that the soil continually drips water into the stream and that a few days without rain will not be enough to stop it.

So the soil is able to soak up water, store it and then let it drip out slowly. This is why rivers can continue to flow in periods of drought. The reason the River Mole nearly stopped flowing in August 1976 is that there had been little rain for over a year. And remember that some of the store of water in a soil is sucked out by plants for transpiration, so it does not all drain to the river.

Sometimes, when rain falls very heavily, the soil sponge cannot take in the remainder fast enough, so some runs over the surface. In Britain such occasions are rare. You can get some idea of how rare they are by thinking of the number of floods you have read about in the news. Reports of floods number only a few a year – and usually these are from different places. The week in November described on page 172 was one such rare event, but notice that rain had been falling for a week before the soil sponge was full and could take no more.

When water runs over the soil surface it moves at about 10 cm a second and this may be ten thousand times faster than through the soil. So when water does flow on the surface it gets to rivers quickly, and floods may result.

Springs

A soggy start

Water prefers to flow straight down a hill slope, using the most direct passages in the soil sponge to get to the river bank. In this way rivers receive most of their water supply through their banks. However, rivers usually start in a hillside hollow: occasionally as a sparking spring bubbling out of the soil; but much more commonly as a seepage which shows itself as a soggy, waterlogged patch of ground.

Springs and seepages are mostly found in hollows (Fig. 8.5). A hollow focusses water to its centre. And, because water from a large region of hillside is gathered together, it is easy to appreciate that the soil in the hollow will not be able to cope. The surplus water flows out at the soil surface and a new river is born. It is interesting to think that both the Mississippi and the Amazon – two of the world's greatest rivers – start as soggy patches of soil in hollows of some remote hillsides.

Figure 8.5 Hillslope hollows
The arrows show how, in a valley, water moving through soil is concentrated towards one place. When water fills the soil, it seeps out over the surface to begin a stream.

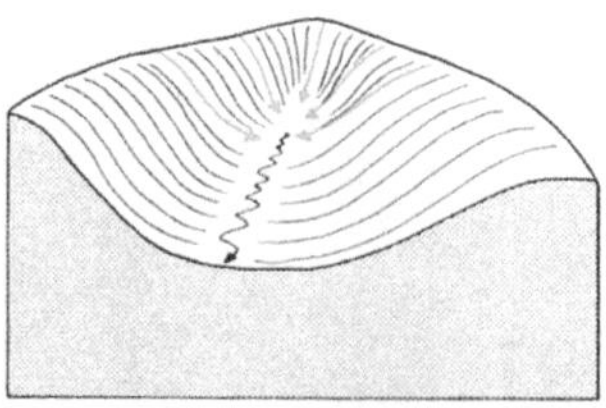

The water beneath your feet

Of all the many pastimes that people enjoy, probably one of the most bizarre is undertaken by people who choose to abandon the sunshine of a warm summer's day, put on a tight rubber suit, a helmet and a lamp, and then squeeze themselves into dark holes in the ground amidst torrents of cold water (see also p. 191). But if these cavers can drag themselves through gaps within the solid rock, what has happened to our soil sponge with its slowly moving water? Well, the sponge is still there, but we shall now examine the rock below.

Most rocks have cracks, but they are too small, too few and too disconnected to permit the fast movement of water. However, some rocks are different. Limestone has sufficiently large and numerous joints in it for water to drip from the soil and deep down into the rock. We shall discover in Chapter 15 how this continual dripping has enlarged the joints to such an extent as to allow whole streams to be swallowed up, but for the moment we should note that water-bearing rocks, called **aquifers,** are quite common. Most of them are made from limestone, chalk or sandstone (Fig. 8.6). Their main function is to store even more water than the soil could alone. Chalks and sandstones are the most efficient, for they contain countless tiny pores and joints very much like a soil, only the rock may be hundreds of metres thick. Limestones, with their larger channels, pass water more quickly and so do not store it as efficiently.

Aquifer storage has a very important effect on the volume of water in rivers. Because water soaks down into the aquifer and not down slopes to the stream bank, streams on water-bearing rock receive their water from the rock rather than the soil. This is sometimes very clear when water can be seen bubbling up in the stream bed. Because the rock can store such large quantities of water, it will usually absorb all rainwater. Water is released only slowly from the tiny pores of an aquifer, giving drainage basins with aquifers an unusually good protection against floods.

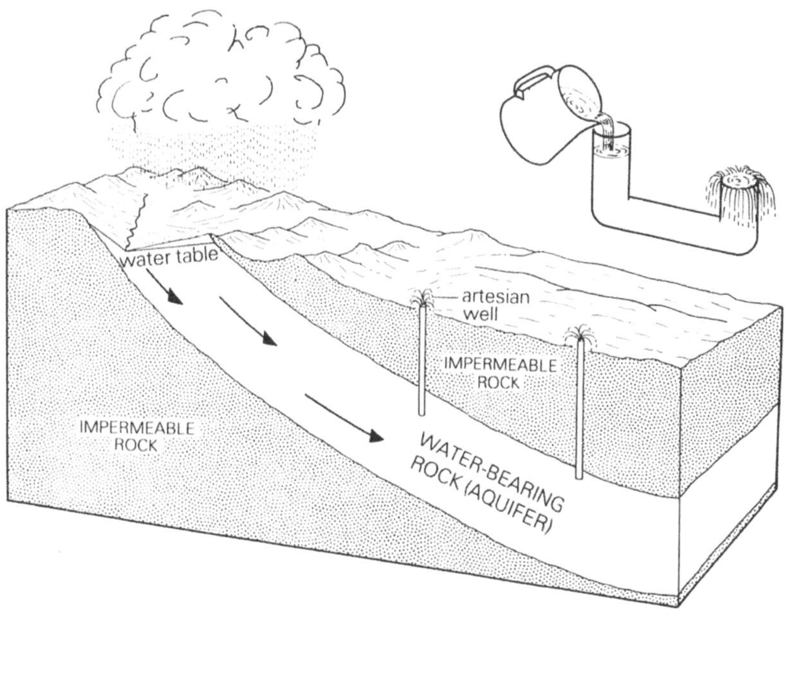

Figure 8.6 Springs The surplus water flows out along a spring line. The remaining water is stored in the rock as artesian water. Wells drilled down to this rock in the nearby lowland will provide water under pressure, much as water poured in at one end of a pipe is forced out at the other end.

Streamflow

Drip, drip, drip

On its own, a drop of water oozing out of a river bank and into the river would be totally unnoticed. If a drop flowed out of a metre-long bank every second, it would take a whole day just to fill an average-sized bucket. Yet somehow these little drops have to combine to yield all the water that flows in a river. The River Mole, for example, as far downstream as Gatwick Airport drains a basin whose longest side is less than 10 km. But in that small area there is about 400 km of channel and therefore 800 km of river bank. Now if we multiply our single drop in each metre of bank by this figure we find we can collect 100 buckets full of water each *second* near Gatwick. So it is quite possible for river flow to be provided by slow 'insignificant' drops of water oozing out of the banks. Most of the drops flow into the river at or below the water line, so water is entering the river even though the bank appears dry.

Following a period of rainfall much more water moves through the soil to the river bank. Now water starts to ooze out of the upper part of the bank as well as below river level and the banks become wet. When the number of drops emerging from the banks reaches the equivalent of about 8 every second, the amount of water reaching the River Mole fills the channel. At this stage it is dealing with 800 buckets worth of water every second, but in the flood of November 1974 at one time there was more than 3500 buckets worth of water flowing. As only 800 could be handled by the river channel, the rest was forced to spill over the surface and flood the surrounding land. Not surprisingly, the flat area on either side of many rivers is called the **floodplain.** Fortunately the amount of rain is not usually great enough to cause flooding and, on average, streams flood only about once in every couple of years.

Flooding is an important natural event, and floodplains are natural temporary reservoirs where water can be held until such times as the river channel can carry it away. Normally this would not matter, but man has a habit of building on floodplains because they are flat. People are not very happy when their living rooms become part of a natural storage system, even for a day or two, so very few river flows are now entirely natural.

Atmosphere and ocean

Connections

There is a continuous cycle of water involved in the stages described above. Clouds produce rain which falls onto the soil and plants. Some is evaporated back to the air and some is used by plants, but most goes on to flow through the soil. This water eventually reaches river banks and causes water to flow in channels. Moving much more quickly now, the river water soon gets back to the sea so that it can be part of the large store of water that is evaporated in order to make clouds. This completes a cycle of water movement (Fig. 8.7).

The water cycle links the big energy flows of the atmosphere with those in the ocean and thus nicely ties up all the loose ends of our present study. Everything is interlinked and is in a long-term balance. However, it is important that we see this as a complete picture. We must realise that if we change one part of the atmosphere – ocean system we are bound to affect others. We need to be sure these effects will not be catastrophic.

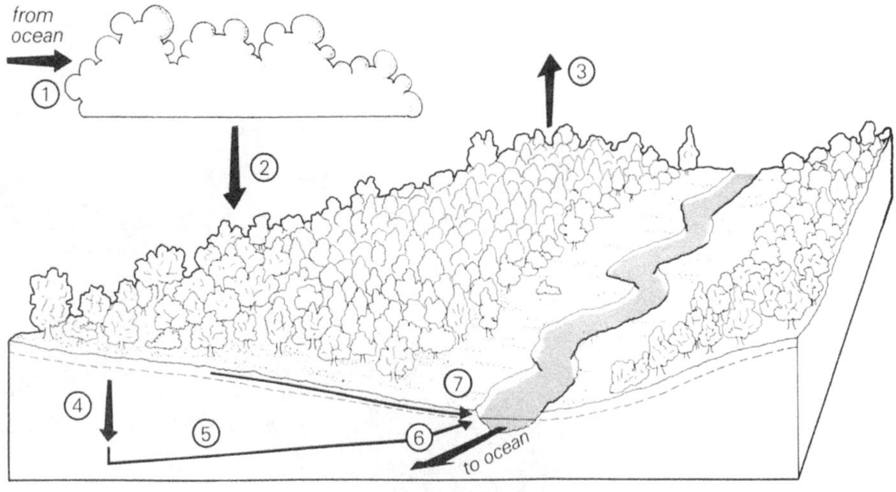

Figure 8.7
The water cycle

SUMMARY: THE WATER CYCLE

Water moves in a never-ending cycle between ocean, atmosphere and land (Fig. 8.7). As air passes over the ocean it picks up moisture. Air and moisture then rise over land to give cloud (1). Clouds release the moisture as either rain or snow (2). This then falls onto the land surface. Here water may be stored on vegetation and later evaporated back into the air (3). The majority passes into the soil (4). Some is then used by plants and transpired back to the air (5). Where there is an aquifer below the soil, water passes down to the rock (6). This delivers water to the streams via springs, otherwise it moves in soil to streams (7). Finally it returns to the ocean.

SUMMARY

Exercise 8.1 The urban water cycle

1 Study Figure 8.8 and compare it with Figure 8.7.
2 List the differences between the routes taken by raindrops in the two diagrams.
3 Explain why the water in Figure 8.8 will get to the river more quickly.
4 Why should the river in Figure 8.8 be more liable to flooding than that in Figure 8.7?
5 Now study Figure 8.9, which is a photograph of a tributary of the River Calder in Halifax (W. Yorks). Firstly, make sure you can find the river! Now describe the changes that have taken place to the channel of the river. What would you call such a river?
6 Describe the routes water follows after falling on the factory roofs. How much water sinks into the soil?
7 The river is used to carry away surplus rainwater. What else has it probably been used for in the past?

Figure 8.8 The urban water cycle

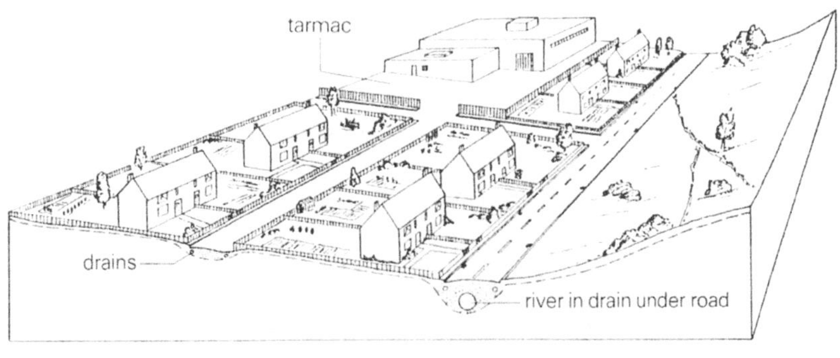

Figure 8.9 Water in an industrial landscape
A tributary of the R. Calder in W. Yorkshire.

Phew, what a scorcher!

On 21 August 1976 *The Times* of London printed the following report on its front page:

'In a move to lessen the risk of fires, many forests were yesterday closed to the public until the end of the drought. Gates will be locked and trails closed. As the water crisis worsened the Welsh National Development Authority became the first undertaking to be granted power to ban the use of water when and where it though fit.

Car washes, fountains, automatic lavatory flushes ... are all to be stopped ... From Monday more than one million people in the south-east (of Wales) will have their water cut off for 17 hours every day ...'

... and in the letters column the inventiveness of the British public included the following suggestion:

'A bucket bath saves a great deal of water as compared to a tub bath, even if only five inches is used. The ratio is one to four, that is water used by Britons for bathing can be saved up to seventy five percent if they abstained from tub bathing and acquired the bucket habit for the duration ...'

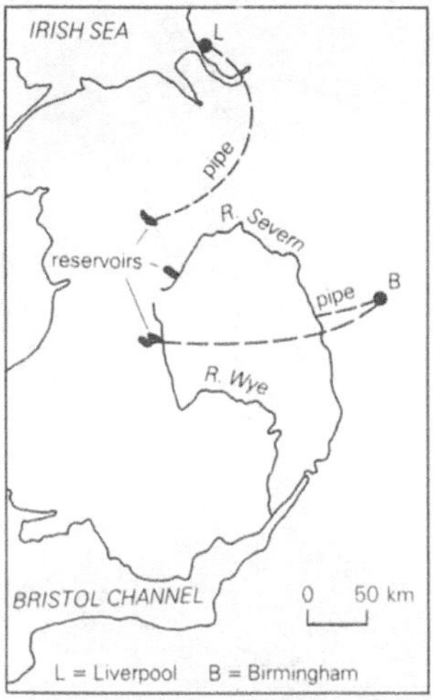

Figure 8.10 Movement of water by pipeline

In the years 1975 and 1976 Europe experienced a dreadful water famine. It seemed that most drastic measures had to be taken as the ground cracked, flowers died in the gardens, and crops wilted in fields. Industry had to be rationed along with households in order to conserve water for as long as possible.

The difficulties experienced at this time show what a delicate balance there is between supply and demand. In areas with a normally regular water supply it is taken for granted that the turn of a tap will produce water, yet this is possible only if rain falls regularly, and even then only because of a vast network of pipes, pumps and reservoirs that move the water from where it falls to where it is needed.

In the past large-diameter pipelines were used to get water from reservoir to town. This became very expensive as distances increased; for example there is a pipeline from Central Wales to Birmingham in central England. However, there are easier and cheaper ways of transferring water. Today new reservoirs are built so that they can let water flow directly into rivers. The river then carries the water naturally to a point nearest to the town that needs it. At this point a pumping station pumps the water out of the river and down a short pipe to the town's purification plant (Fig. 8.10).

Conflict in water demand

Yours or mine?

Water is needed not just for drinking: it is also needed to flush toilets, wash clothes, to cool machinery, as part of many industrial processes, and to irrigate farmland. Moreover, rivers are also used for the transport of goods, recreation and generating electricity. So we have to try to manipulate our water resources in order to serve all of these interests. And while we do this we have to guard against the occasions when we have too much water, or rivers spill over their banks and flood homes and farmland.

You might think that, for man's purposes, the best flow in a river is one that is constant throughout the year. If it were possible to achieve this we would be able to plan for all of our uses to the best advantage. Unfortunately, our demands are not constant. In East Anglia, for example, more water is needed in summer for irrigation while in Devon and Cornwall the huge influx of tourists causes a similar fluctuation in demand. So because rain falls irregularly and our demands change from area to area and from season to season, we need to store water for release at times of need, as well as to prevent flooding. The obvious answer is a **reservoir.** In the past reservoirs were built only to supply water for drinking. They filled up in winter and the surplus just spilled over the top of the dam. Reservoirs like Craig Goch in mid-Wales (Fig. 8.11) are very picturesque but they are of restricted value. Now reservoirs work on a different principle – the outflow is constantly being changed. Take the example of the Clywedog reservoir on the upper River Severn in mid-Wales. In the summer a message from the Midlands might be phoned through to the control at Clywedog 'water level in the Severn getting too low, release more water from the reservoir'. By contrast in winter a prolonged storm over the Welsh mountains might put a lot of water into the river and increase the risk of flooding. A message from Shrewsbury might say 'water is at a dangerously high level here, hold back the water of the Clywedog until water from other streams has passed'.

Figure 8.11 The Craig Goch reservoir in mid-Wales

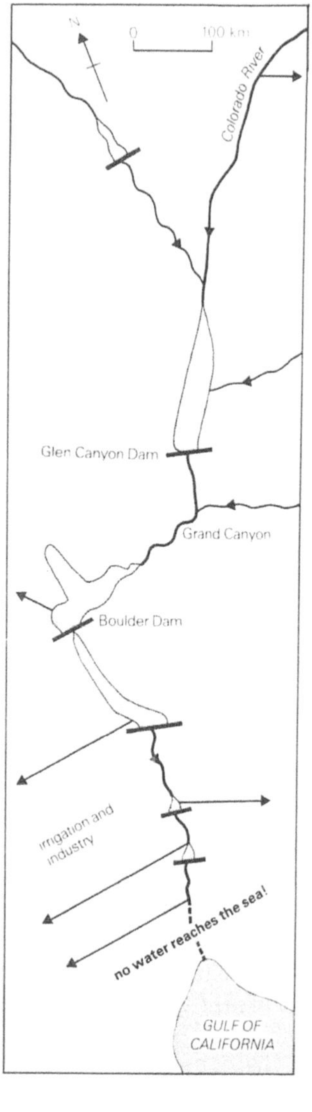

Figure 8.12 The ultimate in river use
The Colorado River is 2000 km long but it never reaches the sea. A series of long lakes store its water for use in irrigating over 1 000 000 hectares of the semi-arid lands of south-western United States and North Mexico and to provide water for drinking and industrial use in the cities of Southern California

In America the ideas behind multiple use of water were employed in many world-famous schemes, including the Colorado River in Arizona (Fig. 8.12) and a whole series of dams to regulate the Tennessee River. The Tennessee is a large river capable of producing much electricity and it was deep enough to be used for navigation. The trouble came because the river flow was very variable. There could be long droughts and then a series of violent storms resulting in flooding. So over a 20-year period 31 dams were built to tame the river. Today it looks like a series of long lakes, but the risk of flooding is small and a large amount of power is constantly available to help boost the economy of the region.

However, there is no point in starting on any regulating programme without a detailed knowledge of the way the water cycle works. You could easily find yourself with a reservoir that never fills, or one that floods after every storm!

Exercise 8.2 Available water

Figure 8.13 shows the pattern of water available for use in Britain for a year. This is the rain that falls, less the amount used by plants and that evaporated directly from the ground. In other words this is the surplus water that goes to fill the streams.

1 Describe the differences you would expect between winter and summer surplus water.
2 Which areas would have little water available in the summer?
3 Sketch a map of Britain from your atlas. Mark up the major cities and urban areas. There is no need to name them.
4 Now compare your map with the map of water available. Into which zone do most of the centres of population fall?

By comparing areas of high demand (centres of population) with available water, you can see the problem facing the water engineer. With much surplus water in the West and many centres of population in the East he is forced to consider ways of moving water large distances.

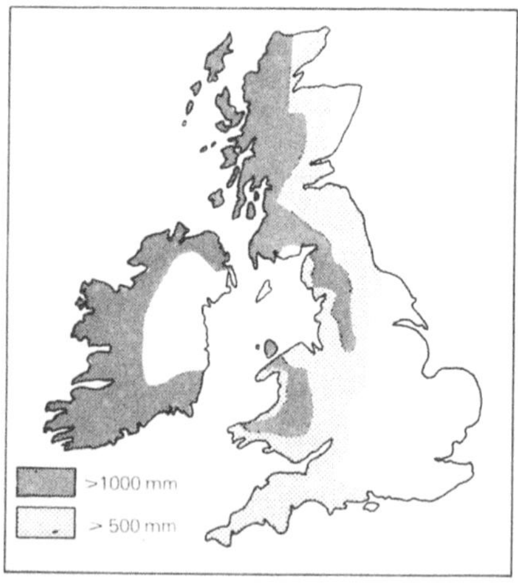

Figure 8.13 The water available for use in Britain

Exercise 8.3 Underground supplies

Figure 8.14 shows the useful water-bearing rocks (aquifers) in Britain.

1 Refer back to the nature of aquifers (Fig. 8.6) and their similarity to a sponge. If a well is put into an aquifer, and water is pumped from it, why will it not flow quickly and why will water be obtainable for a long time?
2 This water – called **groundwater** – is a useful constant supply which is refilled naturally from rainfall. Compare a population map of Britain from an atlas with the map of groundwater supplies. List the areas where groundwater supplies might be the most useful, and name the rocks involved.

Figure 8.14 The distribution of water-bearing rocks in Britain

Exercise 8.4 Surface supplies

Groundwater supplies are rarely sufficient on their own because the rate of supply is low. As a result water has to be taken directly from rivers. Unfortunately we want extra water at just the time the rivers are low – in the summer. So we have to find some means of storing water from winter, when rivers are full, to summer. The usual method is to build reservoirs.

In Figure 8.15 the plans for two proposed reservoirs are shown. The main reason for preferring one site above another are (a) it should not use valuable farmland, (b) it should only require a small dam and (c) it should have a reliable water supply.

State which reservoir site you would use and explain the reasons for your choice.

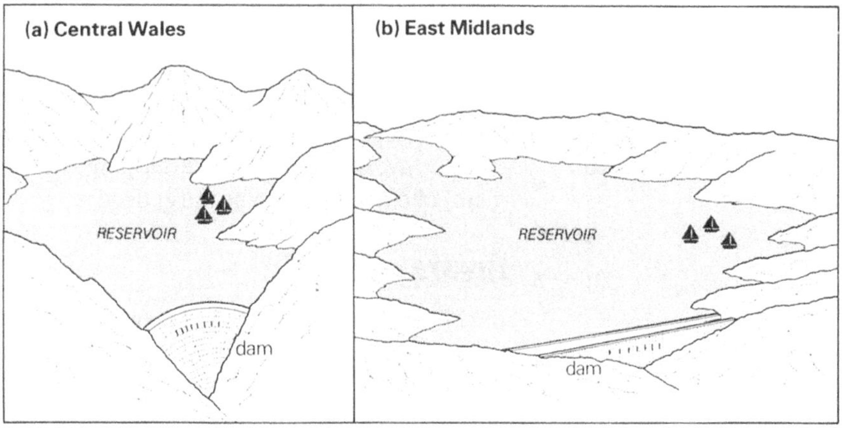

Figure 8.15 Two reservoir designs
(a) Site in Central Wales; dam height 40 m, length 100 m; land to be flooded 1 km², reservoir storage 3 000 000 m³; present land use – rough grazing with sheep; number of houses to be submerged – one.
(b) Site in East Midlands; dam height 15 m, length 250 m; land to be flooded 2 km²; reservoir storage 1 500 000 m³; land use – mixed arable and beef farmland; number of houses to be submerged – ten.

Reservoirs in developing countries

Lessons of Imagi

Engineers in the developed countries such as the United States have a large supply of data to call upon when designing reservoirs. They have river-flow and rainfall data and they also know how much water will be locked up in the soil or used by plants. As a result, if they do their sums correctly, they will end up with a reservoir of the right size for the job. If they understand which rocks are water bearing and which will not let water seep out, then they can also site their reservoir correctly. In developing countries this information is much harder to obtain and many of the early reservoirs were built on guesses. Some guesses worked, but the following is the story of one that didn't.

A town in Tanzania (then Tanganyika) called Dodoma needed water. Before 1910 people depended on an irregular supply from rainponds which dried out in the dry season and after that they had to dig holes in river beds. Around 1910 Dodoma became an administrative centre and the population increased, so a better water supply was needed urgently.

A reservoir was begun in 1929. Unfortunately at this time there were no long-term records of rainfall or evaporation, so the size of the reservoir had to be estimated. Eventually a guess was made and a dam constructed to hold the not very inspiring depth of up to 12 m of water.

The region receives about 500 mm of rain a year. This is the same as parts of South-East England, but Tanzania is a very hot country and evaporation is high. Also much of the rain comes in heavy downpours over just a few months. This means that, while rain is falling, there is a lot of water that runs straight over the surface. The result is that relatively little sinks into the soil to keep streams flowing in the rainless months. So, the rain reaching the reservoir in the rainy season has to last all year and this can be quite a problem with such a high rate of evaporation. In addition, the drainage basin was forested, so much of the water stored in the soil was taken up by tree roots during the dry season.

As a result of all of these factors there was insufficient water to fill the reservoir in years with low rainfall, and constant use by the town caused it to run dry. In an attempt to increase the water reaching the reservoir the trees were cut down. More water did, indeed, subsequently reach the reservoir, but it was full of mud eroded from the now-bare soils. So in a short time the reservoir began to silt up badly and its storage capacity was severely reduced (Fig. 8.16). By 1944 a second reservoir had to be built on a new river. Using lessons derived from mistakes with the first reservoir, the new one was arranged to drain a much larger area, so the reservoir filled even in dry years, without the need to cut down the trees.

Figure 8.16 The effect of sediment accumulating in a small reservoir Large reservoirs also silt up, but more slowly.

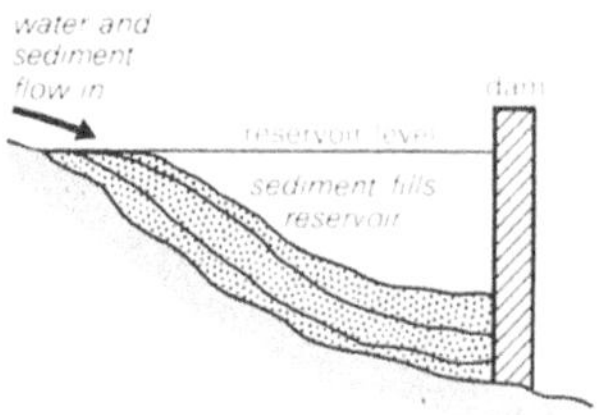

Water supply can be effectively designed only with a full knowledge of how the water cycle works. Unfortunately, the data needed for accurate design are still not available in many developing countries, and without them improvement schemes can never be fully effective.

Water pollution

Rivers – R.I.P.?

In 1775 the English inventor Alexander Cumming patented a flush mechanism for toilets. This was improved a century later by the famous Thomas Crapper. Unfortunately, these actions hastened the decline of many of our rivers into sewers.

In the 19th century the rivers of the developed world provided power to turn the wheels of industry, carried its goods, provided raw material for the chemical

Figure 8.17 The river as a means of disposing of wastes

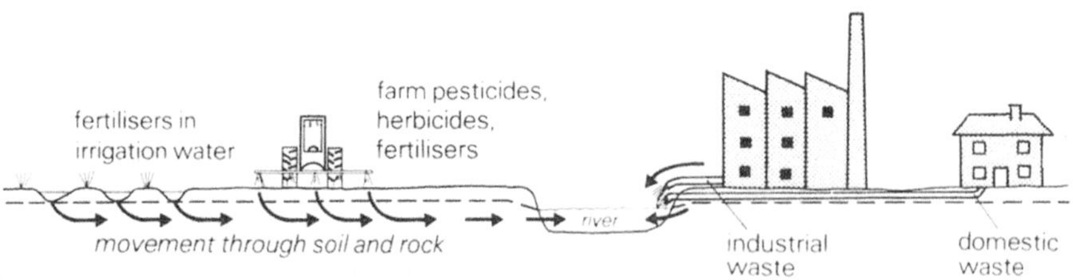

industry, and were a convenient means of disposing of its waste (called **effluents**) (Fig. 8.17). But toxic waste from factories killed off the bacteria that naturally live in a river. It was a disastrous step because it is the bacteria that decompose dead organic matter naturally and safely. So when the Victorians built indoor plumbing systems and flushed human waste into the rivers, there were not enough natural organisms to decompose it. And so the stench rose, disease was spread and the fish died. The rivers had been killed because of greed and a lack of understanding and right through to the middle of this century very little was done about it. By 1958 even vast expanses of water like the 25 000 km^2 Lake Erie in North America were dead round the edge and dying in the middle. And, as if to symbolise the disaster of river pollution, in 1969 a river in America covered with oil and floating debris caught fire!

Of course pollution was worse in the summer when river flows are low, and worse still at the tidal limit where sewage floated out on the ebb tide and back in on the flood tide. There was a vital need to understand how the water cycle worked and what limits could be imposed on it.

It was not just industry and households that were to blame. For over half a century we have been pouring increasing quantities of fertiliser, pesticide and herbicide onto the land. This is washed through the soils by rainfall and, of course, it eventually finds its way to rivers. The fertilisers provide food for algae which then multiply and turn the water green. They use up all the oxygen in the water so there is none left to help in decomposing dead organic matter properly, and sewage just floats in the water. So rivers have been killed by agricultural pollutants in rural areas, industrial pollutants in cities, and human sewage in both.

You can see the geography of pollution by looking at any atlas map. In Europe, America and any other place with industrial development beside rivers, you can almost guarantee that pollution will occur. In the Netherlands the Dutch have stopped taking water from the Rhine: it is so polluted by German and French industry that it has become too costly to filter clean and sterilise for use as drinking water. Britain and the USA now lead in trying to cope with this problem, and the River Thames, for example, has been cleaned up to allow fish to thrive once again even in the City of London itself.

Improvement in water quality has mostly been achieved through laws controlling the pollutants that industry may discharge into a river and by building better sewage works. Our rivers are very forgiving, but they can only cope with a certain load of waste, and no more. Now we understand what these limits are, we are in a better position to allow the rivers to return once again to their natural condition and be of benefit to us at the same time.

Irrigation problems

Water – handle with care!

Throughout the world, more water is used for irrigation than for any other purpose. Just six countries (China, India, USA, Pakistan, USSR and Iran) account for over 70% of irrigated land, 37% (84 million hectares) in China alone. Here, more than anywhere, water is life. Traditional irrigation methods allow water to flow into a field which has a low bank built around it. The water ponds up and sinks slowly into the soil (Fig. 8.18). Very often water is led along trenches

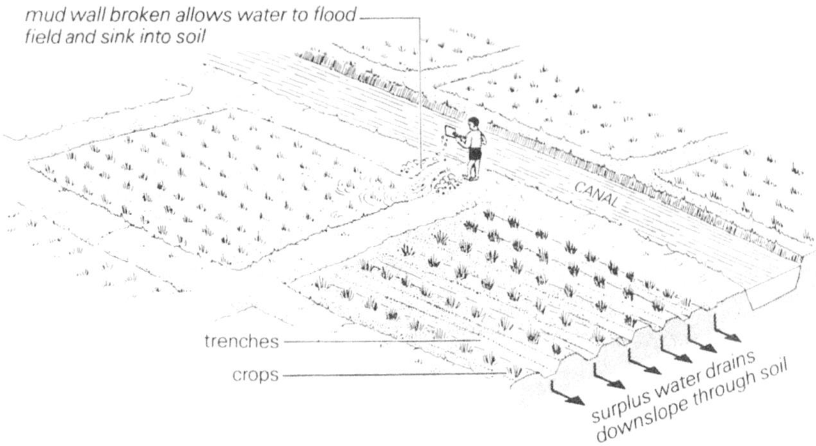

The principle of irrigation in developing countries.

Figure 8.18 Simple irrigation methods

Traditional irrigation uses employ animal power to lift water from a river and pour it into a canal. Fields are irrigated by breaking down the mud walls. This photograph was taken in India.

or furrows dug between growing crops. All irrigation systems need a lot of attention. Usually water is fed into a field by breaking down a mud wall from a nearby canal and the flow is stopped by building up the wall again. In all cases water on the surface evaporates rapidly, so much more needs to be applied to the land than the plants really need. And by diverting all this water from rivers onto the fields, little is available for users downstream.

Some of the world's irrigation schemes are on a massive scale and they represent one of man's greatest influences on the water cycle. In the Indus valley in Pakistan over 100 years of effort has provided water to irrigate more than 10 million hectares of land (Fig. 8.19).

RAYMOND JESSOP

However, irrigation can also bring problems. With large schemes so much water is poured into the soil that it may cause a permanent rise in soil water levels (Fig. 8.20). In downstream regions water can even be forced to the surface. But this water contains many salts that have been washed out of the rocks and are now brought to the surface in such quantities that salt crusts form on top of the soils and plants will not grow.

Great care has to be taken with irrigation to avoid this problem, otherwise as new areas are developed for irrigation, others will be killed off and abandoned.

River control

Taming the Nile

The demands for irrigation can sometimes be so great in arid and semi-arid areas that the water cycle is disrupted completely. In the southwestern United States the famous Colorado River never reaches the sea because all of its waters are diverted along canals to irrigate California and Mexico (Fig. 8.12). Even the African Nile, one of the world's largest rivers, does not reach the sea for three months each year.

The Nile begins its course in the mountains of Central Africa (Fig. 8.21). Here, although rainfall is plentiful, it is seasonal, providing a summer flood to the Nile

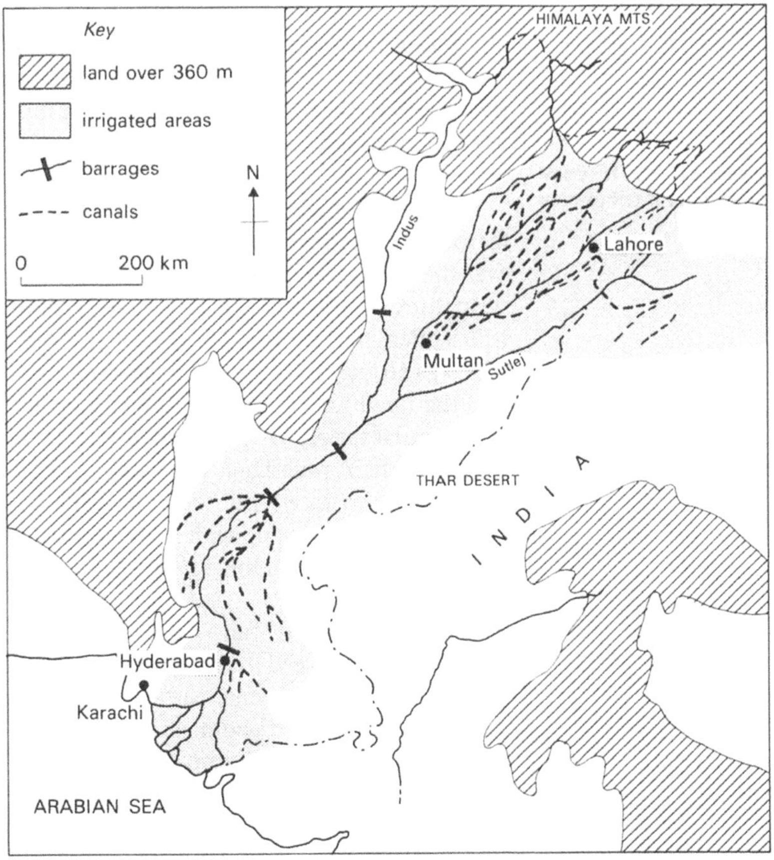

Figure 8.19 Irrigation of the Indus valley
Here 60 000 km of canal irrigates more than 10 000 000 hectares of land.

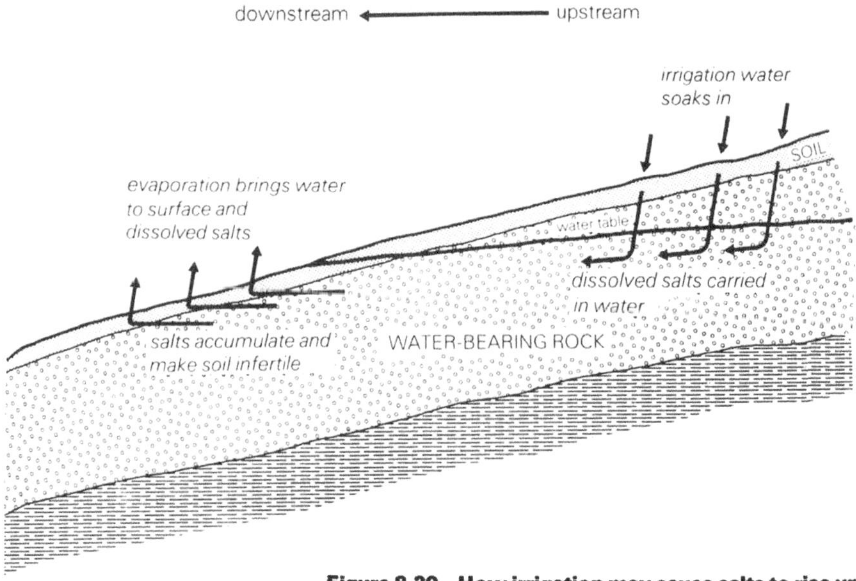

Figure 8.20 How irrigation may cause salts to rise up and pollute agricultural land

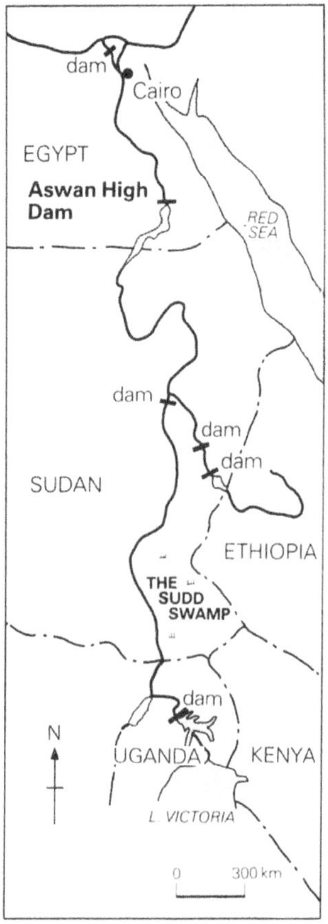

Figure 8.21 The Nile and its international tributaries

but giving much less water in winter. As the Nile flows north towards the Mediterranean Sea, the last 2000 km of its course are through desert. Here the river provides the only means of life to the millions of people that live near to its banks. Irrigation of the lower Nile valley, and especially the Nile delta, goes back over 3000 years.

Today the population in Egypt is over 35 million and still rising, and the need for food is getting ever greater. Traditionally farmers would irrigate fields whenever there was enough water in the river, but the variable supply did not help agriculture to be as efficient as it might. This provided the Egyptian government with a dilemma: how could they provide more water when the natural flow of the Nile was already fully used? The answer seemed to lie in building a big dam to trap the flood waters and to use this 'surplus' to irrigate new lands and allow more regular supply to those areas already being irrigated.

In 1953 the Aswan High Dam was built and the Nile was tamed. Now new lands can be irrigated. But there has been a cost to pay. The reservoir (called Lake Nasser) is filling up with sediment that used to make the fertile soil of the valley. The farmers no longer get free fertiliser and they have to buy expensive artificial substitutes.

Sudan has also built dams on the Nile to irrigate its northern lands. South of the dry lands that need irrigation there is a huge swamp called the Sudd. Now a project is under way to dam part of the Sudd for agriculture and release some of the swamp water for irrigation downstream. To do this the Jonglei Canal is being dug right through the swamp. But again, benefits to some bring hardship to others. The local Dinka tribesmen will have to stop herding cattle near the swamp and try to settle to cultivating land, and the use of yet more water for irrigation in these upstream regions leaves less for those downstream.

EXAMINATION-STYLE QUESTIONS

1 Explain what is meant by (a) a depression and (b) an anticyclone. Draw a diagram of a frontal system in a depression, showing the main types of cloud that are produced at each front. Explain why rain is produced at the frontal regions.

2 Explain why anticyclones usually produce fine warm weather in summer but cold weather, very often with fog, in winter. Explain why areas such as the Sahara experience dry sunny conditions throughout the year.

3 With the aid of maps and diagrams describe and explain three of the following: descending winds, fog and mist, rainshadows, and coastal breezes.

4 Within recent years there has been a marked increase in winter sunshine in most urban areas. State the nature of the pollution that could cause a reduction of visibility and explain what measures were taken to alleviate it. Explain why London, being in a valley, would be particularly susceptible to this kind of pollution.

5 In a series of diagrams describe and explain how (a) cumulus clouds form and (b) stratus clouds form.

6 Name one area which experiences tropical storms (hurricanes) frequently. Draw a diagram to show how a tropical storm develops, and show on a map the typical path of such a storm. What are some of the problems of living in an area with tropical storms?

7 Explain why it is that South-East England experiences the highest temperatures in Britain. Why does snow last longer in the Grampians than it does on the slopes of Ben Nevis? Explain why the coast of Yorkshire receives a lower rainfall than the coast of Lancashire.

8 Using the maps of Figure 5.2 describe and explain the main trend of isotherms for January and July. How far is it true to say that the climate gets cooler from south to north and drier from west to east in the British Isles? Using Figure 6.2 explain why it is less true in winter than in summer.

9 Describe and explain the differences you would expect to find in temperature, rainfall and cloudiness as you went up a mountain from 200 m to 10 000 m. Explain the differences in climate you would expect to find between a station at 200 m on the windward side of a mountain and a station of a similar height on the leeward side of a mountain.

10 Explain what is meant by a maritime climate and a continental climate. Describe the main differences you expect to find between the climate of South-West Ireland and Central Germany.

11 List the main characteristics of an equatorial climate. Explain how the Sun's heating is responsible for the daily pattern of weather within such a climatic region.

12 Name two countries which have a tropical monsoon climate. Explain the difference between an equatorial and a monsoon climate. Describe the sequence of events that leads to the onset of the monsoon.

13 Name three hot deserts. Explain how sinking air is responsible for the presence of large hot deserts such as the Sahara. Explain why there is a marked contrast in day and night temperatures in deserts.

14 Draw a labelled diagram of the water cycle. Explain the differences between the flow of a river on (a) permeable and impermeable rocks, (b) a forested area and a large town.

15 Explain why a river floods more frequently in an area with steep slopes and thin soil than in an area of gentle slopes and permeable rock. Explain how a reservoir may be helpful in preventing floods. Today reservoirs have many uses. Name an example of a modern reservoir that is used for several purposes and explain how it is operated.

16 What is an artesian water supply? Draw a diagram to show how artesian water supplies are obtained. Explain why these supplies are particularly valuable in (a) urban areas such as London and (b) desert areas such as the Sahara.

17 Explain why irrigation is necessary in some areas and drainage necessary in others. Describe one primitive method of irrigation which is still in use today and give one disadvantage of this method. Describe one major modern irrigation scheme and explain how it works.

18 Explain why England still has problems with its water supply despite the fact that it receives about the same amount of rainfall in each month. Describe some of the ways in which the problems of water supply have been tackled in the British Isles. For any one country, explain the methods used for irrigating land, and their advantages and disadvantages. Describe the differences between one large primitive irrigation scheme and one large irrigation scheme in a developed country.

Part three Soils, vegetation and man

All things form delicate links between the Earth and the atmosphere. Each species has slowly adapted itself to occupy a special place within the world's landscapes.

Figure 9.0 Contrasts in ecosystems

Natural (a Joshua tree in a semi-arid region)

Urban (San Francisco)

9 Weathering

Weathering

A grave lesson

When archaeologists dug down under the remains of an Iron Age fort in Hampshire they found much evidence of past occupation. There were bits of broken pottery and a few coins from a long-forgotten civilisation. But most interesting of all was the skull and bones of a man aged about 25 who had been killed by a blow on the head with an axe.

These gruesome finds have important lessons for the geographer as well as the archaeologist. Notice what was found – bones, coins and pottery. Presumably when the body was buried it was complete with flesh and clothes. Since the burial the skin and clothes had rotted away and the coins had lost much of their surface marking. By contrast, the pottery was still intact and had not even lost its decoration. Clearly organic matter – flesh and clothing – very readily decomposes (rots away) whereas bones, which are mostly made from inorganic materials such as calcium carbonate, and calcium phosphate decompose more slowly. Nevertheless the bones were showing signs of considerable surface attack. The coins had started to lose their surface marking, indicating that metal is attacked by chemicals in the soil. The clay pottery was untouched, so clay must be more resistant than all the other materials.

Natural breakdown processes are divided into **physical weathering** (the *disintegration* of rock into smaller fragments) and **chemical weathering** (the *decomposition* of rock or organic material into new materials). In the grave we have noticed only evidence for chemical decay, showing that soil cover prevents physical weathering. Physical weathering is important only in places where bare rock can be found, such as on a mountainside, a cliff, or in a desert.

Figure 9.1 Frost shatter Even a glass bottle cannot withstand the pressure of a liquid freezing. (Reproduced by kind permission of J. Sainsbury Ltd.)

A chip off the old block

Frost shatter

Many parts of the world experience cold, frosty conditions at night. A frosty night will sometimes cause ice to form on wet roads and be a danger to drivers. It will also cause rainwater to freeze in the cracks of a rock.

When water changes to ice it expands by about 10%, creating an extremely powerful and destructive force. Figure 9.1 shows the reason you are told not to put full bottles of liquid in the freezing compartment of a fridge. As water freezes it expands and cracks the bottle. The process is called **frost shattering.**

A bottle is simply a piece of transparent rock, because glass is mostly made from sand. As a result we can compare freezing water in the confined space of a bottle with water freezing in the cracks of a mountainside rock. In nature all that is needed, therefore, is a rock with some cracks in it, rain to fill them, and a large natural freezer. One of the best places to find such a combination is in a mountain area in autumn or spring, when day temperatures can be above freezing but nights are usually still frosty. Any rain falling by day will seep into the cracks and be frozen at night. As the water freezes it expands and it may

Figure 9.2 Frost shatter A block split by freeze/thaw. The black camera case gives the scale.

wedge pieces of surface rock apart (Fig. 9.2). This provides room for more water the next time rain falls. During the day the ice melts and water may also be able to penetrate farther into the rock, before freezing again at night. After many freeze/thaw movements, pieces of rock may well be dislodged so far that they become unstable (Fig. 9.3). The next time the ice melts and releases these pieces from their frozen prison, gravity will cause them to fall away, at the same time exposing new material for shattering.

Frost shatter is most active on mountains and at high latitudes where the temperatures hover around freezing for several weeks at a time. However, the effects of frost shatter will affect any exposed rock. Many quarry faces and coastal cliffs have small piles of shattered rock at their feet after a period of frost. These piles of rock fragments are called **scree.** Frost shatter even affects some roads in a very cold winter. If the water in the soil beneath a road freezes, it will expand and lift up the surface, usually causing the tarmac to crack. In Scandinavia and other northern areas the repair bills for frost-shattered roads is a large part of their total road costs.

Exfoliation

Rockflakes

Much of the income from Middle-Eastern oil is being used for building factories, offices and homes in what were undeveloped countries only 20 years ago. The

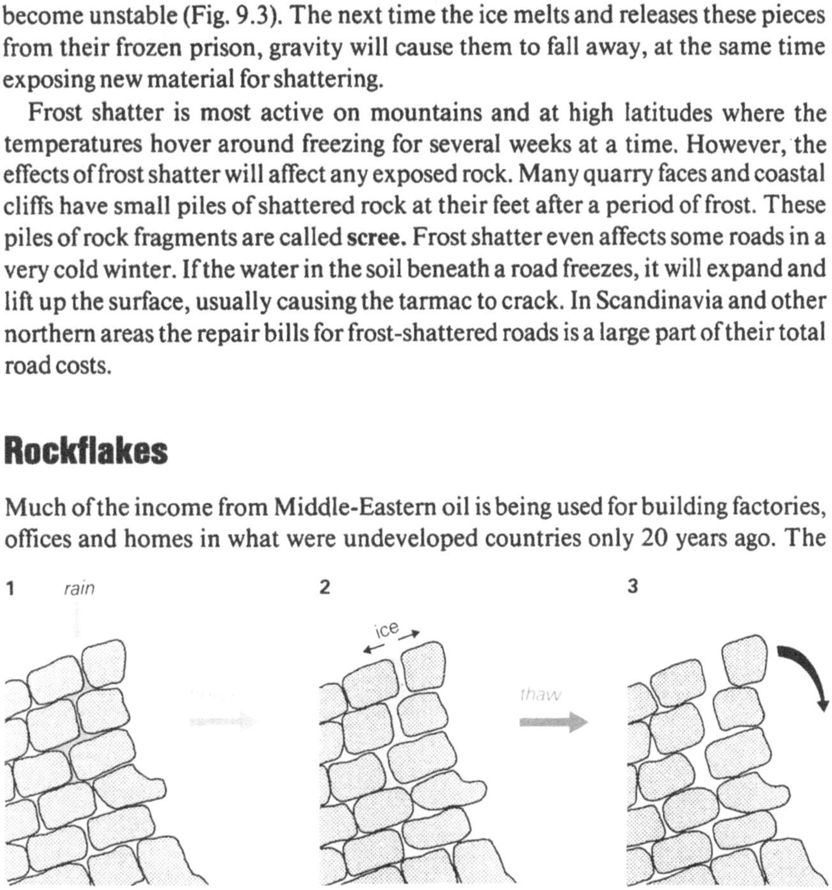

Figure 9.3 Frost shatter (1) This rock has joints which allow water to seep in. During the winter the water expands as it turns into ice. (2) As a result some blocks are prised away from the main mass of rock. (3) Eventually they fall down to accumulate as scree at the base of the slope.

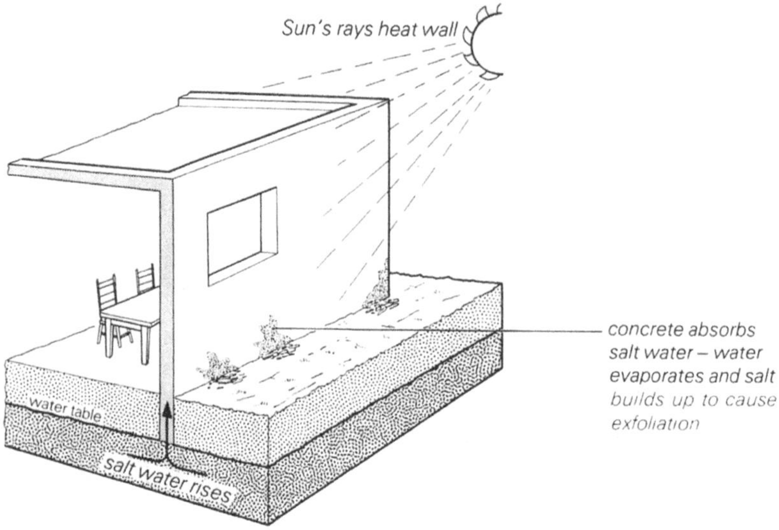

Figure 9.4 Exfoliation Water containing salts has been seeping up these walls for several years. As the water evaporates, the salt crystals force the wall to break off in sheets.

only trouble is that some of the houses are breaking up near their foundations. Great flakes of concrete are being shed from new buildings (Fig. 9.4), causing the people great concern. The process at work in these 'flaking houses' is a speeded up demonstration of a common desert phenomenon called **exfoliation.** It happens whenever water seeps into rocks that are heated and cooled rapidly.

Deserts experience great heating by day and cooling at night. As a result, rocks continually expand and contract. When the rocks are heated they swell, pores and cracks open, and it is easier for rainwater to get inside. Although rain is rare in deserts and does not last long, some water will always soak into the rock. But rainwater is often a mixture of water and sea salt. Soon the water is evaporated away, but the salt is left inside the rock (Fig. 9.5). When the rock cools again at night it cannot contract properly because of the salt, and great stresses are set up within the surface layers. The next time rain falls, the process is repeated and more salt is deposited in the rock. Eventually the rock cannot withstand the forces any longer and it fractures parallel to the surface, producing flakes usually a few centimetres thick (Fig. 9.6). Surprisingly perhaps, heating and cooling without rain in a desert does not cause exfoliation. As a result it is easier to see the effects of this type of physical weathering near desert coasts where there is more salt and more moisture, or near oases and other places where water comes close to the surface. The hottest and driest parts of deserts probably produce very little weathered rock.

Figure 9.5 Exfoliation With repeated showers of rain salt crystals grow and help prise scales of rock from the main mass.

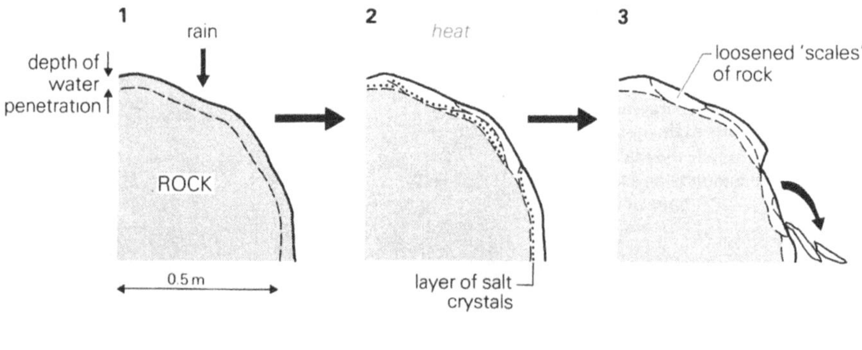

Figure 9.6 Exfoliation on a rock of granite

Chemical weathering

The rot sets in

There is much to be thankful for in the slow rate of physical weathering in a desert. Because of it the pyramids and temples of Egypt still look much as they did 3000 years ago; even their painted surfaces are sometimes fresh and brilliant. In the middle of the 19th century, it was decided to bring one of the Egyptian monuments back to London and to erect it beside the Thames; the people of London nicknamed it 'Cleopatra's Needle'. Within a few years it had become so weathered by chemical rotting that the surface writing could hardly be recognised. In fact it has recently had to be restored. Perhaps it would have been better left beside the Nile! (See Fig. 9.7 for another example.)

Figure 9.7 The result of chemical weathering over the past 100 years

We all know of extreme forms of chemical weathering. In industry, acids are used to 'clean' metals – there has even been a famous case of a murderer who disposed of his victims by dissolving them in a bath of acid. In nature there are many less violent but nonetheless effective forms of chemical weathering. The most important is a reaction between rock and the very weak acid that we call soil water.

If a piece of rock is kept in a tank of rainwater there will be a very slow chemical reaction between the water and the rock surface. After a while – perhaps a few months – although you would probably not see a change, a chemical analysis of the water would show that it contained more soluble materials than at the start of the experiment. These soluble materials were once contained within the structure of the rock and have formed because of chemical weathering. A microscopic examination of the rock would show the surface to be pitted. The pits show where the rock has been weathered. If rainwater were left in contact with the rock for a 100 years, a much more obvious change would

Topic Rocks, blocks and decay

Weathering is a surface reaction, so the greater the surface area of rock, the faster it weathers. Rocks are formed deep below the Earth's surface under great pressure. Often the pressure causes sedimentary rocks to fracture along lines of weakness. A rock made from many bands of sediment will have many weaknesses and will fracture into small blocks (Fig. 9.8). A rock made from a uniform material will have fewer cracks in it. Igneous rocks were once molten and their fractures come from contraction on cooling (Fig. 9.9).

When rocks come to the surface, there is much less pressure on them, and the fracture lines open up, providing places for chemical weathering to work. Rocks that have many factures (**joints** and **bedding planes**) will weather rapidly. The granite has few joints and this helps slow down the rate of weathering.

Figure 9.8 Alternating beds of limestone and shale
Joints and bedding planes are common and weathering is rapid.

Figure 9.9 Granite weathering
Granite has few joints and weathers slowly.

probably occur. The surface of the rock would be covered in a loose, very fine material called **clay**. Clay is the other main product of chemical weathering.

Although pure water will react with rock, usually rainwater can be regarded as a dilute acid, because it has gases such as carbon dioxide dissolved in it from the air. Carbon dioxide gas dissolved in water produces carbonic acid, although it is extremely dilute. Near industrial areas or volcanoes there will also be sulphur dioxide in the air, and when this is dissolved in water it ultimately makes sulphuric acid. These dissolved gases increase the rate of chemical weathering. As you would expect, the stronger the acid, the faster the weathering. Indeed, it has been the high level of pollution from sulphur dioxide that has speeded the decay of Cleopatra's Needle and many other famous buildings.

There is a further source of acidity. On the surface of the soil there is often a layer of partly rotted leaves; this is what gives the 'springy' feel if you walk through a woodland. As these leaves rot they release acids which combine with the rainwater as it soaks through to the soil. They act in the same way as the gases dissolved in water but they are much more powerful. Thus, water moving through a soil is armed with two types of acid. These are the acids that can weather rock and produce new soil.

Chemical weathering affects the *surface* of a rock. Rocks that are well jointed therefore provide many more surfaces to attack than does a rock with few joints. Although all rocks are subject to chemical weathering, its effect on joints can be seen particularly clearly in some limestone areas. The soil and vegetation have been stripped away by glacial erosion and the rock surface exposed to view. Chemical weathering has enlarged every joint, making the area really quite difficult to walk over (Fig. 15.1).

SUMMARY: WEATHERING

Weathering is the natural breakdown of rock. *No movement is involved,* and, unless material is transported away, weathered material will just build up where it is formed. To speak of a rock being eroded it must be both weathered and transported from the place of weathering.

Physical weathering is the disintegration of rock by mechanical force. This can be the result of water freezing and expanding in cracks (Fig. 9.3) – a process called **frost shattering** – or of the deposition of salt in rocks that are expanding and contracting in a desert – a process called **exfoliation** (Fig. 9.5).

Physical weathering can produce only fairly large fragments. Anything smaller than a grain of sand is usually the result of **chemical weathering.** Chemical weathering is the chemical reaction of water or a weak acid with rock. This reaction forms totally new materials. Rainwater is normally a weak acid because it has carbon dioxide gas dissolved in it, making carbonic acid. Water passing over decaying vegetation also produces acids. Both types of acid can be very powerful in rotting rocks, even at depths of many metres. The new products of chemical weathering are mostly very small particles called **clays.** The material left over after the clays have been formed is usually soluble and is carried away in solution with the rainwater.

In places with a vegetation cover, chemical weathering is more important than physical weathering. Soils are largely the result of chemical weathering.

SUMMARY

10 Soil formation

Soil components

A little bit of this and . . .

As weathering occurs all over the Earth's surface, the results of it should be fairly easy to see and have a well known name, such as the scree mentioned in the previous chapter. Scree slopes form when many pieces of physically weathered rock have accumulated at the base of a slope.

Scree slopes are common in mountain areas and at the base of some sea cliffs – but it is unlikely that you pass a scree slope on your way to school! The absence of screes from most areas tells us that physical weathering is *not* the most usual form of weathering. The other form is chemical weathering, but if this is so widespread, where can we see it? The answer to this is 'more or less everywhere', because the usual term for chemically weathered rock is **soil.** This is an important and often overlooked fact, so the next time you go into a garden remember that you are looking at plants growing in rotted rock.

Soil is the normal product of chemical weathering, but as it forms it covers up the solid rock, preventing physical weathering processes from operating. So the rule is: *where you find bare rock, physical processes dominate; where the rock is buried beneath soil, chemical weathering processes have taken over.*

Soil consists of a whole range of 'bits and pieces'. There are some large pieces (which we commonly call **sand, gravel** or **pebbles**) that are made of solid rock which has not yet been completely weathered; there are the very small **clay** pieces that are one of the products of chemical weathering, and there is quite a lot of water (which is actually performing the weathering) in which the soluble products of weathering are found (Fig. 10.1). Then, besides all of these inorganic pieces, there are lots of tiny fragments of dead plant together with earthworms, woodlice, mites, and many tiny organisms. If we use a microscope to look at a piece of soil, we see it crowded with tiny living things. There will probably be millions of white threads of fungus, tens of millions of bacteria, and many other organisms (Fig. 10.2).

Figure 10.3 (Right) Leaf decay
(1) As soon as this leaf begins to die it shrivels, dries up and falls to the ground. (2) The leaf is then decomposed by soil micro-organisms and gradually loses its identity. (3) The decomposed fragments (called humus) are taken into the soil by earthworms digested, coated around other soil particles and thoroughly mixed with the rest of the surface soil, (4).

Figure 10.1 The components of a soil

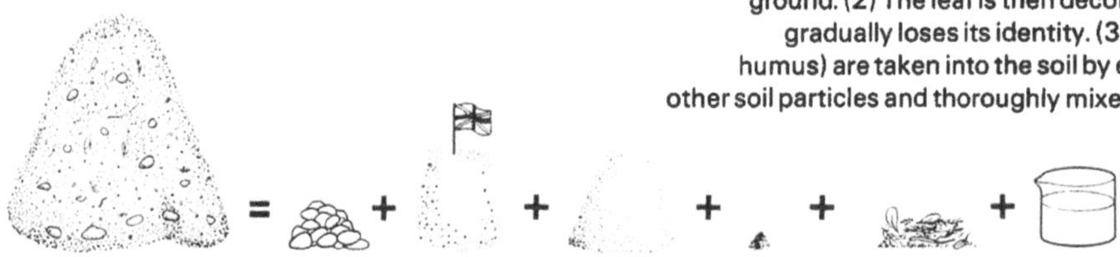

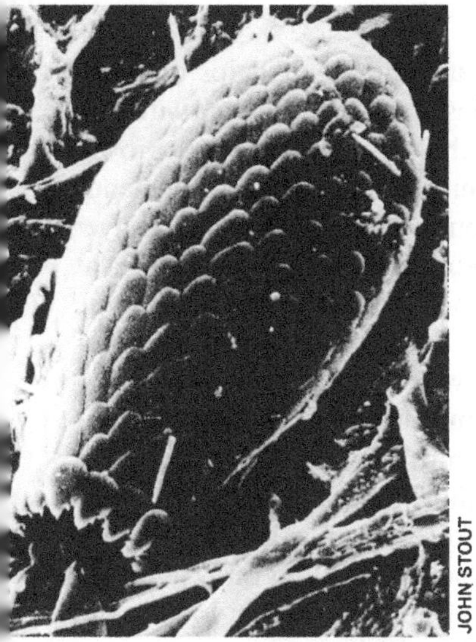

Figure 10.2 Soil animals This strange object looks as though it might have been invented for a science fiction film. It is, however, an extremely common soil organism (a protozoan) enlarged 12 000 times. There are myriads of such beautiful and mysterious objects in every gramme of soil. The twisted threads are pieces of fungus.

It is worth taking a closer look at soil. Imagine a world where you have been shrunk one million times. You will then be the size of a large clay particle! The soil suddenly becomes a world of massive boulders (sand grains), with an intricate network of caverns between. We would call these caverns **soil pores,** and they are important because they allow rainwater to reach down into the soil while allowing the surplus to drain right through. If you were one of the millions of small soil animals you would regard these pores as your home and also your hunting ground for food. If you were a plant root, then they would be easy paths along which to grow as well as sources of water and air.

Both plants and animals need oxygen in the soil pores to survive. Most plants cannot survive complete waterlogging of a soil for long periods because they cannot get enough oxygen to their roots. There are some exceptions, such as rice and the plants that grow by the waterside of most streams and ponds, but a plant needs to be rather well adapted to survive in a very wet place.

From a microscopic viewpoint the soil's inorganic bits and pieces provide a sort of hotel for an amazing variety of small animals and also plant roots. But both plants and animals need to live, so where is the source of food?

Hotel Animalia *Soil life and organic matter*

The soil contains a whole army of small creatures which decompose dead plant material and turn it into a form suitable for plant roots to absorb. Some of the creatures, such as earthworms, are easy to see, but most can only be seen under a hand lens or a microscope. A hand lens focussed on a piece of soil will probably show little orange blobs of 'jelly': these are mites. There will also be little bits of 'animated hair' which are really nematode worms. But these are only the beginning. Every cubic metre of soil contains tens of millions of such creatures, all workers in the soil hotel.

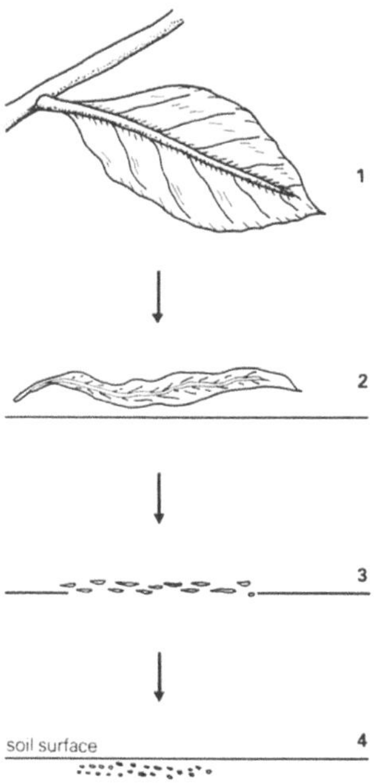

The soil hotel contains various departments. On the surface is the storeroom of dead plant matter such as dead leaves, twigs and tree trunks. There is a continual supply of material to this surface store, but it is quickly removed (Fig. 10.3). Earthworms, ants, mites and other creatures drag plant tissue underground into burrows; if it is too large, they chop it up on the surface. These creatures are like waiters carrying food from the storeroom to the kitchen. Their burrows also help keep the soil ventilated and drained. In the burrows lie millions of bacteria, fungi and other micro-organisms. These are the kitchen staff. They decompose the organic material and produce a sticky substance (**humus**) which stains the soil black. They also produce a range of soluble foods called **nutrients.**

The guests in the hotel are the plant roots. They can take in only soluble foods, so they suck up water which contains nutrients released by decomposition. The plant roots spread out in a very efficient net, delving into a multitude of soil pores especially to find more soluble food. The nutrient foods they find are sent up the plant roots and used to help plant growth. However, eventually these plants also die and their remains are added to the surface storeroom of the soil hotel for future use by other guests.

Completely decomposed organic matter is called **humus.** It performs three vital roles in a soil: first it sticks soil particles together; secondly, it provides an

Figure 10.4 Soil binding
When you pull up a clump of grass, notice how the fine roots intertwine and hold the soil. Notice too how the soil has formed itself into clumps, so making the binding task of roots much easier.

additional supply of nutrients for plant use; and thirdly, it helps to retain moisture. Humus is easy to see in a soil because it is black. When mixed with iron compounds (which are a dull-orange colour) the soil becomes dark brown. Earthworms are mainly responsible for mixing humus, iron and rock fragments. If you look at the sides of a pit dug into a soil, you will see that the surface layer (which has had the 'earthworm mixing treatment') is dark brown, whereas at depth it is much lighter (Fig. 10.7). Earthworms don't like living at depth because their 'restaurant' is on the soil surface. The lighter brown colour of the lower layer is the true colour of the inorganic fragments without humus staining.

The sticky nature of the humus is vital because it prevents individual soil particles being washed or blown away. Also 'clumps' of soil particles do not fit together very well and they provide gaps through which water can drain. Next time you are in a garden, pick up a small piece of soil and look at it carefully. You will see that it is made from a range of different particles, all invisibly gummed together. You can rub the particles apart, but the gum is sufficient to keep the particles stuck together in clumps within the soil (Fig. 10.4).

One thing after another *Soil layers*

As we have seen, soils begin from the weathered products of rock. Sometimes the rock is not solid, such as when it has been deposited in fragments by a glacier, by a river as **alluvium** or by wind as **loess.** Nevertheless, in all cases the material from which a soil forms is unorganised and it doesn't contain organic matter. It is called the soil **parent material.**

Soils are divided into three types of layers: the surface layer containing a mixture of plant and weathered rock material; a middle layer beneath the surface which contains only almost weathered rock; and a bottom layer – the parent material –which has not been weathered at all. We use letters as a form of short-hand for these layers: **A** for the *surface* layer; **B** for the *middle* layer; and **C** for the *parent material* (Fig. 10.5).

Soils form slowly. Physical and chemical weathering may have to break down the parent material if it is solid rock. In any case the soil begins life as only a C layer (Fig. 10.5a). Gradually, perhaps over hundreds of years, a weathered surface layer is formed and this provides the nutrients that plants need in order to live. Soil organisms begin to invade the soil to decompose any dead plant tissue

Figure 10.5 Soil development
A soil gradually develops from unweathered rock.

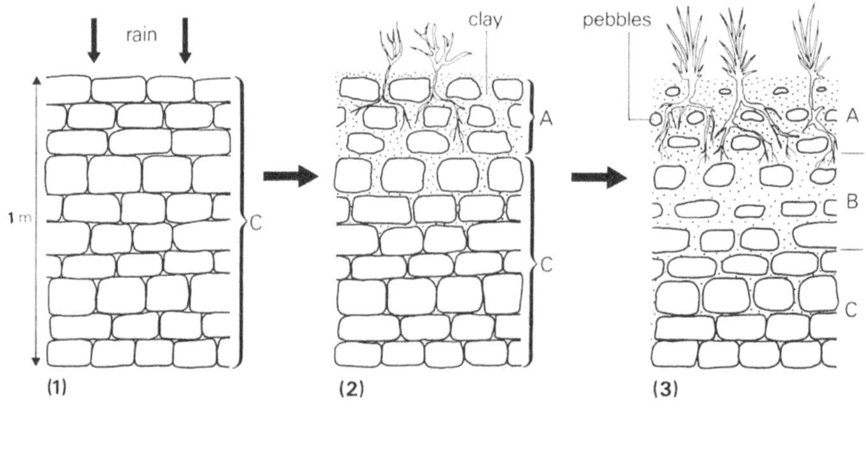

Figure 10.6 A brown earth soil
The (a) mineral and (b) organic parts of a soil combine to make up (c) a complete soil. In diagram (c) the soil layers are a mixture of mineral and organic material (A); weathered mineral matter but very little humus (B); solid rock, yet to be weathered (C).

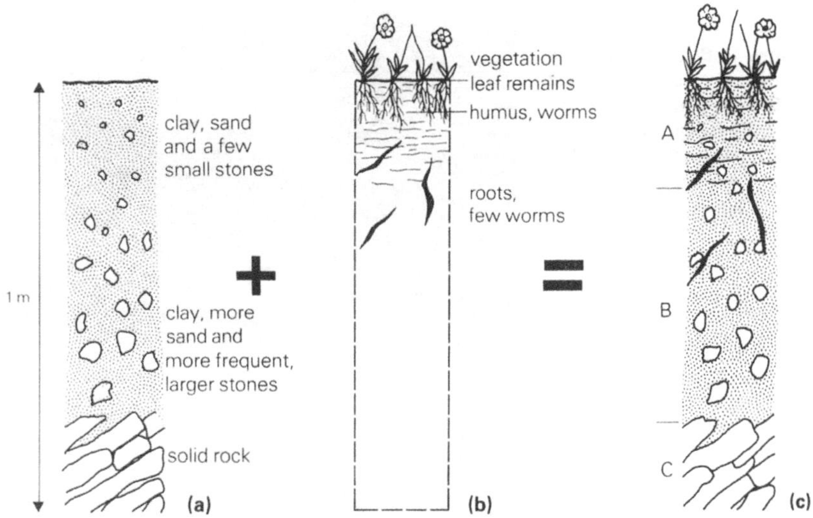

and mix it with the weathered surface rock. This mixed layer of plant and rock material is the A layer and it forms above the C layer (Fig. 10.5b).

As more time passes so the depth of weathering increases, helped by acids produced from decaying plants (Fig. 10.5c). Because plants need light and their roots need air, roots are mostly confined to the surface layer. The deeper part of the weathered rock therefore contains little organic matter and is called the B layer. Slowly the B layer becomes thicker. Most soils contain well developed A, B and C layers (Fig. 10.5c). However, there are many variations to the thickness and properties of these basic layers, as we shall see below.

Figue 10.7 A brown earth soil from a lowland area in Britain
Compare this with Figure 10.6c.

Down in the valley *Lowland soils*

Many lowland soils in Britain are called brown earths because brown colours dominate the soil profile. They typically have A, B and C layers as shown on Figures 10.6 and 10.7. Brown earth soils develop on parent materials that weather to release a large supply of nutrients, making them valuable soils for agriculture. Many of these soils are also deep, providing good anchorage for plant roots and a store of water for plant use between periods of rain.

Exercise 10.1 When is a soil not a soil?

Soil is a mixture of plant and rock fragments, most of which have been decomposed in some way. Soils are most easily distinguished from one another and from other materials by digging pits and looking at the changes that become visible at depth.

1 Study the three sketches of materials found after pits had been dug down to a metre below the surface (Fig. 10.10). State which one of them is (a) a soil, (b) a scree, and (c) a peat. Explain the reasons for your choice.

2 Study Figure 10.6a. This shows the distribution of rock fragments that would be expected in a soil from lowland Britain.
(a) Write down the main sources of acids used in rock weathering.
(b) Why should there be few pebbles (unweathered rock fragments) near the surface?
(c) Why does the proportion of unweathered rock increase with depth?

3 The sketch in Figure 10.6b shows the same soil as in Question 2, only here we see the distribution of organic matter.
(a) Why are most of the decomposed plant remains near the surface?
(b) What do you think the few large roots belong to? Why should they be found at greater depth than others?
(c) Why are worms and microscopic soil animals concentrated near the surface?
(d) What do we call decomposed plant tissue?
(e) What effect does decomposed plant tissue have on soil colour?

RAYMOND JESSOP

Figure 10.8 A tropical red soil, Malaysia
The trees on top of the cliff are 30 m high and there are steps cut up to the house! The rock of the C layer is a light-coloured granite.

Figure 10.9 A chernozem or black earth from Canada
Notice the deep organic-rich surface layer.

Lowland soils in tropical areas such as Malaysia and Nigeria also have soils with A, B and C layers (Fig. 10.8). The main difference is that chemical weathering under tropical conditions releases red-coloured iron oxides into the soil instead of the brown or dull orange that occurs in Britain. Because of their colour these soils are called **tropical red earths**. Another difference is that soils in Britain may only be 1 m deep, whereas those in the Tropics may be many times this value. (These soils used to be called laterites.)

In lowland areas with much less rainfall, such as the American Prairies or the Russian Steppes, soils are not brown or red but black at the surface. The Russian word **chernozem** is used for these black soils (Fig. 10.9). The black layer may be 1 m deep and it is mostly coloured by humus. These deep A layers develop because of the way soil animals are influenced by climate. Summers are scorching hot and quite dry, whereas winters are very cold and the upper part of the soil is frozen solid. By contrast, spring and early summer are moist and warm, and soil animals thrive near the surface. To escape the extremes many soil animals make their way to lower levels of the soil and take their food supplies with them from the surface. As they digest their plant food they release the humus that provides the black colour. Although these soils are remote from us, they are extremely important to the world because they are the soils of some of the world's chief grain-growing regions.

Over the hills . . . *Upland soils*

In uplands soils experience higher rainfall and lower temperatures than in lowlands. As a result much more water soaks through the soil and carries away vital nutrients. Low temperatures do not suit soil animals very well and, with fewer of them to decompose the dead plant leaves, vegetation tends to accumulate on the

Figure 10.10 Some surface materials

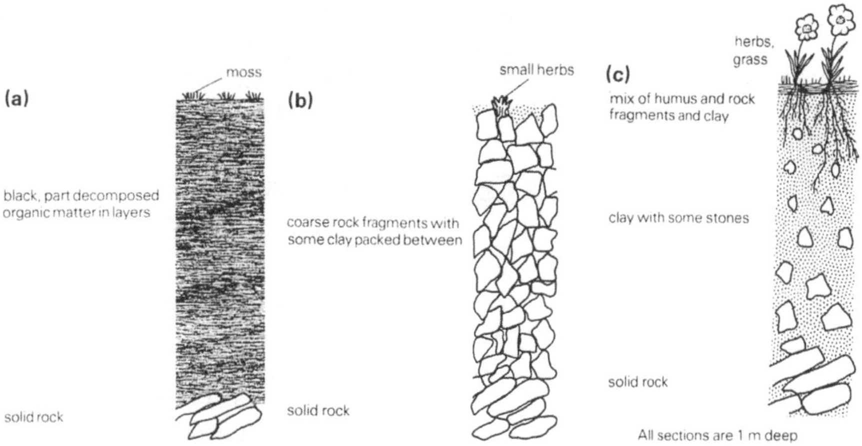

Figure 10.11 Profile of a podzol soil This is a podzol soil from an upland area in Britain. The clays and nutrients have been washed out of the upper layer, leaving only ash-grey sand. The organic material rests *on top of* this and it is *not* mixed with it. Conditions are too acidic for earthworms and most other soil organisms to survive.

soil surface. In extreme cases such as on flat moorland or very wet valley bottoms the accumulation gets so thick that we use the term **peat** to describe it.

Once the number of soil animals decreases, a vicious circle of events begins and a new type of soil called a **podzol** develops (Fig. 10.11). The word podzol comes from the Russian word for ash, and an ash-coloured layer is very clear.

With leaves building up on the surface, the nutrients needed for further plant growth are not returned to the soil very quickly. Those nutrients that are released are often washed through the soil by the high rainfall. Most plants cannot tolerate cool wet conditions and a lack of nutrients. Only woody plants such as heather and coniferous trees survive because they can store nutrients very effectively. However, the woody nature of these plants makes them hard to decompose by the few soil animals that remain, so the dead conifer needles tend to build up on the soil surface even more. Rainwater falling on this poorly rotting vegetation becomes much more acid and this passes through the soil, removing even more nutrients. Soil animals do not like acid conditions, so fewer soil animals inhabit the soil. In this way the vicious cycle of soil decline continues.

The extreme acidity of these soils is clearly seen in the colours of the layers (Fig. 10.12). The surface layer is black because it consists of an accumulation of poorly rotting leaves. Rainwater passing through this surface layer becomes very acid and it **leaches** away (washes out) the nutrients and the iron from the layer

Figure 10.12 A podzol soil

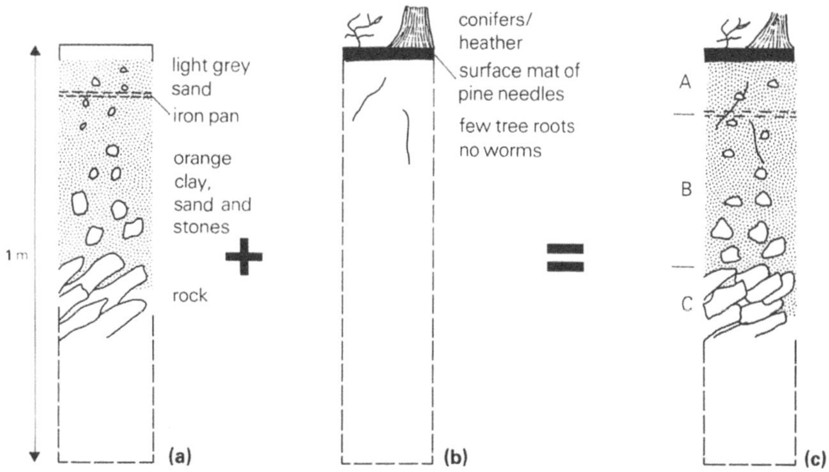

below. Without the iron to stain the soil it becomes a distinctive ash-grey colour. Below this layer the iron is redeposited because conditions are less acid. The extra supply of iron turns this layer a rusty orange colour. Podzols are easily recognised by their coloured layers: black, ash-grey and orange.

The role of relief

Getting it together

Soils are closely influenced by climate, drainage, vegetation, and the time during which they have been forming. On a worldwide scale, climate may be the most important factor influencing the type of soil that develops, but on a local scale this is likely to be less true.

A valley often contains a sequence of soils, even if it has been eroded in only one type of rock (Fig. 10.13). On the upper part of the valley, rainwater may soak through the soil and leach away nutrients. In this way a podzol soil may develop. Further down the valleysides, nutrients are also leached away, but they may be replaced by those from up slope. Slopes are also more sheltered, temperatures higher, plants grow more quickly, and more soil organisms are present to decompose and recycle plant tissue. Brown earths are more common in these lower areas. Finally, at the bottom of the valleyside, near the river, the water that has soaked through the soils builds up before entering the river. Here there is very little leaching, but the water may remain near the surface for longer periods. Water-logging turns these riverside soils dark blue-grey in colour.

Figure 10.13 A sequence of soils in a valley

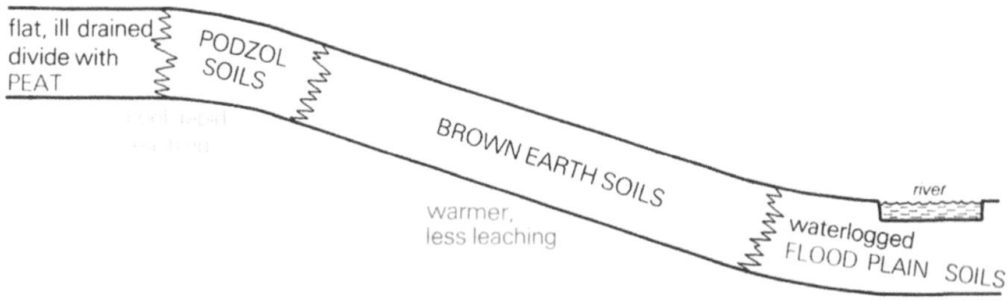

SUMMARY: SOIL FORMATION AND MAJOR SOIL GROUPS

Soils develop on unweathered rock. This may be solid rock, or transported fragments such as river alluvium or deposits of ice sheets. Chemical weathering attacks this parent material (Fig. 10.5a) and forms a weathered surface layer of clay particles. Weathering also releases the nutrients that plants need. Plants and soil creatures now invade the soil. Micro-organisms decompose any dead plant tissue, producing humus and releasing nutrients so that more plants can grow.

Slowly the incorporation of organic matter into the surface layer turns it a darker colour (Fig. 10.5b). Chemical weathering proceeds below this A layer, helped by acids released as plants are decomposed. Soil micro-organisms only live in the upper soil where air and new plant material are available. The deeper weathered layer therefore contains little organic matter. It is called the B layer or subsoil (Fig. 10.5c).

Brown earth soils have a dark brown A layer and a medium-brown B layer, with a gradual transition between them. **Podzols** have been heavily leached. They are acid soils and they contain few soil creatures to mix organic and mineral material. As a result they develop sharply contrasting layers. The A layer contains only organic matter and rests directly on the surface; the layer below is ash grey in colour due to extensive leaching; the B layer is orange where iron has been redeposited. Some podzols have all the iron redeposited as a hard **iron pan** at the top of the B layer. **Chernozems** are deep brown earth type soils of the plains. Their deep black A layer is partly the result of soil creatures migrating up and down the profile during the year because of extremes of heat and cold. **Tropical red soils** (formerly often called laterites) are deep brown earth type soils where the iron has turned a red colour under tropical conditions. **Rendzina** soils are thin brown earth type soils found on chalk and limestone rocks.

11 Clothing the surface

Ecosystem principles

The delicate web

The Earth has several 'skins': below our feet is a skin of crustal rocks from which the landscape has been carved; above our heads is a skin of atmosphere which provides our weather, and the water and wind that help erode the land; but between them there is a delicate and beautiful skin of green – the vegetation that clothes most of the Earth's surface. This web of plants is vital to our well being for it provides food for man and animals. It also holds the soil in place and, in so doing, exercises a control on the landscape far out of proportion to its apparent size and strength.

As we shall see, man has had more influence on the distribution and character of plants and animals than on any other part of the Earth. Consequently, we have to make even greater efforts to sort out the natural order of things so that we may best learn how to preserve some of the remaining natural environment while adapting much of the rest for our long-term survival.

The nature of vegetation

Dust to dust

The patterns of plant and animal life on the Earth's surface are the result of long periods of evolution and adaptation to the natural environment. Some rocks in Canada, for example, show fossil forms of life over 3000 million years old whose descendants can still be found building the limestone of the Great Barrier Reef of Australia.

Most scientists think that life probably formed at a very early stage in the history of the Earth when the atmosphere was very different from that of today. At this early stage there was a lack of oxygen, but the atmosphere was just right for the most primitive forms of life to be created. In a research institute in England scientists are beginning to succeed in recreating these conditions in a laboratory. Using a mud 'soup' of the kind that may have existed on Earth near volcanoes, these basic building blocks of life can be formed. This may show the way that evolution began. Once started from these simple substances, the long path of evolution has been able to produce the millions of different life forms that we see today.

SUMMARY

All living things are made from very simple building blocks organised into very complicated arrangements. The basic chemicals that make up our bodies, for example, could be separated and put into just a few jars on a shelf! One jar would be labelled calcium, another iron, and so on. Below the jars would be a large tub of water. As with all living things, human bodies are mostly water.

But why do we need to understand that our origins may go back to simple forms created in hot mud pools? What has this to do with geography? First, it allows us to see that only simple materials are needed for life. Secondly, we can see that a vast length of time has elapsed since the starting date. In this huge time span there should have been plenty of opportunity for development, specialisation and adaptation to produce a wide range of life forms, each adapted to its own environment. It is the geographer's task to explain the present distribution of plants and animals over the Earth's surface, so clearly he needs to know something of the past events first.

Habitats

Home, sweet home

Every plant and animal has a **habitat** – a home – that is suitable for its survival. This is the fundamental idea that underlies biogeography. But plants and animals do not live in isolation: one may provide a shelter for another; some may provide the food for others; and so on. An environment in which a group of animals and plants all live together in a stable and mutually dependent way is called an **ecosystem.** Ecosystems (Fig. 11.1) provide frameworks for studying communities of plants and animals along with the local climate and soils on which they depend.

When we look around we see that most places provide a habitat for some living things. Even the top of a mountain, a desert or a piece of wasteland in a city all have their own distinctive ecosystems. This tells us that life has become very well adapted to a whole range of environments. Now we need to know just how this has become possible.

Natural selection

Little by little

It is probably just as well that we do not all think, feel and look alike, but we do still have much in common. Together, humans make up a **species,** a biological group that is capable of interbreeding.

Over the past half century, archaeologists have excavated many sites where our ancestors lay buried and they have found considerable differences between

Figure 11.1 Ecosystems A natural ecosystem provides a stable habitat for many species.

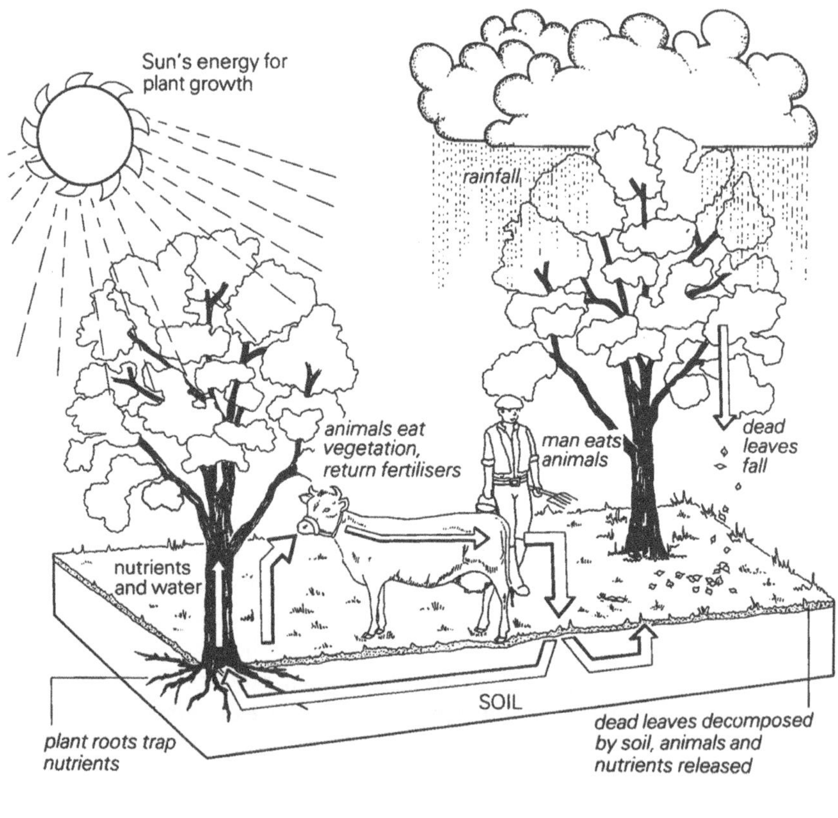

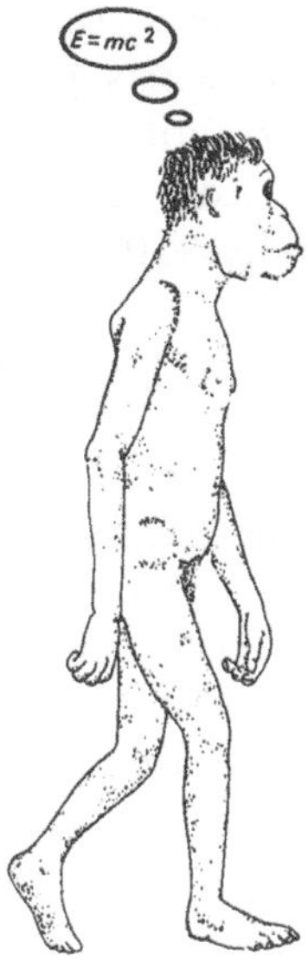

Figure 11.2 Your classmates 3 million years ago?

these people and ourselves. Several million years ago man had to have teeth that would crush food rather than bite, but as this became less important the large teeth in the front of the mouth became smaller and developed sharper edges. Many other changes can also be recognised: the shape of the head, the height of the body and the amount of hair have all changed. Reconstructed models of prehistoric man show that he looked very different from ourselves and yet the changes by which he evolved came about only by generation after generation of tiny, repeated differences (Fig. 11.2). If the differences between father and son were an advantage, then they helped him to survive and to be more successful in his hunt for food, and increased his resistance to disease. If the differences were a disadvantage, then he would be less likely to survive. In this way the more successful trends in evolution have been preserved while the others have died out. This is the way that **natural selection** works to help species adapt to their environments. It is often called the survival of the fittest.

Man is the most highly evolved of all species and he can survive in the most varied of environments. However, most living things are not as mobile, nor can they go in search of food and shelter. Plants cannot move with the changing seasons and may have to brave harsh, frosty winters as well as hot dry summers. So, in order to understand the distribution of a species we must find out what it needs to survive.

Food chains

Then we shall cum an' eat thee up

One of the most basic needs for most living things is sunlight. Plants, for example, use water, oxygen, carbon dioxide and nutrients as they grow, but they need the energy of the Sun to combine these substances into new tissue. Animals eat plant tissue in order to convert this stored energy into forms they can use in their turn. Thus one creature is dependent on another for survival. Cattle, for example, are plant eaters (herbivores) but they in turn provide food for man who is a meat eater (carnivore). Thus the plants, cattle and man are linked together in a food chain. At each stage along the chain only a small part of the stored energy is passed on, so man needs to eat a large number of cattle in his lifetime and the cattle need to eat a large amount of plant tissue (Fig. 11.3). However, if there were

Figure 11.3 The food chain

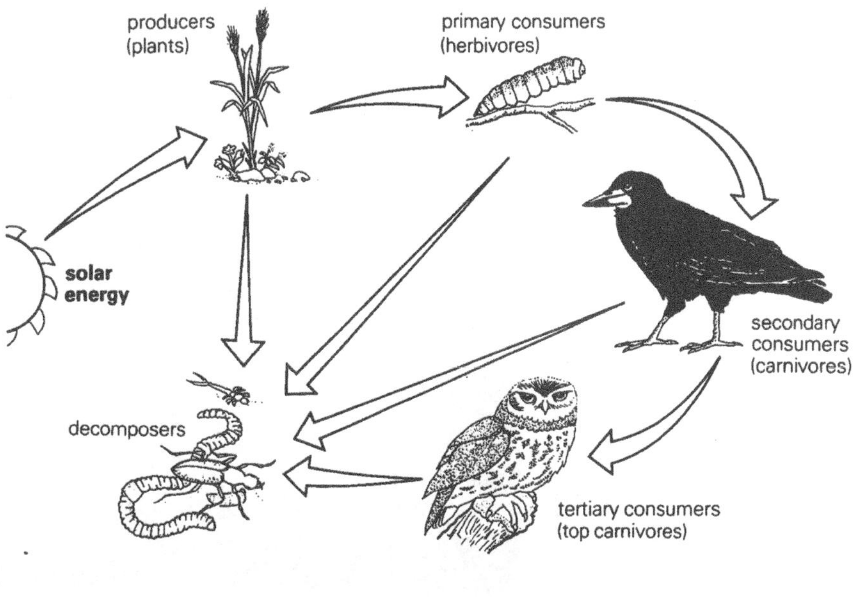

too many cattle in a region, the plants would not be able to grow fast enough to replace what was eaten, overgrazing would occur, and the plant supply would run out. Eventually the cattle would starve and some die until their numbers were in balance with the rate of plant growth. In a stable ecosystem, therefore, there must be the correct balance of carnivores and herbivores so that links in the food chain do not get out of step.

Each ecosystem must be adjusted so that every part of it is in equilibrium in other ways too. For example, some plants can flourish in the presence of one another because they do not compete for the same amount of light or the same type of nutrient. Direct competitors are never found together in equilibrium.

Many types of stable ecosystem are possible, depending upon the climate, the slope of the land, and the soil. On a worldwide scale such ecosystems often give rise to distinctive types of vegetation. Many of these have been given common names such as the prairie grassland and tropical rainforest. However, within these broad groups there are smaller, but distinctive groups that adapt to, for example, a wet riverside or a sandy soil.

The range of species

Far and wide

We call plant and animal tissue food, but to plants 'food' comes as a drink sucked through straws (the roots). An appetising drink to a plant is one containing all sorts of 'goodies' such as calcium, nitrogen and sulphur. Except for some dissolved metals that are poisonous, the general rule is the more dissolved substances from rock in the water, the better. In fact, plants are mostly indiscriminate feeders – they suck up nutrients more or less in the proportions that are available in the soil water, and what controls the plant growth is the nutrient that is in shortest supply. This is the growth 'bottleneck'. So here we find a potential source of limitation for some species that may require a copious supply of one particular nutrient or loathe the presence of another. Rhododendrons and heathers, for example, just will not grow well in calcium-rich water, which is why they are found on sandy acid soils. Hydrangeas, on the other hand, will not grow without a lot of calcium!

Figure 11.4 Colonisation These ferns are growing in a pool of sunlight which has filtered to the ground because a tree has recently fallen over.

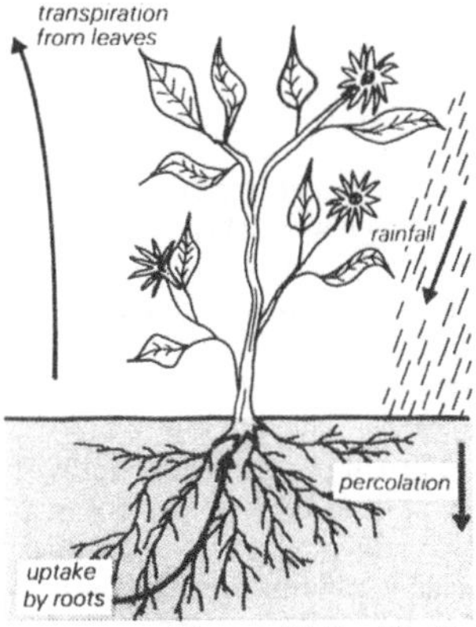

Figure 11.5 Limits to growth
Many plants need continuous water supplies to provide enough nutrients for fast growth. Where water is limited, plants tend to grow more slowly and adapt to conserve their supplies.

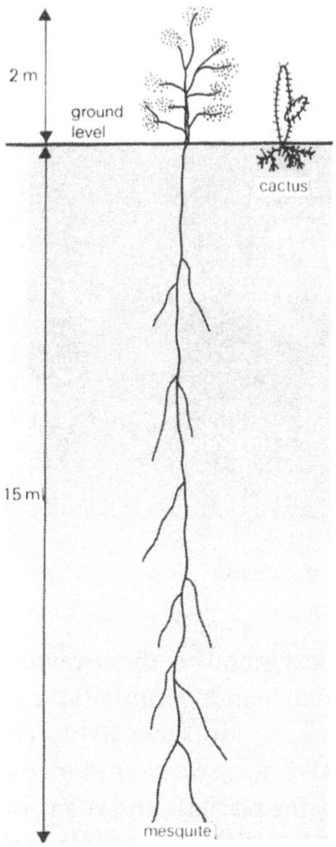

Figure 11.6 Adaptations
The mesquite bush survives during a drought by tapping reserves of water up to 15 m below the surface. The cactus, without deep roots, has had to adapt in a different way. (See also Fig. 9.0a.)

Very few of us prefer winter to summer, and most of us feel better on a sunny day than on a dull one. So it is with plants. Indeed, plants need a certain amount of sunlight to grow at all. In many cases where there is too little light, either because of distance from the Equator or a canopy of tall trees, the seeds of some species simply do not ripen. As a result their reproductive cycle is not completed (Fig. 11.4). Remember that, even with 24 hours of sunlight in the arctic summer, this low-angled radiation cannot provide the same energy as just 12 hours of tropical sunlight. This does not mean that shady or polar areas are without vegetation, but rather that different plants need different amounts of energy, and only plants tolerant of low-energy levels can thrive in these difficult places – an example of 'adapt or die'.

Both sunlight and nutrient supplies are important in controlling the distribution of species. But looking back to our list of plant needs we see that the other vital requirement is for water. So what happens if the water supply is reduced? In fact water does much more than just provide a means by which nutrients can reach plants: it takes part in the formation of tissue and, just as air fills out the shape of a balloon, it maintains the shape of non-woody parts of tissue. A glance at a dried plant will soon confirm this last point.

Unfortunately, plants cannot just blow themselves up with water and then remain inactive. They need to keep pumping water so that the 'food' of nutrients and dissolved gases can be fed in. This constant need for more water means that the 'used' water has to be removed. This usually happens as a chain reaction started by sunshine or air drying the surface of leaves. More water flows to the surface of the leaves and this in turn pulls water up from the roots (Fig. 11.5). And if there is a limited supply of water, some rather special adaptation is needed, otherwise the continued removal of water from the leaves will simply cause the plant to wilt and die. Small woody plants or those with thick rubbery leaves are typical adaptations (e.g. the olive, cactus (Fig. 11.6), acacia tree, and mesquite bush (Figs 11.6, 11.7)).

Figure 11.7 Plants of the desert

Topic Plant resistance to fire, drought and pests

In the deserts of America, one of the most commonly found bushes is a grey-green woody plant with small leaves, called mesquite (Fig. 11.6). Rain falls only a few times a year and there are long periods of drought. At the same time temperatures can reach well above 30°C – enough to make most plants wilt in minutes! Yet the mesquite grows on and survives because it has adapted to its environment in a remarkable way. Although woody stems and small leathery leaves are important in reducing water loss, the key to the mesquite's survival is seen if one is dug up. In fact, to dig up a mesquite complete with roots may require a mechanical digger and pneumatic drill! It is normal to find the tap roots sinking down to 15 m – more than the height of a house – just to find a water supply.

When the famous wooden horse was being wheeled through the gates of Troy over 3000 years ago, a small sapling had just begun to rise above the forest floor in California. The seed from which it grew weighed less than 4/1000th of a gramme. When Christ was born, the tree had grown to over 30 m in height. Today it is 90 m tall and weighs 1000 tonnes (Fig. 11.8). Such is the history of the giant redwood and, with a 3000-year growth period to its credit, it is surely one of the best survivors. However, its means of survival are quite different from the mesquite. Because it takes so long to grow it must be very resistant to disease and for this purpose it has grown a bark up to 60 cm thick. The same bark has enabled it to resist fires that have swept through the forests over the centuries. Indeed the redwood has benefited from forest fires because other species have been burned away, reducing competition for growing space.

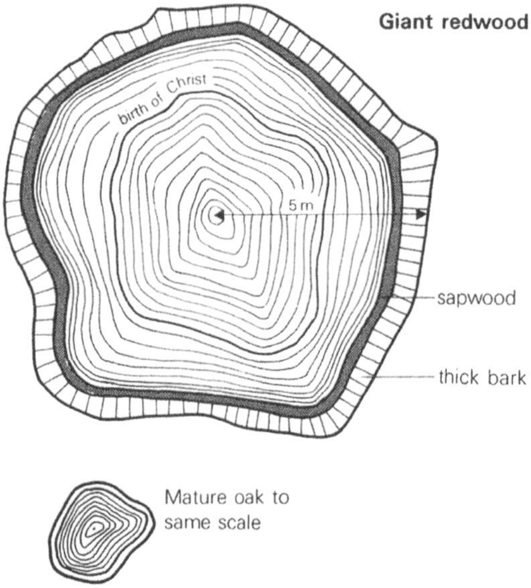

Figure 11.8 Adaptations
The giant redwood has survived by a combination of fire and insect resistance, using its very thick bark.

Sunlight, water and other nutrients are all fundamental to the growth of plants. Of these, water and sunlight are the most critical and, because these are also features of climate, people have long related vegetation zones to climatic zones. In a broad sense there is obviously logic in this but, as we have seen, on a more local scale there can be many other controls on the plants found at any one place.

Let us leave this idea of broad controls by re-emphasising the effects of water restriction. When water is limited, the number and size of plants that can be supported in a given area decrease both in volume and range of species. Eventually the plants may have to be widely spaced to obtain enough water for themselves. Water has also had such a powerful influence on evolution that widely separated areas containing unrelated species can take on much the same appearance and pattern.

Exercise 11.1 Desert survival

Figure 11.7 is a photograph of an area where rainfall is very low and irregular.

1 Explain why plants have to make special adaptations to hot, dry conditions.
2 Choose two different types of adaptation and describe how each plant type survives.
3 From the photograph, list any other adaptations that you can see.
4 Why do the plants not cover the ground completely?

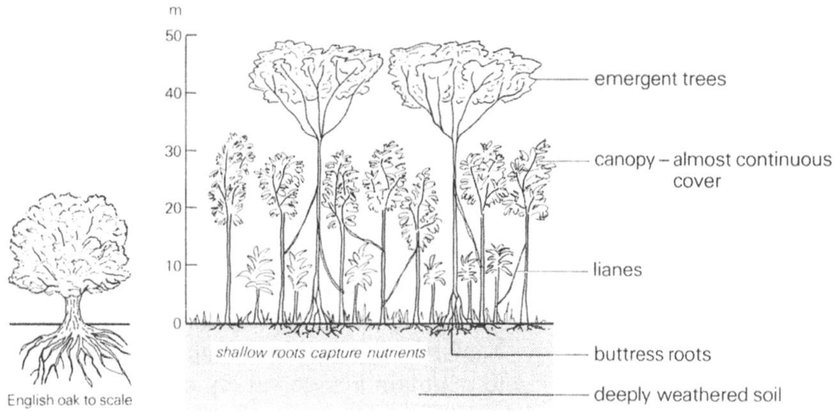

Figure 11.9 The vegetation pattern of the tropical rainforest

Climax vegetation

Survival of the fittest

TIMOTHY GOULTER

Figure 11.10 Tropical rainforest buttress tree

Let us start an examination of vegetation by investigating the rain forests of the Amazon, Zaïre or South-East Asia where there is much warmth and rainfall. In these places we can try to see what the results of a long-term battle between species are like.

A **tropical rainforest** is dominated by giant trees, some over 45 m high (Fig. 11.9) whose roots are even adapted into 'buttresses' (Fig. 11.10). These trees emerge above the main tree species which provide an almost complete canopy of leaves at about 30 m and block out much of the sunlight from the layers below. Hanging from the trees are the lianes – creeping plants, which like the ivy, live on other trees. These are the natural ropes of the jungle that some animals use for swinging from branch to branch.

Below the tree leaf canopy there are smaller trees that can grow with less light, and then below them, on the forest floor, are the shade-tolerant plants that need a low light, high humidity and constant temperatures to thrive. Here is a balanced and interdependent community which has developed to its most mature and complete stage – a **climax vegetation.** In a tropical rainforest there are more species of plant and animal than anywhere else in the world. There are evergreen trees and some deciduous trees, shrubs, herbs, and mosses all surviving together.

However, the equilibrium is a continual battle for survival. You can see how fierce the competition is in such a forest when a giant tree dies. Often it will fall over and tear out other trees as well so that a broad swathe of forest is exposed to the full light of day. This is the time when the vegetation changes rapidly and the battle for colonisation of this new empty space begins in earnest. First, grasses and other low, sun-loving plants colonise the ground. This light-tolerant ecosystem is not stable, however, and it is soon overwhelmed by a scrubby bush-dominated ecosystem. But even this dense thicket-type of vegetation only survives until the saplings of larger trees grow to a sufficient height to shade the bushes. As the shade increases, the light-loving plants are deprived of their source of energy and they die back, and their places are taken by shade-tolerant plants. In this environment of ample heat, sunlight and moisture, the sequence of changing ecosystems is very rapid and it may only take a few years before large trees dominate.

The dense vegetation that builds up in these early stages has led to the term 'jungle' being used for parts of the rainforest. In such areas it may not be possible to see ahead for more than a few metres, and progress on foot requires constant cutting of trails. Indeed, people often describe the nature of the rainforest by the distance that can be seen ahead, e.g. a 10 m jungle or even a 2 m jungle! However, gradually the largest trees grow through the newly colonised jungle and then provide shade, so that many low tree species die back and the forest becomes more open again near the ground.

So the key to all climax vegetation patterns is the survival of the most suitable *in the long-term.* Where heat, light and moisture are abundant, as in the tropical rainforest, it is possible to support a great bulk of plant matter at every level and to maintain a wide variety of species. However, away from these regions of plenty water limitation and/or temperature limitation are encountered and the plant bulk is reduced. Under these conditions different species come into their own and dominate the ecosystem.

Topic Plant colonisation in a garden

Suppose a piece of land is cleared by a landslide or a volcanic eruption, or as a building site. Initially the ground is laid bare, but it doesn't remain that way for long. If the piece of bare ground were a cleared building plot in Britain, the first colonising plants would probably be called weeds. Our gardens are also cleared ground and a battle is engaged through the summer between weed and gardener. This is the main reason gardening takes up so much time.

Suppose a garden were abandoned and left to these weeds. Dandelion is a typical example of a British weed, which will rapidly grow over abandoned flower beds. It is soon accompanied by many other colonisers such as plantains, buttercups and grasses, until the whole surface of the ground is covered once again. After a while the odd shrub, often some form of willow, may invade, although eventually trees such as birch will grow above this ground cover.

Trees create shade. Some of the initial ground cover may not be able to live without direct sunlight and so it will die. In this way competition ensures that the composition of the vegetation changes. It may take up to 100 years for a proper forest to develop over the garden, but even then a stable plant community will not have been achieved. The 'pioneer' trees like birch only have a short life and after a couple of hundred years they will be replaced by longer-living species such as oak or elm. Only then will a more or less stable or climax vegetation be attained over what used to be a garden. You only have to look at disused railway lines, factories or airstrips to see colonisation in action.

Man's use of plants

Plants rule – OK?

A climax vegetation is broadly dependent on climate in most cases. For example climate usually determines whether the vegetation will be of tundra shrubs and mosses, coniferous forest, savanna parklands, and so on. Nevertheless, within this broad framework the detail is largely under the guiding influences of soil, drainage and slope. There are, of course, some instances where climate may not be the main control. For example, a closed depression from which the water cannot escape may well develop into a bog, and permit only water-tolerant plants such as rushes where otherwise there might be forest.

Bearing the local differences in mind, in broad terms similar climates around the world will encourage similar types of climax vegetation (Fig. 11.11). Not that the same species will be present in the Zaïre jungles of Africa as in the Amazon jungles of South America, but the overall tropical rainforest will look the same. This is important for man because it means that he can often take a plant from one region of the world and grow it successfully in a different region with a similar climate. There are numerous examples of the benefits this has brought: vines from America have been grown in France to improve the disease resistance of European stock, and coffee trees are now grown in Asia, West and East Africa and South America although they originally came from Ethiopia in North Africa.

Nowadays man mostly ignores or fights the resilience of the biological world in its attempts to restore itself. Farmers resort to extensive use of herbicides every year in order to prevent 'weeds' colonising their fields. Yet some people still use nature to advantage. The foresters of Scandinavia and Canada encourage natural regrowth of trees in their lumbering areas as a means of renewing their reserves. Woodsmen cut back trees for cricket bats, hockey sticks and fence posts knowing that in the struggle toward a climax vegetation the trees will grow up again. Primitive farmers in tropical forests know that they can cultivate land for a while, then leave the forest to grow anew, gradually replacing the needed soil nutrients as it does so (p. 151).

Figure 11.11 The world's major natural vegetation zones

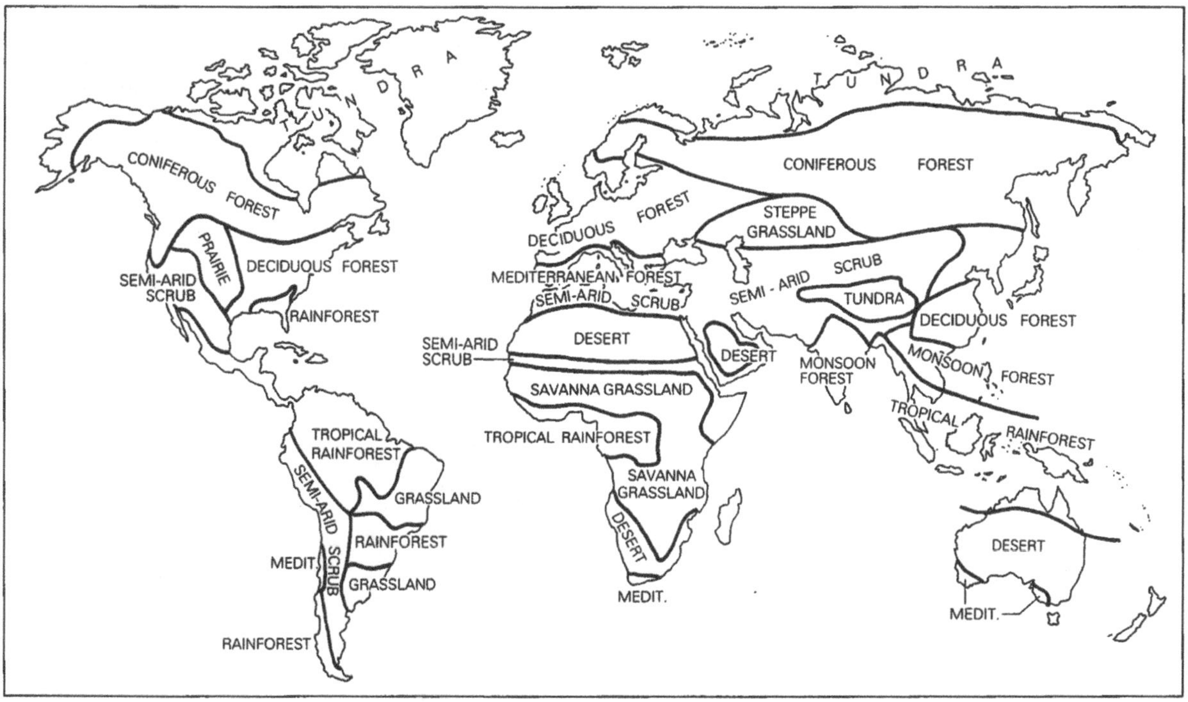

Tropical rainforest. Tropical rainforests occupy about one sixth of all the land surface covered by broad-leaved trees. As described on p. 127, rainforests may contain hundreds of different tree species, sometimes occurring in patches, sometimes singly. Because there is no cold season there are trees in all stages of their reproductive cycles at all times of the year. Intermixed with a bewildering variety of evergreens (Fig. 11.12), there will be some deciduous trees coming into leaf at the same time as leaves are falling from others, and yet others are in flower.

Each type of tree grows to a different height, so that a forest has many tree canopy layers (Fig. 11.9). This layering has given the rainforest the name 'gallery forest'. Together these tree layers shade the ground so that near the forest floor there is little light. Few plants will grow in deep shade, so it is often possible to walk about quite easily.

The tropical rainforest is a closed system – an ecosystem with a very tight control over its nutrients. Luxuriant forests can be found even on soil that is not very fertile because they recycle nutrients so effectively. The trees are shallow rooted and they trap nearly all the nutrients as they are released by rock weathering and the decay of dead leaves. These are used *immediately* to make new growth. It is a never-ending cycle but *it relies on the ecosystem remaining undisturbed.* So delicate is the balance within the cycle that, once disturbed by clearance, rain soon washes nutrients from the soil, and the re-establishment of a new forest at a later stage may prove impossible. Rainforests occur in the Amazon Basin, Mexico, Zaïre and South-East Asia, all regions which are being exploited at a rapidly increasing rate. The possible effects of this are causing great concern (see p. 153).

RAYMOND JESSOP

Figure 11.12 Rainforests
The canopy of mature tropical rainforest can be so dense that the ground is totally obscured.

Savanna. Savannas are tropical grasslands with a small proportion of trees. Together, grass and trees make up a sort of parkland, covering some 20% of the world's land surface. Most savannas occur between tropical rainforests and semi-arid scrub that border deserts. Because they are so widely distributed, the type of grass varies considerably from the African elephant grass, which is several metres high, down to low sparse grasses on the borders of the deserts. Trees in the savanna all have deep strong roots because the rainfall of these regions is seasonal and there are long dry periods. Trees become specially adapted to these conditions in other ways as well. Many, such as the acacia (Fig. 11.13) are deciduous, shedding their leaves in the dry season to conserve water. They also often have a thick bark and scaly buds to protect themselves from the fires that become a major hazard in the dry seasons.

It is difficult to know how natural the savanna parkland really is, because man has cultivated these areas for so long. However, in some regions woods are more common, so it is possible that the natural vegetation of the wetter savanna should really be deciduous woodland (Fig. 11.14).

Steppe and prairie grasslands. Steppe and prairie grasslands occur in temperate regions mostly on the poleward side of the great deserts. Like the savannas they may owe their vegetation to the low rainfall and a long dry season which restricts most tree growth. Steppe grasslands have an additional problem to overcome because they experience very cold, snowy winters.

JOHN MACK

Figure 11.13 Savanna parkland, Sudan
Above. In the dry season and, *below,* a little after the start of the wet season.

Figure 11.14 Savanna
This is a semi-natural area in Zimbabwe, where low-growing acacia trees make a woodland with grass patches. The vegetation and soils are developed on a granite batholith. The upstanding piles of blocks are called **tors**.

JOHN BURN

Figue 11.15 Deciduous woodlands of Britain

Summer

Winter

However, the grassland that we see today may be as much an effect of fire burning off any trees, together with the effects of an enormous population of grazing animals. For example, 50 million buffalo used to roam the prairies of the USA, each eating as much as a cow does. Again we do not know how 'natural' the steppe, pampas or prairie grassland really is. Much of it anyway has long since disappeared under the plough to become some of the world's great wheatlands (see p. 143).

Temperate deciduous forests. As rainfall becomes more regular and plentiful, the steppe and prairie grasslands sometimes merge into deciduous forests. Deciduous forests of the mid-latitudes (Fig. 11.15) consist of only a few dominant trees, such as oak, beech, elm and hornbeam, in contrast to the amazing variety of the tropical rainforests. Light often filters through the tree canopy and allows an undergrowth of smaller species such as holly and a ground vegetation including bracken and blackberries. Temperate areas receive less energy from the Sun than do areas nearer the Equator, so growth is slower and trees less densely packed. Nutrients are also recycled more slowly and they may accumulate on the soil surface for many months before being decomposed.

The American and Asian deciduous forests have many more species in them than do the European forests. This is because trees in Europe had to migrate from Turkey around the difficult barrier of the Alps since the ice retreated only 12 000 years ago. There is no such barrier in America or Asia (Fig. 11.21).

Coniferous forest. Conifers have adapted to a wide range of harsh environments, including dry and wet sites. The trees are all very efficient in their use and storage of nutrients, because in these areas nutrients are in short supply. Spruce is prominent in areas where rainfall is heavy or where soils are wet. Pine, with a shallow rooting system, tends to be found more often in drier sites. Together with other species the coniferous forest stretches in a broad belt polewards of the deciduous forests in Europe. However, in the drier areas of America and Asia there are no deciduous areas, and the steppe grassland merges directly with the conifers. Conifers also occur widely on mountain sites as far apart as the Rocky mountains of America and the Alps of Europe (Fig. 11.16).

These forests are made of even fewer species than the deciduous forests and they often appear dark and monotonous (Fig. 11.17). Here, recycling of dead plant matter is very slow. The floor of a coniferous forest is always littered with a thick layer of needles that have not yet been decomposed. The soils are almost always poor in nutrients and the number of soil creatures is low. These forests have rarely been cleared for agriculture because of the cold climate, and poor, pebbly soils make them unsuitable for most other plants. They do, however, provide the world's largest timber reserves and are widely exploited in Canada, Scandinavia and the Soviet Union.

Figure 11.16 Coniferous forests
Coniferous forests extend in a wide belt polewards of the broad-leaved trees and up mountainsides. Their downward-sloping branches help shed snow in winter.

Tundra. The poleward or altitudinal limit of the coniferous forest is rarely sharp. As the boundary is reached, trees become more stunted and eventually only reach the height of shrubs. The problem that all plants face in these areas is survival under extreme climatic conditions. Arctic plants face long winters and poor summer heating and so are very restricted in size and variety. Many seek the shelter of rocks and hollows for their survival. Most alpine mountain tundra is more varied because it has a shorter winter and greater summer heating. Therefore, don't expect the variety of pretty flowers found in the Alps to carpet the regions of Arctic Canada or Siberia. The most common arctic plants are cotton grass, sedges, mosses, and lichens which together provide a grey-green cloak of just a few centimetres in height to these exposed regions of usually waterlogged soils (Fig. 11.18).

Figure 11.18 Tundra vegetation
Tundra plants are low shrubs, mosses and lichens, and they give a grey-green look to these otherwise barren lands.

Figure 11.17 Coniferous forest
Coniferous forest can often be monotonously regular.

HARRY WALKLAND

Figure 11.19 Juniper trees in semi-arid hills

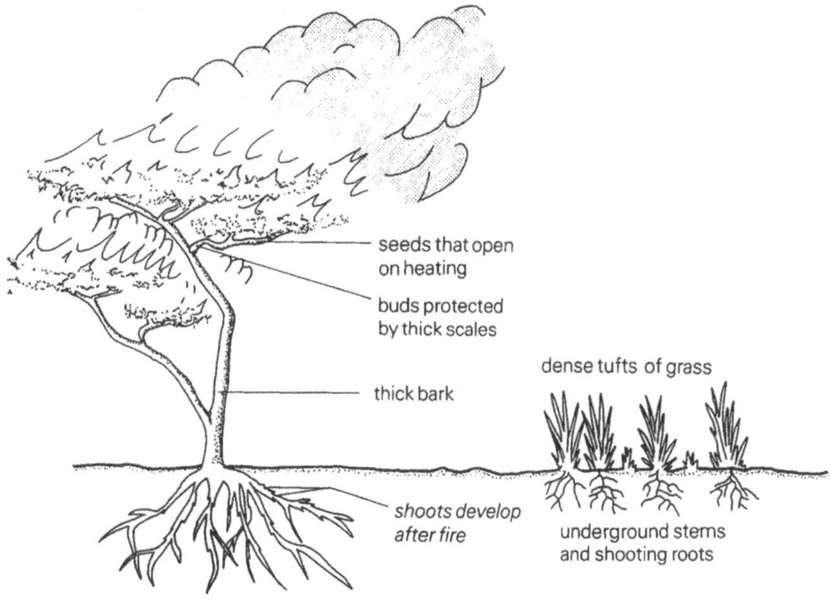

Figure 11.20 Fire-resistant plant features

Exercise 11.2 Drought and fire

1 Figure 11.19 shows juniper bushes growing semi-arid region. Suggest how they may have adapted to survival in low rainfall regions.

2 Figure 11.20 shows some of the features of vegetation that might help preserve it from fire. Explain how each of these is useful. Are fires likely to be important in determining the nature of (a) evergreen rainforest, (b) steppe, (c) savanna, (d) temperate woodland and (e) desert?

The effects of continental drift

Drifting apart

In earlier sections we have seen how plants have a particular set of environmental needs and how they fit into a community of other plants to make up an ecosystem. But we have also seen that continents have drifted in the past and are still drifting today. Only 50 million years ago Britain was experiencing a tropical climate. It has only been 12 000 years since an ice age left the land barren, and yet it is now clothed in vegetation. These two examples show that there have been some very remarkable changes in vegetation in the past.

However, plants cannot simply pull up their roots and migrate to more suitable areas. Change happens because conditions become too harsh and species die off in the direction of climatic worsening, and at the other end of the range the environment becomes better and seeds will grow for the first time. This change of distribution is called **migration.** It must have happened in the past and, of course, it is still happening today.

There is one other important factor to be remembered when thinking of migration. As continents drifted across the Earth they did not necessarily stay in one piece. Continents that have split in the past have been out of contact and have thus evolved separately. Given a long enough period of time new species will evolve. This is why much African and American **flora** (plants) and **fauna** (animals) are distinctively different in detail despite having the same basic shapes and functions. Africa and South America both have rainforest and grassland belts and they look alike from an aeroplane, yet from ground level it is clear that there are considerable differences in the species present.

Migration patterns

Silent waves

Figure 11.21 The spread of vegetation at the end of the Ice Age

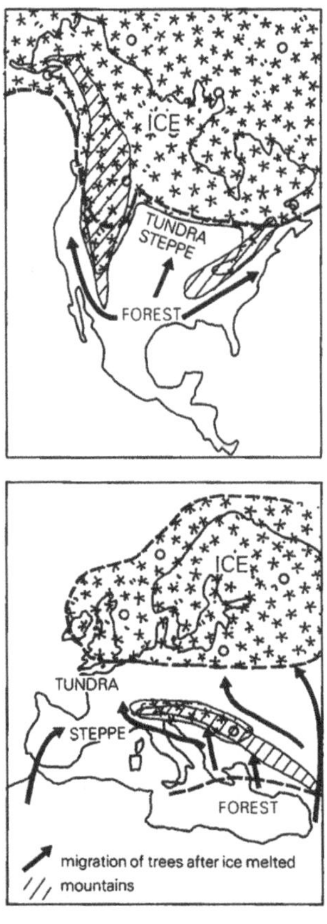

We can use an example to demonstrate that the pattern of plants and animals is always on the move if we ask the question 'What happens as the climate improves over a previously inhospitable area? To answer this question we can look at Britain's vegetation over the past 10 000 years.

As the ice retreated from the land it revealed a barren wilderness of crushed and shattered rock. This was the foundation for our soils, and the home of colonising plants and animals as they made the first tentative steps across a then dry English Channel. Watching the colonisation of Britain must have been much the same as observing the changing pattern of vegetation from the North Pole to France would be today. However, instead of making a journey across a quarter of the globe, you could have stood still and watched the different forms of vegetation move past you (Fig. 11.21).

Today, the vegetation that first grew on the barren landscape can be found only on high mountains or within the Arctic Circle. In these areas an ecosystem including mosses, lichens, small flowering plants and dwarf shrubs make up the pattern called tundra. This survives as a major vegetation form in these areas because the plants can endure the rigours of Arctic winters with their long periods of snow cover and also waterlogged soil in summer. No other vegetation forms can withstand this type of climate, but in warmer regions what chance would such low lying vegetation have against the trees of a forest? Our first wave of vegetation may have been of tundra plants, but they were soon overwhelmed by the advancing armies of birch, pine and willow trees in all but a few special places, such as the highest mountains of Wales and Scotland.

For example, in the craggy backwall of a glacially eroded hollow (corrie) called Cwm Idwal, there is a steep, narrow ascent that climbers call the 'Devil's Kitchen' (Fig. 11.22). This area of Snowdonia in North Wales is so precipitous that the only places where soil can remain are in the chance cracks left after thousands of years of frost shattering. Near the bottom of the 'Devil's Kitchen' trail there is a notice with the request: 'Please do not pick or damage the wild

Figure 11.22 The craggy walls of the 'Devil's kitchen' overlook Cwm Idwal (N. Wales)

flowers'. There is a particularly good reason for this request because these inhospitable cliffs are the last refuge of some of Britain's rarest plants. The delicate flowers that are so much admired by climber and botanist alike are irreplaceable – once destroyed, plants such as the purple mountain saxifrage and mountain sorrel will not grow in Cwm Idwal again because there are no nearby plants to allow recolonisation. The nearest plants of the same species grow in the Alps and northern Scandinavia!

Coniferous trees such as pine are still a common sight in Britain, but these are nearly all plantation trees which rarely survive naturally here. However, as warmer conditions spread over Europe, Asia and America after the end of the last cold period, seeds from the hardy birch and willow and coniferous trees such as spruce and pine were able to grow amongst the tundra mosses and lichens. Soon they developed into mature trees and shaded the ground so that the tundra vegetation died off. As these trees matured so they scattered their seeds farther north and continued to migrate.

In amongst the conifers, seeds of deciduous trees started to grow, but they matured more slowly and therefore spread less quickly. However, eventually the new arrivals of elm, oak and, later, beech became more successful in their turn and replaced the conifers.

Over Britain we can therefore envisage a series of waves of vegetation advance as conditions became warmer. First came the tundra plants, then the conifers, and finally the deciduous trees. Eventually the tundra and conifer vegetations were driven out of Britain in all but isolated places such as Cwm Idwal. Today Britain's natural vegetation over all but the highest hills should be deciduous forest. Only in the higher regions would conifers be a natural vegetation form. Nowhere should there naturally be moorland. This is clearly shown in two ways. First, there are new plantations of conifers being grown at up to 700 m without any trouble; secondly, old pine stumps are regularly dug out of the peat bogs and heather moors at heights of 900 m. All the changes from this pattern are the work of man.

Exercise 11.3 Mountain climates and mountain vegetations

1 Figure 11.23 shows a profile from a lowland plain in a warm dry region, such as California, to a mountainous peak. Using your knowledge of the effects of mountains on rainfall (from Ch. 5, p. 60) describe how the rainfall would vary between the plain and the mountain peak.

2 Explain how the rainfall changes might account for the changes in vegetation.

3 Describe the vegetation you would expect to find between the tree line and the snowline, and explain the special characteristics of the vegetation that allow it to survive.

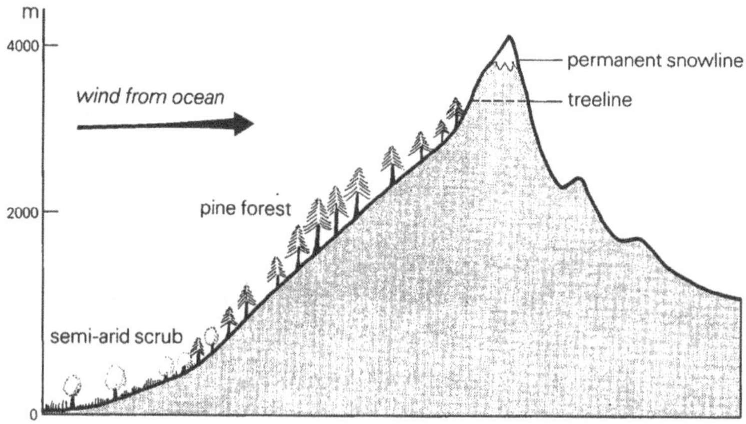

Figure 11.23 A profile from a lowland plain to a mountainous peak (latitude approx. 35°N)

Topic Changes in vegetation with altitude

Mountains show very clearly the way vegetation responds to changes in environment. A walk from an Alpine valley up to the summit of a nearby peak provides, for example, a change of altitude of more than 2000 m (Fig. 11.24). The change of climate observed in this climb is the same as a change of over 25° of latitude – the distance from the Alps to the Arctic shores of Russia.

In the valley there would have been a natural forest of deciduous trees such as beech and oak intermixed with some conifers. Much of this has been replaced by agriculture, so cereals and permanent pasture are now more common. But as slopes steepen and there is less farming, so the natural forest becomes more dominant. Within 200 m of the valley floor, coniferous forest has completely replaced the deciduous woodland. The main reason for the change is the decreasing temperature and a longer period of snow: this allows a shorter growing season. As the growing season shortens, deciduous trees cannot complete their reproductive cycle – fruits do not ripen and therefore trees do not perpetuate themselves.

As we walk up through the conifers it is much darker than the deciduous woodland. Trees are closely packed and little light penetrates. On the ground there is a dense carpet of fallen pine needles but no ground plants. Climbing higher we find a shorter and shorter growing season and lower summer temperatures until even the conifers become stunted, thin out and then disappear altogether. We have reached the tree line, which in our Alpine valley is 1200 m above the valley floor.

Above the tree line we are in open country for the first time, walking across a carpet of grasses and small alpine flowers that have adapted to survive in a cold windswept climate with a short growing season. However, as we climb to 2000 m only the hardiest forms of vegetation are left: mosses and lichen clinging to rocks in screes.

At 2600 m we reach the snowfield and even in summer all vegetation has now ceased and we are left looking across frost-riven peaks down to the forested valley below.

Figure 11.24 Vegetation changes in the Alps

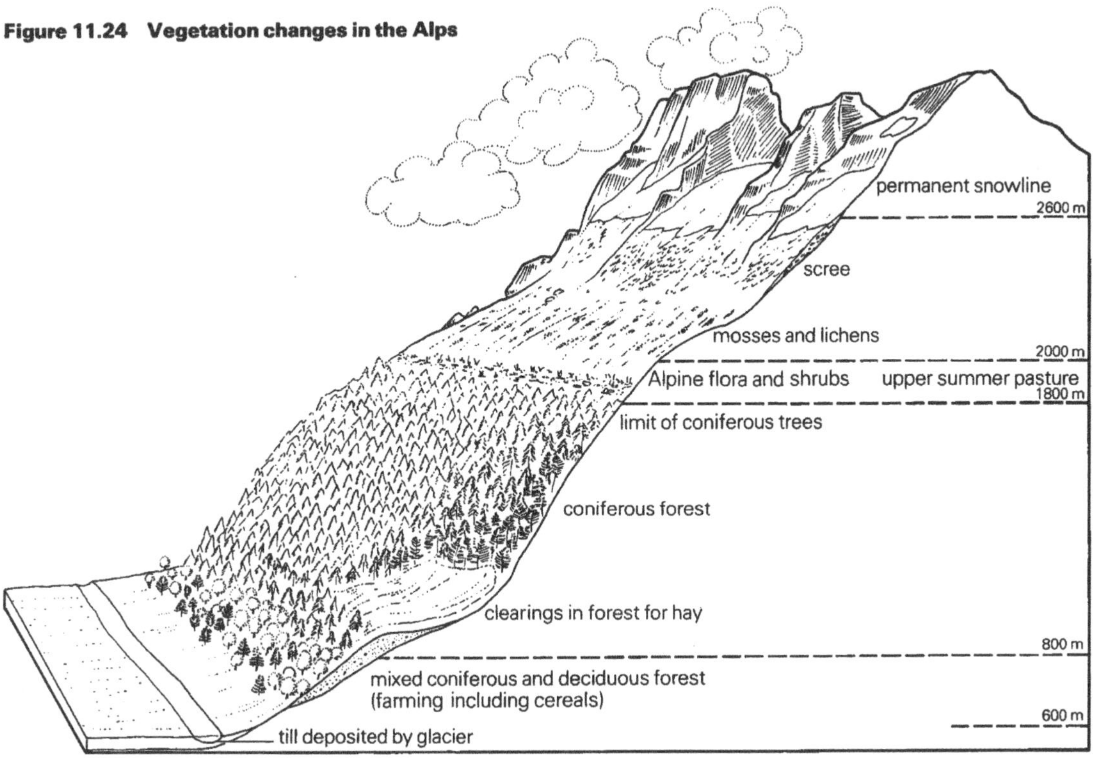

12 Man-made ecosystems

Man's impact on ecosystems

. . . and then came man

JOHN MACK

Figure 12.1 Primitive survival
Man still hunts his food in some parts of the world, but he could not survive in large numbers without more control over his food supply.

Look through the window of the room you are sitting in. How much of the vegetation that you see is really natural and not influenced by man? Can you see any vegetation at all?

Natural vegetation is a tangled mass of plants that might almost have been designed to prevent the development of man. For example, forests make the pursuit of agriculture and the construction of buildings and roads difficult. Similarly, if a man wants to build a house or a factory he does not usually want a tree growing through the middle of it! Further, most wild plants do not produce much that man now wants to eat, and even if they do their productivity is low. Man has to put up with an inconveniently placed mountain or unpleasant weather, but today he does not have to suffer the competition of natural vegetation.

Weathering, soil formation and plant growth are all closely connected. For example, rock weathering is speeded up by acids from rotting vegetation; plants need weathered rocks to provide new supplies of nutrients; and soil type is influenced by the plants from above and the rock below. But soils and plants are influenced by man more than any other parts of the physical environment.

Man and animals need to eat plants to live. Without cultivation and animal husbandry there would only be enough food to support a tiny fraction of the Earth's population (Fig. 12.1). So, unwanted plants are uprooted or burned away so that large areas can be cultivated. In this way man does not simply destroy – he replaces. But of course the new plants and cultivation methods will also cause the soils to change because the plants with which they were in balance no longer exist.

Clearly, without a careful understanding of the nature of plant–soil relationships, man may eventually destroy the very stuff on which his survival depends. This chapter shows some of the ways man has changed the natural order and created his own new ecosystems. We call these ecosystems 'farmland', 'pastureland', 'moorland' and 'woodland', but because they are all man-made they are very unstable and fragile.

Man-made changes through time

Fire!

One of the world's greatest problems is to feed its ever increasing population. Lack of food is closely associated with the spread of disease and poor resistance to infections, and even today much of the developing world must still face the stark challenge of survival. Although Europeans no longer have these immediate problems, in the past they too were often at starvation level because they just couldn't produce enough food with the farming techniques available. For example, in 1348 a plague called the Black Death swept across Britain. At the time of its arrival the population was about $4\frac{1}{2}$ million. Two years later the population had fallen to about 3 million – a third of the people had died. The disastrous effect of the Black Death was mainly because many people were on the verge of starvation. Strange as it may seem, 4½ million people were just too many to be supported on the land with only simple cultivation techniques and low-yielding crops. Being on the verge of starvation meant that many of the people were prone to disease, just as they are in Kampuchia, Uganda and other places in the developing world today.

Starving people do not sit around when there is further land to be cleared and cultivated. We can, therefore, be sure that virtually all possible land had been cleared and put under the plough. But as a consequence, the natural vegetation in Britain disappeared fast.

As we have seen this struggle for survival reached a crucial stage in the 14th century, but the natural vegetation had already suffered a long line of blows. Large areas of forest had been cleared as soon as man learned the use of fire. Indeed, we still get frequent reminders of the way fire spreads rapidly. For example, in 1978 someone carelessly flicked a smouldering cigarette end out of their car window and onto the verge of a road in southern France. Nearby was a large coniferous forest. It had been a dry summer and the grass, shrubs and leaves near the road smouldered and then caught fire. With a stiff breeze to fan it, the fire swept forward quickly and small burning embers were whisked up into the air. Soon these embers became lodged in the nearby trees and it was not long before the fire was sweeping through the branches of the forest. Unfortunately, nobody saw these early events, and people became aware of the tragedy only when smoke was seen billowing up from within the trees. By then it was too late. An emergency telephone call soon brought firemen to the scene, but there was little they could do. It took several days to get the fire under control and in the end tens of square kilometres of forest were destroyed – all accomplished by one person dropping a smouldering cigarette. So, it does not take very much imagination to understand that even prehistoric man could have cleared large areas of forest. Indeed, over many generations, just a few thousand people probably cleared the forests of Dartmoor, Exmoor, the North York Moors and many other British upland sites. Of course, once the trees were destroyed they stood little chance of growing again because any seeds that germinated and reached the sapling stage would have been eaten by domesticated animals.

In the ways described above the natural vegetation was under attack by man from a very early stage. Although the easiest areas to clear and work were the thinly forested uplands and chalk down, succeeding generations cleared more and more land so that by the time the Normans arrived in 1066 the natural forest was already severely depleted even in the lowlands. Only in the Royal Parks did it survive intact. After continued felling, in 1700 it was reported that hardly a tree was left standing within 25 miles of Southampton. If this is typical of other areas the removal of natural vegetation must have been more or less complete by the 18th century. Instead of forest, the land was laid out mostly in large arable fields.

Slowly, farming methods changed, but this time they accidentally allowed the trees to revive. The newer methods required land to be divided into rectangular fields, much after the fashion that we see today. To keep animals within the fields, quick growing trees such as hawthorn were planted and trained into hedges. Inevitably seeds of other trees and plants settled in these newly made hedgerows. Slowly our native trees – oak and elm – rose above the hedges to provide variety in the landscape once again.

JOHN MACK

IAN FENWICK

IAN FENWICK

Figure 12.2 The bush fallow system
Left. Rainforest is cleared by fire. *Below left.* A village is established and a variety of crops grown. *Right.* Burned tree ashes are piled up to provide fertiliser. *Below.* The area is abandoned and allowed to revegetate. In the foreground a variety of cultivated plants can be seen amongst recent scrub. Virgin rainforest is in the background.

RAYMOND JESSOP

If you look across the hedgerows in lowland Britain today, you can sometimes get an impression of a largely forested land. This is because the eye tends to merge all the hedgerow trees. In this way we are able to gain a glimpse back in time to the age before man began to destroy with fire and then cultivate with the plough. Today, with Dutch elm disease taking a heavy toll of one of our main hedgerow trees, and farmers grubbing up hedges completely to take advantage of bigger machines, we may be getting our last glimpses ever of this type of scenery. But wherever you look out over a British landscape, the chances are that very, very little of it is natural.

Increasing soil productivity

Out of the ashes

The history of our changing vegetation is not unusual. Man has adapted virtually all of the areas where plants grow naturally, and even some, such as deserts, where they do not. But all new types of land use have one basic need in common: they must have a **fertile** soil. If the soil is not fertile the cost of adding artificial fertilisers may make farming unprofitable.

A soil is naturally fertile if it can provide as many nutrients as are needed for the healthy growth of plants. As we have seen, a natural vegetation slowly gathers nutrients released by rock weathering and then traps most of them in a continuous growth and decay cycle. In this way the demands on the soil may be quite small. Tropical rainforests and northern coniferous forests, for example, grow on soils with very few nutrients.

The difference between farming and nature is that whereas nature can cope with a small reserve in the soil by continuous recycling, the farmer cannot. A farmer needs to remove the natural vegetation in order to grow his crops. He cannot easily use the nutrients stored up in the tree trunks, so he burns them. After burning, only a small proportion of nutrients are left in the ashes, the rest literally going up in smoke. As a result, the farmer begins with a loss of nutrients and by harvesting his crop, he takes away even more: *the nutrient cycle has been broken*. Consequently the soil become less fertile, slowly in temperate areas and very rapidly in the Tropics.

Loss of fertility in a soil has always been a major problem to the farmer. Traditionally he has had to leave part of the land unused (in fallow) for a few years until it has recovered naturally (Fig. 12.2). Then he ploughs up or burns off the vegetation that has recolonised the fallow site and he uses the land until it is again exhausted. This type of inefficient system is still widely found in the tropical areas of the world today and it is called the **bush fallow system.**

Fertilisers and rotations

Fertile minds

In the 18th century, European farmers started to look for ways of ridding themselves of the wasteful fallow land. European populations were rising to a level where production had to be increased. Gradually and by continued experiment they discovered crops need different nutrients and that the soil could be used all the time provided a **rotation** of crops was used. For example, wheat and clover can be rotated without wasting land or reducing fertility, because clover adds extra nitrogen to the soil to replace that used by wheat.

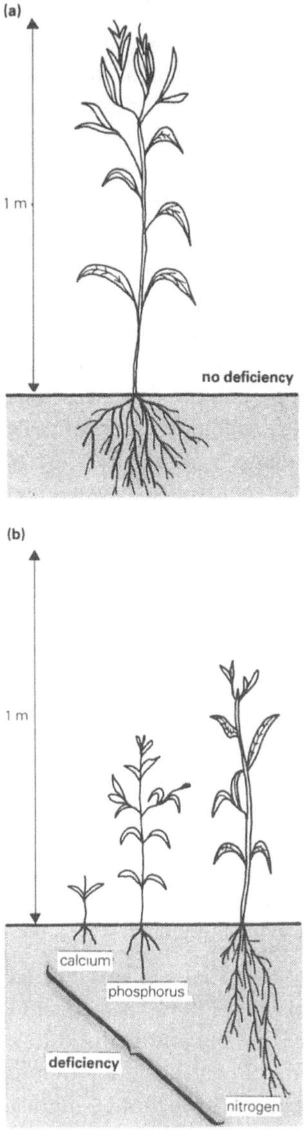

Figure 12.3 Fertilisers Even tiny deficiencies can dramatically affect plant growth and yield. Most fertilisers contain a range of plant foods designed to overcome the loss of growth shown here.

Further experiments showed that if baked limestones (usually called 'lime') were scattered on the fields, an increase in yield could be achieved. The age of the artificial fertiliser had arrived. A **fertiliser** is any material that is spread over the soil to restore or improve the nutrient content. Today, farms in developed countries make use of a whole range of fertilisers so that their yields are tens of times greater than those of their ancestors, or even of their contemporaries in the developing world. Each fertiliser has its own effect and the absence of just one can have a serious effect on plant growth (Fig. 12.3).

Falling apart *Causes of soil erosion*

Artificial fertilisers replace only one of the losses from a soil. They cannot replace the sticky gums that come from humus and which hold the soil particles together. Humus can only come from rotted plants or animal dung.

When soil particles start to fall apart the results can be disastrous. Initially the small particles fall into the soil pores and block them up. Without pores, water cannot drain through the soil properly and it is forced to run over the surface carrying soil particles and starting erosion. When wind blows, it also whisks soil away from bare field with little humus.

The unnaturally fast removal of soil by water and wind is called **accelerated soil erosion.** In the past, Europeans have not suffered from this problem because frequent and plentiful rain keeps the soil moist and prevents blowing. Usually rain falls quite lightly and it can even soak into bare soils without running over the surface. However, in other areas to which the Europeans took their farming methods this was not the case. America provides some clear lessons on the need to observe and try to imitate the way nature cares for its soil. The great soil erosion disasters in the United States tell one story of how man tried to work against nature – and failed.

The Dust Bowl *Erosion in the USA*

The Great Plains are a flat expanse of land which occupies a north-south belt of country just east of the Rocky mountains in North America (Fig. 12.4). Because

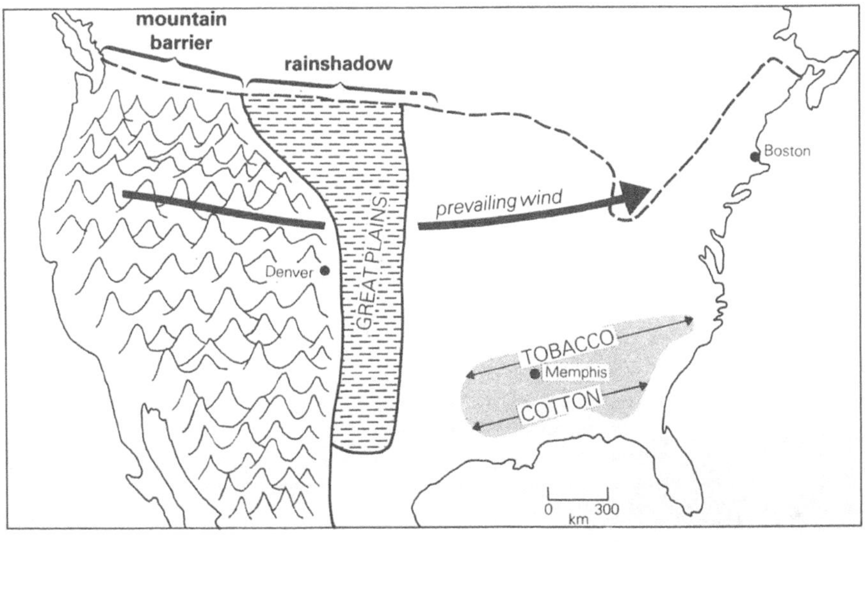

Figure 12.4 Areas most subject to soil erosion in the USA

they lie in the rainshadow of the Rockies, rainfall is less than 500 mm a year. Indeed, water on the Plains is worth more than property. In summer the rain falls only as occasional isolated storms, so that in any particular part of the Plains there may be enough rain for good grass and wheat to grow for two or three years; but then there may be a year when virtually no rain falls, when the streams from the mountains dry up and fill with sand. In fact the area used to be called the 'Great American Desert' on early maps. In the years with little rain, strong winds dry up anything that is exposed on the surface, crops fail, and despair comes to the farmer.

Before the European settlers arrived, this was the grazing land for vast herds of buffalo whose numbers were controlled naturally by the amount of grass and water available. When the railway reached the Plains in the 1860s, the Indians were chased off their land, the buffalo were slaughtered, and the new features of the land were the cowboy and the cattle ranch. Cattle ranching was profitable because cows could be transported by rail to the densely populated east of America. Now, although the numbers of cattle were still controlled by water and grass in the long term, just as the buffalo before them, the chance of quick profits led to overgrazing and exposure of bare soil. The problem of accelerated erosion had begun. Worse still, farmers arrived in the Great Plains to *cultivate* the hard, dry soil. Unfortunately, these homesteaders were mostly immigrants from Northern Europe and they were accustomed to plentiful rain. Their farming methods, their seed, their ideas – all were geared to moist climates. But by practising these techniques in the dry Plains they doomed their land to disaster. Right from the start they had to face the lack of water. In times of severe drought only windmills saved them. With these they could pump artesian water out of subsurface rocks, but this was only enough for human use; there was little they could do to solve the water shortage on the fields. These farmers, known to the cattlemen as 'sodbusters', ploughed up the grass and left the soil bare while they waited for their seed to germinate. But with no protection during this vital period, slowly the topsoil began to blow away (Fig. 12.5).

By the 1930s wind erosion was so severe on some days that soil was carried right across America. For example, on the morning of May 11 1934, the people of Boston stopped in the streets to look at the dirty yellow sky and realised that millions of tonnes of Great Plains topsoil were being blown into the Atlantic Ocean. After episodes such as this it became clear to people throughout America that something had to be done. But wind erosion in the area that became known as the 'Dust Bowl' was not the only problem. When the occasional rains did come, they were downpours of such intensity that the soil could not absorb the water. Consequently, the surplus water ran over the surface into the furrows and it carried even more topsoil away.

Figure 12.5 Wind erosion
In areas with low rainfall, soil is soon blown away even while it is being cultivated.

SUMMARY: SOIL EROSION BY WIND AND WATER

Wind erosion occurs when a soil is left without a complete vegetation cover. It is most effective with:

(a) strong winds with sufficient energy to pick up soil (usually on plains);
(b) soil particles that are poorly stuck together with organic gums (usually soils that have not been manured);
(c) ploughing parallel to the wind direction which results in wind being funnelled into the furrows and speeded up;
(d) a long period of drought so that the soil can dry out and become dusty.

The areas of greatest risk are plainlands used for arable cultivation, e.g. the Prairies and Great Plains of the USA and Canada; the Steppes of Russia; the Sahel region of Africa; the eastern plains of Australia; and, on a smaller scale, the Fens of East Anglia, the whole of Denmark, and the new polders of the Netherlands.

Water erosion occurs when a soil is left without a complete vegetation cover. It is most effective with:

(a) rainfall greater than can be absorbed by the soil (usually in hot regions where heavy thunderstorms are common);
(b) rainfall with large droplets that can smash up soil clods and splatter small soil particles down slope or into soil pores (again most likely with thunderstorms);
(c) steep slopes where water can run quickly over the surface, form small streamlets and erode shallow channels called **rills.** Rills are only a few centimetres deep, but they can merge into bigger channels called **gullies** which can be several metres deep.

The areas of greatest risk from water erosion are those that receive thunderstorms. They include all of the hot regions of the world where the land is under arable cultivation or where the soil has been exposed by overgrazing.

The Sahel region has one of the largest gullied areas in the world.

SUMMARY

Soil exhaustion

Too much of a good thing

The problem of water erosion that helped spell disaster in the Plains also plagued the farmers of the south-eastern USA. They had a hot climate with plenty of rain and very fertile soils. Indeed the soils and climate were so good that high-yielding crops such as cotton and tobacco could be grown year after year. However, inevitably the nutrients in the soil were used up by the cotton and tobacco plants faster than they could be replaced by natural weathering. Slowly the yields went down, but there was plenty more land left, so the farmers just moved on.

Cotton and tobacco grow best in well spaced rows. Between the rows the ground was kept hoed and bare (Fig. 12.6). In these areas the rain falls in heavy

Figure 12.6 Cotton growing in Nigeria

IAN FENWICK

downpours, and as it fell on the bare soil – a soil depleted in humus and nutrients by continuous cropping – it ran over the surface and washed the topsoil away, million tonne after million tonne.

When, after many years of such mistreatment there was no more new land to use, the farmers looked around and saw the results of their abuse of the soil. They had ruined thousands of square kilometres of land.

Understanding soil erosion

Lessons to be learned

It is a hard job trying to repair the destruction of so many years. The core of the trouble in all of these cases of erosion was that the farmers simply did not understand the delicate balance between plants and soil. It was difficult for them to see what they were doing wrong because it can take years of mismanagement for a soil to deteriorate so badly that it will not produce a crop and become vulnerable to erosion. During this time the soil is losing its humus, its gums and its nutrients, but for a long time this does not show. Then one year a farmer will be faced with a soil blowing away, or a crop with a very low yield, even though he has done nothing special that year. As a result he is left puzzled. Of course we now know that plant leaves protect the soil from winds and plant roots bind it together. We now know that as leaves die they are incorporated into the soil to provide humus, gums and nutrients. Further, we now understand that humus retains moisture and that the gums stick soil particles together. In this way easy routes for water movement are provided and clumps of soil are created that are too big to be blown away.

With a hundred years of research behind us it seems obvious that if you take the natural plants away in order to grow crops you must somehow put back into the soil the nutrients and organic matter that it needs. But the first research was done by the farmers the hard way – by trial and error, and a willingness to learn.

Soil conservation

Repairing the damage

There were many lessons to be learned by decades of soil erosion and most were learned too late. By the end of the 1930s even winter wheat would not grow in the Great Plains. The soil was deprived of humus, nutrients and a plant cover, and it had simply dried up. It was no idle jest that the area was called the 'Dust Bowl'.

A soil needs years of tender care to rebuild. In the Dust Bowl the topsoil had been blown completely away from some areas. Other places were so badly gullied by water running over the surface that the land could not be worked with machines. The most severeley gullied land had to be abandoned, although it was planted with grass and trees whose roots would reduce further erosion. The remaining soil was protected from wind and running water by planting clover or alfalfa. These crops have broad leaves that absorb the force of heavy rain, they have complex trailing roots which bind the soil, and they also produce a surplus of nitrogen. When these crops were ploughed into the soil they added both organic matter and the valuable nitrogen.

However, farmers needed a lot more help than this. They had to be taught to return wheat stubble and straw after harvesting, both to put back some organic matter and also because straw, spread over the soil, helps prevent wind erosion (Fig. 12.7). On the steeper slopes of the south-eastern USA they learned to

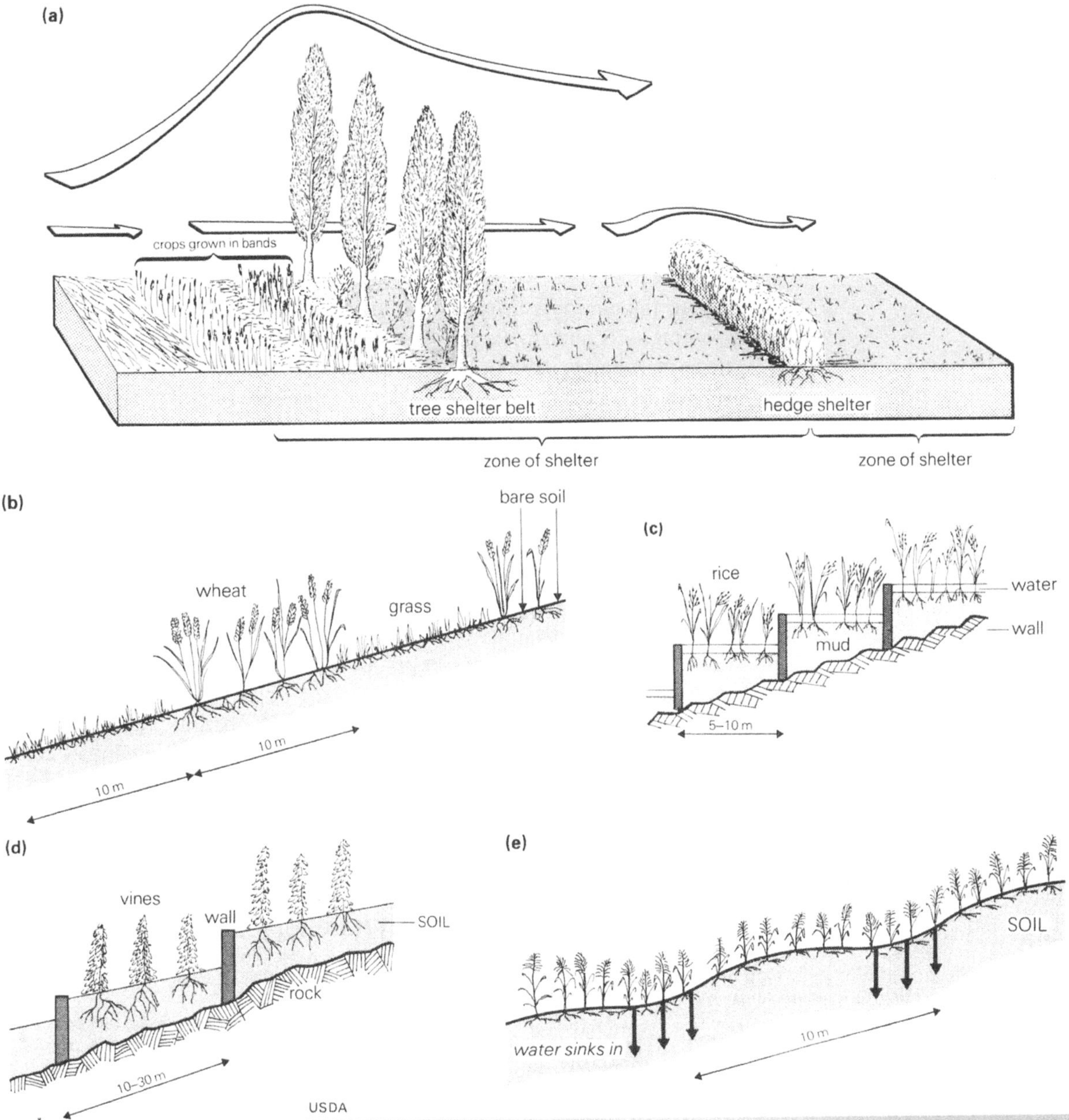

USDA

Figure 12.7 Soil conservation methods
(a) Protection from wind using shelter belts; (b) alternate crops across a slope; (c) rice grown in water ponded up behind terrace walls (see also Fig. 1.23); (d) vines on terraces (see also Fig. 5.4); (e) slope ploughed to produce soil terraces; (f) contour ploughing in the USA.

Exercise 12.1 Soil erosion

An experiment was conducted to find out how much soil was being lost from land by erosion in East Africa. For an experimental site the results were as in this table.

Land use	*Soil loss in tonnes/hectare*
grain	80
forest	0
bare soil	150
grass under controlled grazing	1
grass being overgrazed	120

1 Write down these figures in order of size and place first the land use suffering least erosion.
2 Plot a bar graph of the data.
3 Explain the lessons to be learned from these results which might be of benefit to the farmers.

plough with the contours instead of up and down the slope. This prevented furrows acting as river channels and causing erosion after rainfall.

Cattle farmers had to learn to control the numbers of animals on their land so that overgrazing was prevented; cotton farmers had to be taught that there were often better crops that could be grown on their land and that fertiliser application was essential. They had to be taught how a layer of rotting vegetation (called a **mulch**) laid over the bare soil between the rows would not only stop erosion but also add humus to the soil. In a little town in Alabama there is even a monument erected to yet another lesson. The monument commemorates the time when an insect called the boll weevil destroyed the cotton crop and forced the farmers to change from **monoculture** (growing only one crop) to growing a variety of crops such as peanuts and keeping dairy cattle. With a wider range of crops these farmers are now more prosperous than they have ever been, because if one crop fails they have others to rely on.

Today most of the Great Plains and the Cotton Belt are once again thriving agricultural regions. Although the farmers suffered badly from mistakes and many had to abandon their land and look for work elsewhere, those that remained were able to call on the resources of a prosperous country and high technology to help them. The farmers were able to obtain bank loans and the

SUMMARY: SOIL CONSERVATION

Soils at risk from erosion must be protected from strong winds or heavy rain. The main methods for preventing wind erosion are:

(a) planting shelterbelts of trees across the path of the wind (Fig. 12.7); this is not widely used because trees are effective shelter only for short distances, and they use up space and soil water, can shade crops and thereby prevent crops ripening:
(b) planting crops that ripen at different times of the year in alternate bands across the fields; only small strips of soil are left bare at a time;
(c) spreading straw over the soil as soon as ploughing is over (Fig. 12.7a); this prevents the wind reaching the soil and drying it out, makes the ground surface rough, and slows down the wind;
(d) drilling the seeds of the next crop through the remains of the harvested crop using special machinery; with this method the soil is never laid bare.

The main methods for preventing water erosion are:

(a) drilling the seeds for the next crop through the remains of the harvested crop; no soil is then exposed to beating rain, and even if water runs over the surface the soil will be held in place by plant roots;
(b) ploughing across the slope (contour ploughing) rather than up and down; surplus water then collects in the furrows and slowly soaks away;
(c) planting grass strips across the slope parallel to arable crops (Fig. 12.7b). Any soil lost from the bare soil will be trapped by the grass strips below;
(d) building terraces to reduce the slope of the soil and to encourage water to sink in; terraces can use brick, mud or stone walls (Fig. 12.7 c & d), or they can be made by ploughing (Fig. 12.7 e).

government was able to give grants and subsidies until the problems were over. However, today the problems of soil erosion have spread to the developing world, as we shall see later in the chapter. In these areas, although we now know how to combat erosion, there is often not enough money available to tackle the tasks, so erosion increases year by year.

Farming in Florida

Farming in the Sun

In the earlier part of this chapter we saw how poor land management caused a loss of fertility and catastrophic erosion in parts of America. Today in these areas farming methods make full use of the lessons learned. In this part of the chapter we shall look at some examples of artificial ecosystems as they are managed today both in the developed world and the developing world. To make the contrast more realistic we will compare these methods within the tropical and near-tropical world.

Florida – known as the sunshine state – is an example of a region where artificial ecosystems are kept in balance by the use of high technology. Because Florida has a mean annual temperature of over 21°C and a high rainfall (1400 mm per yr) it is well situated to grow citrus fruits such as oranges (Fig. 12.8). It can also grow vegetables for the markets in the northern USA and Europe during winter when prices are highest. The land, however, is low lying and often swampy. Using large machines the swamps have now been cleared away and kept dry by means of large numbers of ditches. On the cleared land oranges are grown in large numbers each tree uniformly spaced to allow easy access by machinery. Of course, drainage has changed the ecosystem: what was once a dense swamp and waterlogged soils with many plants competing for space has become a single species of tree grown in a regimented pattern. Estates of single species crops are called **plantations** and are often run by large companies with access to large distant markets. They do not provide food directly for the local community.

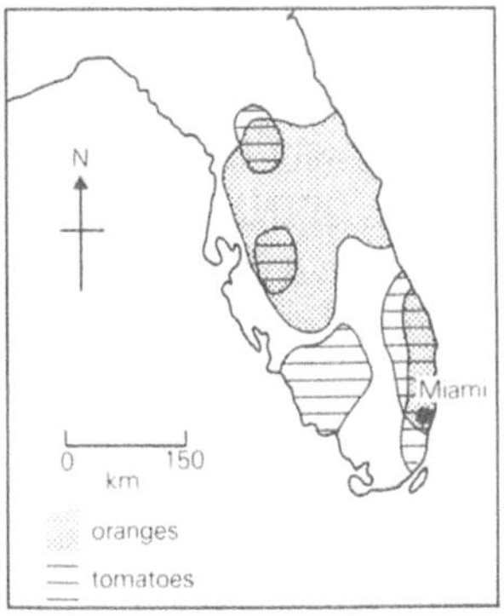

Figure 12.8 Areas of citrus fruit and tomato cultivation in Florida

Because this is an artificial system, farmers have to work hard to control weeds and pests with the widespread use of herbicides and pesticides. Occasional frosts are even countered by large oil-burning stoves which are lit in the orange groves when needed.

Without constant attention, vegetables such as tomatoes and cucumbers are even less able to survive than citrus trees. Daily doses of water, fertiliser and pesticide are provided by overhead sprinklers in a pampered form of agriculture that survives only because of the high prices that can be obtained for the high-quality crops.

SUMMARY

Farming in Israel

The green desert

In Chapter 16 there is a description of the normal nomadic way of life of the desert peoples of Egypt. Their survival depends on finding small plant shoots amongst the pebbles of the desert to provide grazing for their goats. For water they are limited to scooping up from shallow wells using a leather bucket and a long piece of rope. Understandably, with this system the desert will only support a very small number of people.

In the Negev Desert of Israel things are very different. Here water is brought to

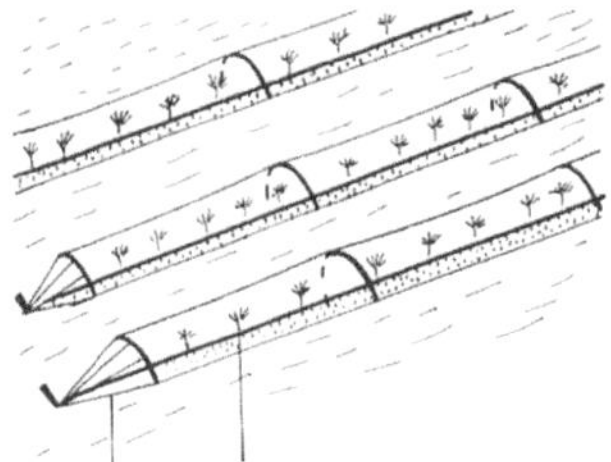

Figure 12.9 Water conservation
Individual treatment is given to plants in countries such as Israel where water is limited.

the area by canal and pipeline so that cereal crops can be grown. The animals are sheep, not goats, and they are fattened for market using cereals rather than being allowed to grub about in the desert. Citrus fruits and vegetables are also grown, again using irrigation water, each plant being drip fed from plastic hose pipes. In this totally artificial system a large pumped water supply is the vital means of success, although even so the vegetables are often grown under plastic sheets (called cloches) especially to help reduce moisture loss (Fig. 12.9).

To a limited extent the high-technology desert farming is practised in other parts of the Middle East. For example, is southeastern Libya, some of the oil wealth has been used to change the Kufra Oasis from a region of marginal subsistence agriculture to one where a large surplus can be grown. In this case the vital water is pumped out of an artesian supply 200 m below the surface and used to irrigate fields, so that clover, wheat and barley now grow within sight of sand dunes.

Coffee production

Iced coffee

In the previous examples much *money* and *energy* have been used to grow very *special* crops. However, only small amounts of food can be grown in this way. Most of the world's food has to be grown without such excessive attention. As a result, crops usually have to be grown within the limits of their tolerance to drought, cold or heat. Nevertheless, there are great pressures on people to try to grow more and more of a crop if it sells well on the world market. To do this they have to grow the crop in more and more marginal regions. Coffee growing is a very good example of some of the results of such expansion.

Coffee comes from a tree which first grew naturally in Ethiopia. It is a very adaptable tree and it will grow in a wide range of tropical and subtropical climates and soils. Gradually, as demand rose, it was introduced into India, Java and then South and Central America (Fig. 12.10). This was relatively easy because its simple cultivation is suited to less-developed agricultural methods as well as on large plantations. Above all, coffee needs little attention and it produces a higher return in money than any alternative crops grown in the fields.

Frost is the biggest enemy of the coffee tree. But despite this limitation, high value on world markets has meant that coffee trees have been planted further and further polewards and at higher altitudes. Similarly, coffee growing has been extended equatorwards (e.g. into the Caribbean islands) where the climate is really too damp and the trees become very prone to disease.

Figure 12.10 The world's major coffee-producing areas

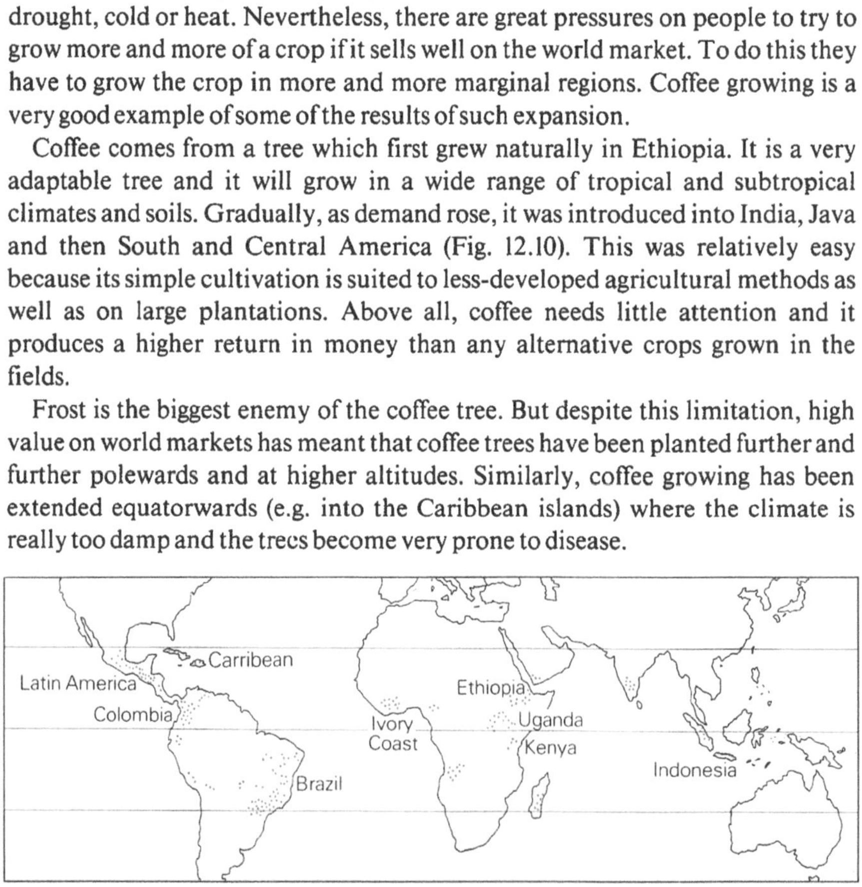

RAYMOND JESSOP

Figure 12.11 A palm oil plantation in Malaysia

Exercise 12.2 Plantation farming

Figure 12.11 shows part of a palm oil plantation in Malaysia.

1 Explain the advantages of a plantation system for farming.
2 The bare soil is a plantation road. There is also bare soil around the base of each tree, but elsewhere low-growing vegetation is allowed. Explain the advantage of this.
3 What must be added to the soil to replace nutrients lost by crop harvesting?

Not surprisingly, therefore, the marginal areas at risk from frost have increased greatly in recent years. When a bad frost occurs it can kill off much of the crop. This in turn causes hardship to the growers and a rapid rise in the price of coffee on the world market. Something like a fifth of the coffee crop is now regularly damaged by frost, but the continuing demand from the developed countries will mean that the marginal areas are still kept in use.

Farming in equatorial rainforests

The Indians know best

The Trans-Amazonian Highway strikes like an arrow right through the heart of the Brazilian rainforest. The road has not been blasted out of solid rock but bulldozed through the deep red soil that has supported a luxuriant forest for millions of years.

You will remember from Chapter 11 that everything recycles fast in the rainforest. Teaming multitudes of insects are organised into search parties that continually gather every fragment of dead organic matter. If you were to leave a newspaper on the ground it would be cut up into tiny pieces and carried away within a couple of hours. Wooden furniture in houses will even be eaten where it stands – from the legs up!

Because recyling is so rapid, the rainforest is in a more delicate balance than anywhere on earth. The native indians know this well. They also know that although the vegetation grows so well, the soil in which it grows is rarely fertile. As a result the indians only cut and burn temporary small patches called 'gardens' out of the forest. They make little attempt at weeding and they leave the charred stumps of trees lying where they fall. It all looks very disorganised

Exercise 12.3 A rubber estate

1 The map and section (Fig. 12.12) show a rubber plantation set among hills in Malaysia. Why does the estate occupy the area indicated?
2 It is intended to expand the estate into undeveloped areas. Which area do you think would be exploited first? What erosion control measures would be needed if prolonged cultivation of rubber trees is to be successful?
3 If the main estate road were driven up into the hills of area B, how would this encourage erosion?

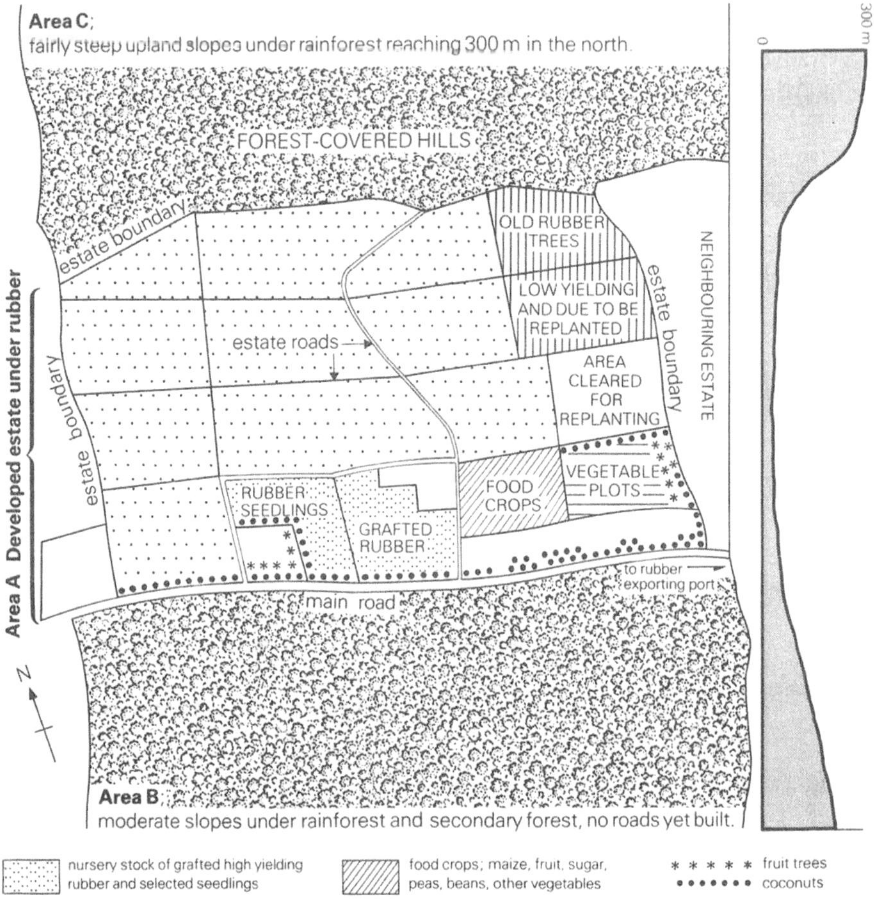

Figure 12.12 A rubber estate

but there is a reason for the lack of weeding. Any bare soil is liable to erosion and leaching of nutrients. It is better to keep the soil and its nutrients and tolerate some competition from weeds (see Fig. 12.2 b & c).

Overall, the indians only scrape a subsistence living from these gardens. Even with these 'conservation' methods, the soil will only grow crops for two years before it is exhausted and they have to cut and burn a new area. *But the method is adapted to the environment.* With small numbers of people the vast Amazon lands will provide food indefinitely.

The low real fertility of the tropical rainforest was not understood by many. At the end of the 19th century a city called Belem grew rapidly on the banks of the River Para in Brazil. Here people from the drought-stricken northeastern coast of Brazil came in their thousands, cut down large tracts of forest, and planted corn, beans, rice and tobacco.

Soon they found out how fragile the land really was. The soil was exhausted within a couple of years and they had to move to new areas . . . and new areas . . . and new areas, until there were no more areas and they had to return to the original cleared lands which had still not recovered. But they had ruined 30 000 km^2 in the process. Today there is no longer enough food grown on these impoverished soils to support the 800 000 population of the city and the government is trying to evacuate them to new areas.

More recently in the southern rainforest an area called Mato Grosso has been cleared and planted to grass. Over 10 million beef cattle graze on these lands, but not for much longer – soil fertility is already low and erosion has begun. But still the lesson has not been learned. Now the government is attracting farmers to the newly cleared areas beside the Trans-Amazonian Highway. The talk is again of the old mistaken idea of 'fertile valleys' and 'a thick layer of humus accumulated during centuries'. So far, more than 130 000 km^2 (nearly the size of England and Wales) have been cut down and burned. The new farmers to this area are going to be the latest in a long line who have all discovered that the fertility of these rain-forest soils just cannot be maintained without a ruinously expensive use of fertilisers. And meanwhile the forest devastation goes on . . .

Desertification

Man made the desert?

14 % (630 million) of the world's population live in arid or semi-arid areas. One of the main problem regions in the semi-arid zone is called the Sahel (Fig. 6.11). Sahel means 'the border' in Arabic and it is the name used to describe the semi-arid grass, bush and occasional tree vegetation that forms the dry scrub land zone next to the Sahara desert. Here men have followed a yearly pattern of activity controlled by the seasons. The longer season is hot and almost completely dry, the shorter is hot and humid with rain falling in heavy showers. However, being on the edge of the desert, the rainfall is not reliable and sometimes the land may not receive the 400 mm or so of rainfall needed for cultivation.

Traditionally people from this region were herders of cattle, sheep, goats and camels (Fig. 12.13). They were nomadic and they moved south in the dry season

Figure 12.13 Cattle herding in the Sahel of Africa

JOHN MACK

ANDERS RAPP

Figure 12.14 Gulley erosion in Africa following overgrazing

and then returned north when the rains came. Others cultivated the land in small patches which they clawed from the savanna by burning (Fig. 12.2b). When the fertility of the land was lost they burned a new small area and left the used area to recover naturally. In the areas of Central Africa with the poorest soils, fallow could take up to 35 years to revive. Six to ten years is needed even in the better regions, but such long resting periods did not matter as there was plenty of land to go round.

But the population of Africa is rising fast. Without the sophisticated farming methods to keep pace, the population is exceeding the productivity of the land. Land that was left in fallow for long periods is now used every two to three years, but it does not recover in such a short time and so its yield is very low. The productivity of the fields has actually gone down under this pressure despite the hard work of the farmers.

The herdsmen have fared just as badly. National boundaries have stopped most of their nomadic wandering in search of good grazing. In the Sudan, for example, 3–4 km^2 of savanna are needed to support a family herd of 30–40

Figure 12.15 Isolated village in the Sudan
The difficulties of applying modern techniques in this environment are clear. There is not even a road. Note the lack of grass is due to overgrazing.

JOHN MACK

cattle. In 1945 the stock numbers were just equal to the land available. Since then the human and animal population has doubled. It is no surprise to find that overgrazing is widespred and the soil left open to erosion (Fig. 12.14).

The clear answer to this problem is in the use of fertilisers and farming methods that use soil moisture more efficiently. The problem is that the people do not understand how to use fertiliser, they have never had any surplus of money to buy it and the isolation of many areas makes for difficulties of supply (Fig. 12.15).

At the moment because of overuse of land in the Sahel, soil erosion is turning more and more land into an unusable desert. A new word has been coined especially to describe this disastrous situation – **desertification.** Already desertification has claimed more than 30 million km² of the Earth's agricultural land, not because of any change in climate but solely due to man's destruction of the natural vegetation and his inability to stop the resulting soil erosion. It is not that he is being intentionally careless with the land, but desperation for food makes long-term goals difficult to see.

We finish this chapter on a note of despair. We now understand much about the way the environment works and the results of changing the naturally balanced ecosystem. Agricultural advisers mostly now understand the correct management methods. It is vital that this understanding is used, but so far little has been achieved.

Figure 12.16 Future land-use possibilities
(See page 156)

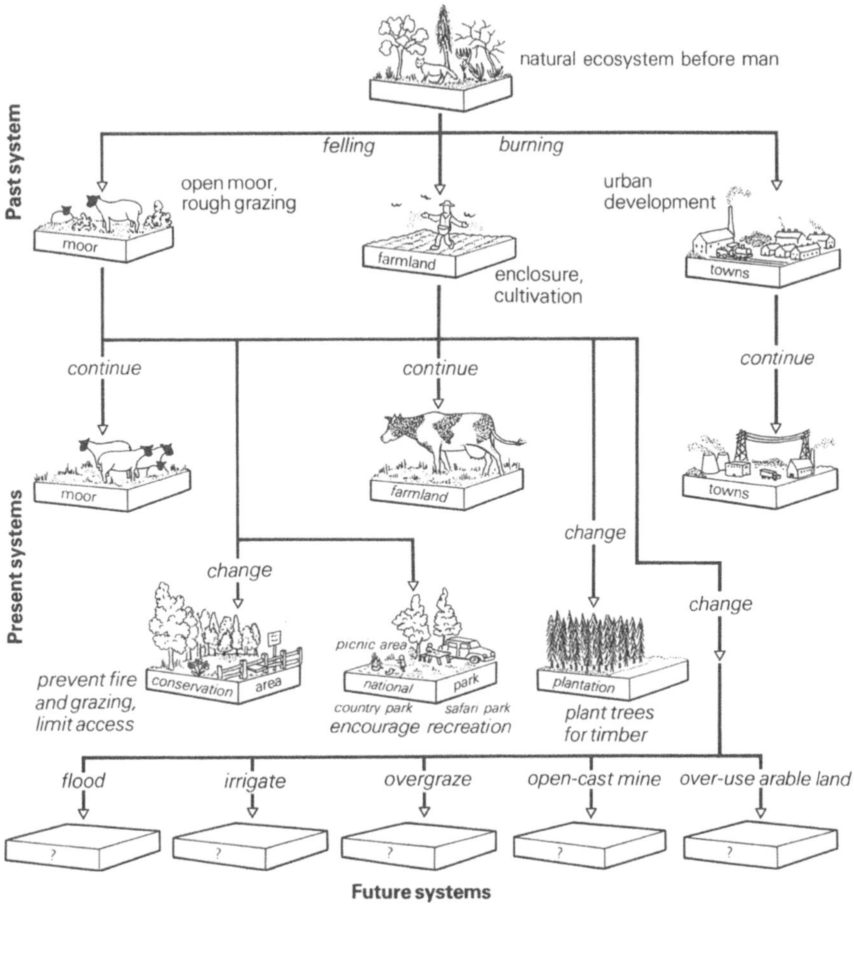

Exercise 12.4 The future of our land use

Man has many alternative ways in which he can use the land. Some are destructive, others creative. Some of the choices are shown in Figure 12.16. Five boxes have been left unlabelled. For each box describe and explain the result of the land use shown.

EXAMINATION-STYLE QUESTIONS

1 State the two main components of a soil. Explain how a soil forms, illustrating your answer with labelled diagrams.

2 What is the name of a soil in upland areas which experience cool climates and where annual rainfall exceeds evaporation. Draw a labelled diagram of such a soil and contrast it with a soil from a humid tropical region.

3 Explain what is meant by the terms **texture, humus** and **soil profile.** Choose one type of soil which you have studied in detail and describe, with the aid of a labelled diagram, its main characteristics. Explain how this soil may have been formed.

4 Explain how natural vegetation is adapted to survive in areas which regularly receive (a) drought, (b) long cold winters, (c) extremely high temperatures, and (d) shade.

5 Name an example of an area with (a) tropical rain forest, (b) tropical grassland (savanna), (c) temperate grassland (prairie or steppe), and (d) tundra. Explain how man has changed the pattern of natural vegetation in the tropical rainforest area you described above.

6 Explain what is meant by natural vegetation and explain why it is becoming increasingly rare in the world. Name one savanna region and describe how man has altered its vegetation.

7 Describe the main characteristics of desert vegetation and explain how each type of vegetation is adapted to its environment.

8 Explain why there is a much greater variety of trees and plants grown much higher in equatorial regions than in regions of coniferous forests in cool temperate areas. Explain why it is difficult for plants and trees to grow in the tundra and desert regions.

9 Do you think it is true to say that soil erosion is the result of careless cultivation? Describe one area in a developing country which illustrates your answer. Describe some of the methods being used to control soil erosion.

10 Name two types of soil erosion found in the United States. Name two areas where soil has been a problem and explain some of the means that have been used to combat soil erosion. Explain why some of these methods cannot be used to combat soil erosion in developing countries such as those in Africa.

11 Name and locate one example of small-scale subsistence farming you have studied; describe the methods used, the crops produced, and explain how these are influenced by physical, social and economic conditions.

12 Describe and explain some of the main features of plantation cultivation giving specific examples to illustrate your argument. Describe what methods of soil erosion prevention must be employed in a plantation.

13 Explain why a tropical rainforest area has the greatest variety of trees and the deepest soils of anywhere on Earth. Explain why it is impossible for plantation agriculture to survive for long periods in such areas without extensive use of fertilisers. Explain how primitive cultivation methods have managed to allow survival of small groups of people in these forests.

14 Referring to specific examples and named areas show how natural vegetation is affected by climate, soil and man.

Part four **Shaping the land**

Moving water, ice and wind continually remove soils and rock, exposing new rock to weathering and changing the landscape. They also scratch and tear at the rock, producing a set of distinctive landforms.

DENYS BRUNSDEN

Figure 13.0 Landslide! The roads at the top and in the middle indicate the scale. The photograph was taken in the foothills of the Himalayan mountains, Nepal.

13 Mass movement processes

Soil movement

On the move

Soil is the result of rock weathering. It appears to be an unchanging part of the landscape because we rarely see it move. We tend to think of a landslide as something unusual and unnatural. In fact, except on flat land, soil is always on the move, creeping downhill a few millimetres a year – an unnoticed part of landscape erosion. The soil appears unchanging because new rock weathers to replace that being moved away and the change is so slow that there is plenty of time for well developed soil layers to form.

The steeper the slope, the faster soil moves downhill. You may have noticed that steep slopes have thin soils. This is because weathering of new rock can only just keep pace with movement. Another sign of rapid movement is that soils on slopes contain many pebbles of unweathered rock. Sometimes boulders move so often that chemical weathering never has any real chance to act. In these cases, movement is faster than soil formation and a scree slope of boulders is produced.

Scree formation

Block by block

On very steep slopes, where soils do not have the time to form, bare rock is exposed and mechanical weathering becomes important. Blocks are slowly prised loose and they fall, bouncing and rolling until they come to rest. Mountaineers on frosty rocks are well aware of the importance of frost shatter. As the morning Sun warms the rocks, falling blocks prised from the rock face overnight can be a major hazard. These blocks are mostly large and sharp-edged and so do not travel far, soon building up as piles of loose rock called **screes** (Fig. 13.1).

Figure 13.1 Scree slopes in Sutherland, Scotland

ROGER JONES

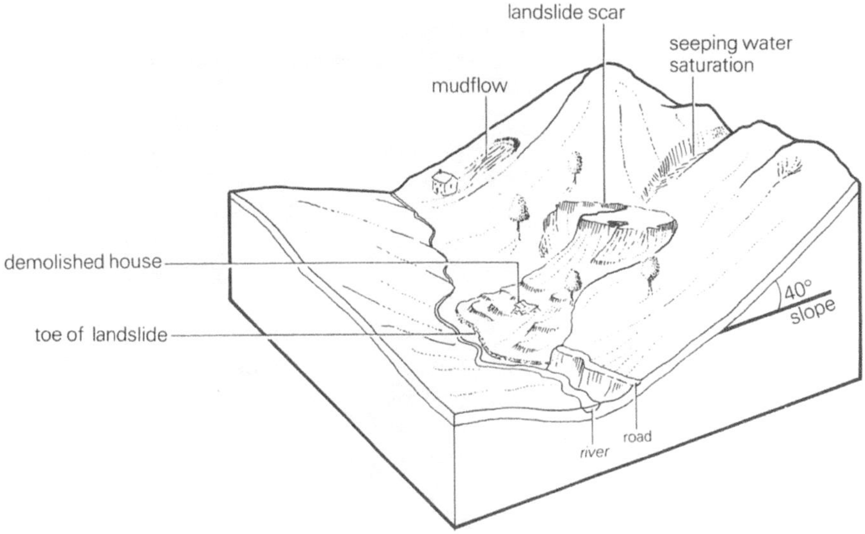

Figure 13.2 The events of South Topanga Canyon, Los Angeles, USA

Landslides

Slipping and sliding

When slopes are less steep, screes do not form and weathered blocks remain in place, eventually making a soil. Nevertheless soil can be very unstable.

We often hear of landslides in the newspapers, especially when houses are destroyed or people hurt. In South Topanga Canyon near Los Angeles (California), for example, a man was woken up at 3 am one morning by a loud rumbling sound outside. He later told local newspaper men that just as he opened his eyes the house collapsed and the next thing he knew was that he was in hospital. A near neighbour was also awakened by the noise only to find a mudflow oozing in through his back window (Fig. 13.2)!

What had happened to allow gravity to work so rapidly on the slopes in California? It was as though the weight of soil had been too great for the angle of the valley slope and the whole mass had simply slid downwards. Yet nearby slopes of similar angle had not failed. This is because there is a fragile balance on all slopes between the weight of the soil and the amount of friction opposing movement. The presence of large amounts of water in a soil is one very good way of tipping this balance towards disaster.

The landslide and mudflow in South Topanga Canyon occurred partly because much rain had fallen the day before. Water soaked into the soil and caused some of it to become saturated. The soil was so rainsoaked that the water supported part of the soil weight, much as a person is supported by water in a swimming bath. With less pressure on the slope there was no longer enough friction to hold the soil on the rock, so it broke loose in 'slabs' and slid quickly down the valley side and into the houses below.

Soil landslides sometimes turn into mudflows, depending on the amount of water present and the type of soil. Sandy soils tend to flow; clay soils tend to slide as slabs.

Landslides mostly occur when soil on steep slopes becomes saturated. This does not happen very often and most of the time water can drain out of the soil

Figure 13.3 Landslide sites This meandering river is undercutting a slope at the bends, causing landslides to occur.

as fast as it soaks in. So it takes a really long period of heavy rain to cause water to build up in a soil and allow a landslide to occur. Landslides also occur if valley slopes are made steeper, such as when they are undercut by a river (Fig. 13.3).

Although landslides occur only occasionally, they are always dramatic. Landslides near rivers often dam up water temporarily. They are also probably the most important way in which soil moves on steep slopes.

Figure 13.4 Terracettes on a steep coastal slope (Sutherland, Scotland)

ROGER JONES

Figure 13.5 A landslide at Mam Tor (Peak District)

Exercise 13.1 Landslides

1 Figure 13.5 is a photograph of a landslide near Mam Tor (Derbyshire). Describe what seems to have happened.
2 On the hillslope in the background is there any other evidence of instability?
3 Describe any evidence you see on the hillslope of previous attempts to drain the soil and prevent saturation.
4 This road has not been rebuilt. Why do you think rebuilding would be costly and only temporarily successful?

Clay soils are most at risk from landslides because water drains away so slowly through the tiny pores and they are easily saturated. On slopes that are at the limit of steepness for soil formation, only a thin layer of weathered material exists (Fig. 13.4). Under these conditions normal landslides are not possible and the soil and grass tears away in strips and slips a few centimetres downslope. **Terracettes,** as these strips are called, seem to move quite slowly. A slope with terracettes can clearly be seen by the thin ridges that are produced.

Figure 13.6 A panoramic view of the extent of the mudflow which developed from a rock and ice flow in the Andes of South America

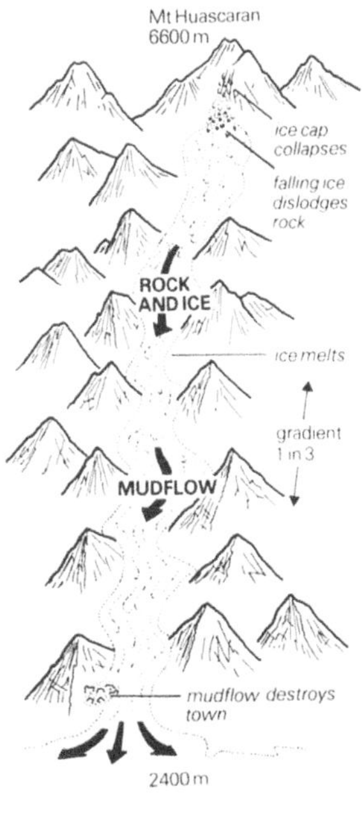

Catastrophe! *Severe landslides*

Very occasionally much larger landslides occur, involving rock as well as soil. In these rare cases rock and soil can move at speeds up to 100 km/hr and cause disasters. There are many reasons for large landslides, but they are important because they may move as much material in a few minutes as a river can move in a thousand years. It is also important to realise that one type of movement can often trigger off a whole series of others.

The following example describes the largest single movement of debris in recent times. It started when an earthquake caused the collapse of part of the summit ice cap of Mount Huascaran in the Peruvian Andes of South America. As the ice cap fell away it dislodged thousands of tonnes of rock which, together with ice, went hurtling down a nearby valley. The valleys of the Andes are so steep that the mass of rock and ice, instead of coming to rest, gathered speed, and as it burst into the valley below it picked up thousands more tonnes of rock, water and soil (Fig. 13.6). Some of the pieces of rock being carried were bigger than a house! As it went down the valley the ice in amongst the rock fragments melted and the mass of material turned into a mudflow. 15 km down the valley, the people of Yungay heard a distant roar. One of the few eyewitnesses to survive the next few minutes described the scene:

‘The crest of the wave [of mud] had a curl, like a huge breaker coming in from the ocean. I estimated the wave to be at least 80 m high. I observed hundreds of people running in all directions [and] all the while there was a continuous loud roar and rumble. I reached . . . the cemetery near the top just as the mudflow struck the base of the hill and I was probably only ten seconds ahead of it. It was the most horrible thing I have ever experienced and I will never forget it.’

Soil creep

But quietly creep the hills

We have started our description of hillslope processes by looking at the most spectacular. They are important processes because, although rare, they cause much erosion. But there probably isn't a landslide or an avalanche of rock and ice outside your home and you are unlikely to live on a steep slope, so what is happening everywhere else?

Here we meet one of the difficulties of landscape study. Some of the processes are spectacular and get all the news; others work all the time but in such small steps that only delicate instruments will detect that they are even working at all. The most important slow process is called **soil creep.**

Most soil creeps down hill very slightly every time rain falls. This is because clay particles swell as they get wet and shrink again as they dry out. During this change of size the soil is being pulled down slope by gravity and the swelling and shrinking with each storm help to ease the soil down slope. This is an extraordinarily slow process and at the end of a year a piece of soil might have moved only one millimetre downslope. Soil creep works fastest at the surface where wetting and drying occurs most often. As a result sometimes the process can be seen indirectly. A telegraph pole or a wall may lean over as soil creep pushes against it. Even trees may be pushed over (Fig. 13.7). As they continue to grow, their trunks bend to regain an upright position. During the winter, freezing and thawing of water in the soil also causes swelling and shrinking and this adds to soil creep. Thus soil creep may move a piece of soil perhaps only 10 cm down a

Figure 13.7 The supposed effects of soil creep

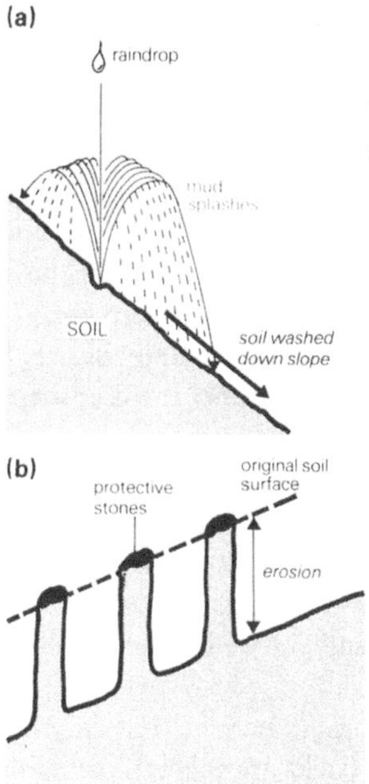

Figure 13.8 Surface wash
(a) Heavy beating rain splashes the soil away. (b & c (*right*)) If large pebbles protect the surface, then earth pillars will survive for a while.

RAYMOND JESSOP

slope in 100 years; hardly worth considering you might think. But every particle on the soil surface moves by similar amounts all over the hillside. So if you add all of these movements up they can be as important as a localised landslide. Scientists have still not made enough measurements to be sure of the answer to this, but we do know that landslides can only work on steep slopes and so soil creep is possibly the most important way of moving soil particles on gentle slopes. On the other hand, on steep slopes soil creep is probably less important than landslides.

We tend to take grass and other vegetation cover for granted, even on steep slopes. The roots of plants bind themselves and the soil together and so reduce soil movement. Hillslopes change very slowly under these conditions. However, if we lived in a low rainfall or low temperature area we would not be so used to vegetation. In these areas both vegetation and soil are uncommon. On steep slopes in well jointed rocks, screes occur, whereas weak rocks are extensively cut into by channels – called **gullies** (Fig. 13.8d). These features are common any place where water can flow over the surface. They are especially common in semi-arid regions, but where man has cleared land in the Tropics the effects of beating rain are also clear to see (Fig. 13.8a–d).

(d) Eventually the land will become heavily gullied.

Soil solution

Before your very eyes

In the previous sections we have looked at the movements of soil particles. However, there is also a very important way in which soil material moves completely invisibly. This is part of the soil taken away in solution. We should not be surprised to discover that solution is important because we know that soil is formed by chemical weathering. Soluble materials remain after rock has been chemically weathered to clay, and they are easily transported away in the soil water. Even less noticeable than soil creep, solution, and consequently the removal of rock constituents **(leaching),** is happening all of the time. Slowly but surely, on steep slopes, on gentle slopes and even on flat ground, the landscape is dissolving away. The ground is dissolving away beneath you right now!

Of all the hillslope processes, leaching is probably the most difficult to understand and measure. There is much work still to be done, but scientists now think leaching may sometimes account for half of all the material lost from a hillslope in areas which have adequate rainfall. So *leaching can be as important as soil creep, landslides and gulleying all put together* – and yet we can never see it at work.

One final point is worth emphasising. It is common to believe that only limestone rock is affected by chemical action, but other rocks are subject to chemical attack just as much as limestone. This means that *all* landscapes are affected by solutional erosion.

The nature of avalanches

Avalanche!

Cold regions experience types of movement that are not found anywhere else. Taken together they make up a very special sequence which is described below.

In Antarctica there are some places where temperatures remain below freezing throughout the year. Here frost shatter does not break off pieces of rock and the landscape ceases to evolve. In most other cold regions, however, spring arrives by about June or July and lasts only a few weeks. There is no 'summer' and by September conditions are wintry again. But although spring, with its temperatures creeping above freezing, does not last long, thawing allows a whole range of activities to occur quickly. Earlier in this chapter we mentioned how blocks of rock can be dislodged by freeze/thaw processes to produce scree slopes. But on slopes below about 40° loosened debris tends to remain in place. At the same time the slope becomes gentle enough for snow to accumulate, allowing a new process to take over: we all know its name well – **avalanche!**

Winter snow can be the delicate snowflakes that clothe mountain tops like the icing on a Christmas cake. It can also provide the material that can come crashing down on a town and blow its houses to bits as effectively as a bomb.

Throughout winter snow falls and settles on mountain slopes. Because it is light it is whisked up by the winds – cleared from some places but piled up in others. And this is where the trouble starts. A few metres depth of snow clothing a mountainside may weigh a million tonnes. On some slopes this enormous weight builds up an intolerable stress, the delicately interlocking snowflakes are torn apart, and avalanches begin.

For a moment consider the way snow accumulates on a steep mountain slope in winter. Every few days a few centimetres or more of snow settle or are blown onto the slope. It is cold. A snowy day may be followed by a clear freezing night. The snow settles and compacts; the snowflakes freeze together more tightly, lose their angular form and become more rounded. It snows again and again building up a thick layer of snow, which is powdery and made of finely shaped crystals at the surface, and more compact with old, more rounded flakes below. Some of these crystals deep in the snow eventually behave almost like ball bearings, enabling the whole snow pack to move more easily. In the spring, snow and mountain rock are warmed by the Sun. But rock and snow warm at different rates and, as the rock faces that surround the snowpatch warm up, so they gently heat the snow from below. By the spring only a very small amount of heating may be all that is needed to cause instability.

Whether it be from a heavy snowfall or spring thaw, the start of an avalanche can hardly be detected: in a small patch of snow perhaps just a few metres across, delicate snowflakes tear apart and allow the snow to slump forward onto the untouched slope below. With the whole snow mass delicately balanced, the new weight causes the snow on this part of the slope to slip forward as well. Soon a chain of action and reaction is started which grows every second. What was only a small slip ten seconds earlier is now a plunging mass 100 m across (Fig. 13.9).

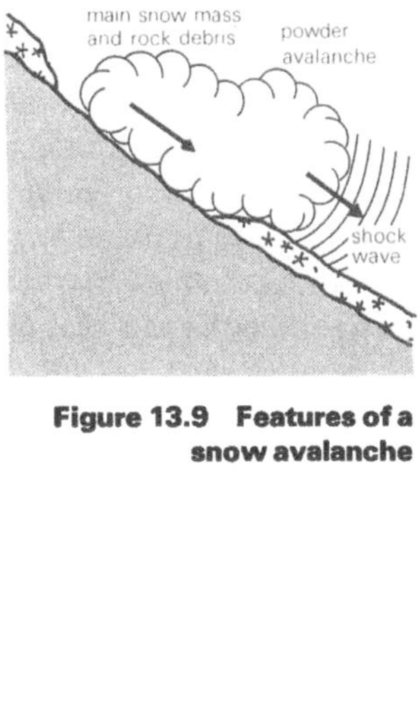

Figure 13.9 Features of a snow avalanche

At the start, the slipping snow moves slowly – a few metres a minute. But as it grows bigger, so its speed increases until, half a minute from the start of the avalanche, the bulk of the snow is moving at 80 km/hr. By now nobody can out-run it.

As the mass gathers speed it shakes off more and more powdery surface snow into the air. This powder becomes mixed with the air to form a high density 'gas', ten times as dense as air. This, too, is pulled down slope by gravity, but is not subject to the frictional resistance of the slipping mass. As a result it starts to move ahead of the main avalanche, picking up more and more powdery snow so that it is soon hurtling down at over 150 km/hr and dragging in eddies of air whose speed exceeds 300 km/hr.

At this stage the powder avalanche hides the main sliding avalanche in a bubbling mist. The powder is now moving so fast that the air in front does not have time to get out of the way. Pushed downslope by the powder avalanche, the air becomes a compressed shock wave with its own devastating power.

In the valley below the small plume of white snow first seen high on the slope becomes a long column of bubbling whiteness whose approach can be heard as a low rumbling note, vibrating the rock of the mountain. Less than a minute after the first snow flakes have torn apart and caused that slip on the mountain, a million tonnes of snow and a hundred thousand tonnes of air may have hit the valley bottom at over 300 km/hr.

For buildings in the path of the avalanche, there is little hope unless they have been specially built. But the importance of an avalanche for landscape formation is that it tears away frost-shattered rock from slopes, and hurtles boulders down to the valley floor. Thus it is a major source of erosion in mountains and it must, in the past, have propelled millions of tonnes of rock fragments onto the surfaces of glaciers to add to their debris load.

Humpty Dumpty

We can now move on to discuss the way in which soil movements affect man's activities. No movement can be helpful to a road, a house or a dam. In some cases it might just be possible to put up with the effects of soil creep, but even this is unlikely. Clearly, landslides cannot be tolerated at any time. It is therefore vital not to build in an area of slope instability. The lesson of Humpty Dumpty is one that is all too real to planners and engineers because many lives could be sacrificed if a mistake is made.

Hong Kong regularly receives large amounts of rain. So in June 1972, when 200 mm fell in one day, there was little surprise. However, this deluge was followed by similar amounts on both of the following days. In total the rain that fell was equivalent to a whole year's rainfall over London. Towards the end of this period a giant landslide started to form on a steep slope above one of Hong Kong's most densely populated areas of high-rise flats. Within a space of just five minutes the slide had swept across two roads and completely demolished a 12-storey block of flats. Within those same five minutes 67 people living in the flats had been killed.

Now from our hillslope studies we can see why this should have happened. In this humid tropical region chemical weathering quickly rots solid rock to great depths, so deep soils cloak even steep slopes. We also know that steep slopes and deep soil are an unstable combination. We can now see how man turned an unstable situation into a recipe for disaster. The trouble stems from the large population of the colony and the small area of land they have to be packed into. The gently sloping sites have long ago been used up and the planners were forced to turn to areas of steep slopes. As a result we have a conflict between human

Figure 13.10 Man-induced hazards on slopes
(a) This slope is stable because the vegetation helps to anchor the soil to the rock. (b) This slope is less stable. The cutting made for the road gives a short, steep, unprotected slope above the road. Part of the road is built on 'fill' (soil and rock dug out of the cutting). This material is very liable to being washed away. (c) This slope is highly unstable. Although the building is securely anchored to the underlying rock, the soil at the top of the slope is unsupported and liable to slip – for this reason, a retaining wall often has to be constructed.

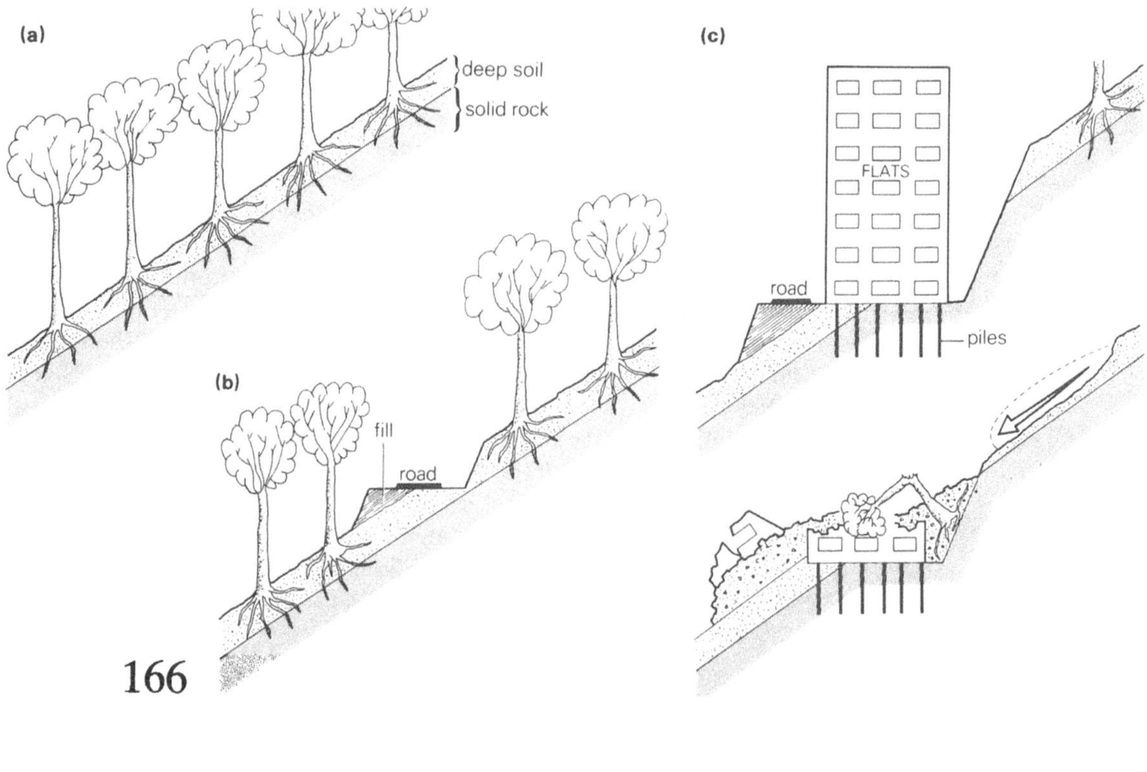

pressure for building and dangers from unstable land. The results are shown in Figure 13.10.

Finally, you will remember that water moving through a soil does not always travel in straight lines. Sometimes it can converge, even on a uniform slope, and when it does, the soil gets ever closer to saturation. Above that Hong Kong block of flats, water converged and saturation was reached on 16 June 1972.

Building on frozen ground

I thawt this was rock!

Cold conditions affect not only mountainous areas covered by ice. Much of the plains of Siberia and Arctic Canada are extremely cold places but they do not receive enough snowfall each year to compact into ice. Nevertheless these areas are underlain by frozen ground called **permafrost.** Soil and rock, and the water contained within them, are all frozen by the long, cold winters down to the depths of hundreds of metres. But the frozen ground is very fragile: warm it ever so slightly and the ice turns back into water and the top metre or so of ground melts, turning the soil into a sort of sludge. Normally this is of little consequence to man because such regions have traditionally been sparsely populated. However, in recent years the discovery of minerals – especially oil – under these frozen wastes has changed all that. Now there is an urgent need to understand the way permafrost behaves because vast unspoilt lands and hundreds of millions of pounds depend on it.

For example, about half of Canada and all of Alaska are underlain by permafrost. The discovery of the Alaskan oil field and the need to build a 1300 km-long pipeline, oil rigs, drilling wells, storage tanks and homes has recently posed a whole variety of new problems. The trouble comes from trying to build on the 'skin' of the permafrost that turns into a waterlogged sludge as soon as it is heated. Thus a pipeline carrying hot oil would melt the surrounding permafrost.

SUMMARY: TRANSPORT DOWN SLOPES

Soil and rock debris move down a hillside or cliff because they become unstable. There are many forms of instability: very fast sudden movements that we call rockfalls, landslides or mudflows; and more gentle, long-term movements called soil creep, slope wash or solution.

On very steep slopes soil or vegetation are missing and the main processes at work are mechanical – blocks are prised away by heat or cold (exfoliation and frost shatter) giving **rock falls** which accumulate at the slope bottom as a **scree** (Fig. 13.1). Screes are not usually very thick but they can cover slopes and prevent further mechanical weathering. Unless streams continually wash the scree away from the bottom of a slope the debris will gradually rot where it lies and form a soil.

Moderate slopes are too gentle for scree material to move and so they have developed soils. The movement of soil is different from scree because the particles are small and can be moved with the aid of water. **Landslides** and **mudflows** are typical fast soil movements on slopes. After heavy rain, water may build up in the soil and help to reduce the pressure of one particle on another. In clay soils a whole slab of material slides downwards, leaving a crescent-shaped scar behind. When the soil is sandy, water build-up often causes the soil to flow downhill. Where soils are very thin, soil slips in strips called **terracettes.**

On more gentle slopes **soil creep, solution, rainsplash** and **surface wash** are important (Fig. 13.8). Soil creep results from swelling of soils as they are wetted by rain, and shrinking as they dry out again. It can also be produced by freezing and thawing of water in the soil pores. Surface wash occurs with very heavy rainfall in the semi-arid areas where there is not proper vegetation cover to hold the soil in place. It often produces **gullies.**

Finally, although we can never see solution and **leaching** at work, we know that it is the main way of forming a soil and so must be important on most slopes. Leaching may remove as much material as all the other movement processes combined in lowland areas.

SUMMARY

With the solid support removed, the pipe would settle into the sludge irregularly and it is likely that somewhere along its 1300 km it would crack (Fig. 13.11). The oil spillage that could result is too serious to risk. Therefore, the pipe now runs above ground in an insulated jacket.

In the past, oil storage tanks have been built with their foundations in permafrost, but over a long period of time heat from the oil has melted the permafrost below the concrete support raft, so that the tanks have settled and are starting to rupture.

Figure 13.11 Permafrost hazards

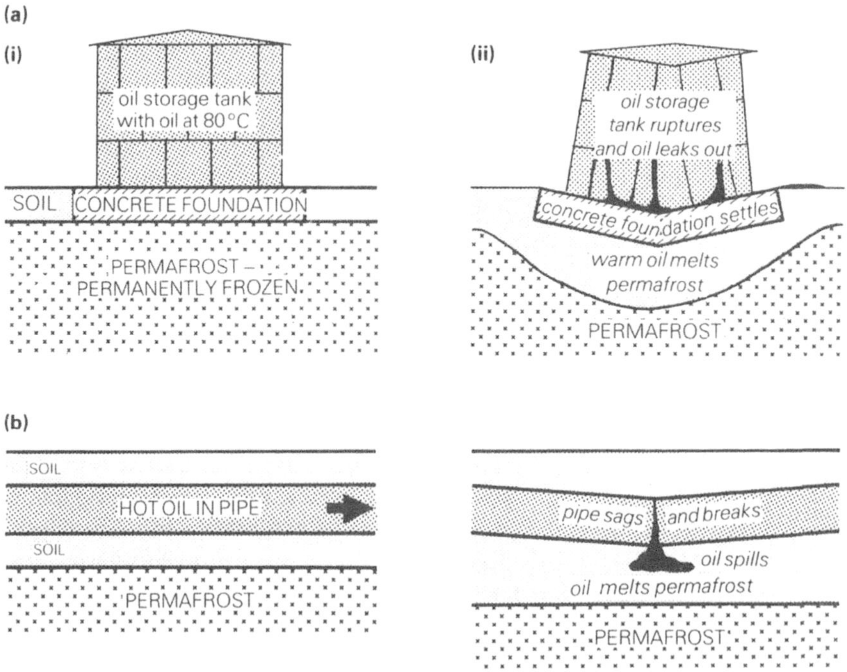

Exercise 13.2 Problems of waste disposal

Waste from man comes in a variety of forms. Industrial waste can be solids such as coal, rock or ash from a furnace; it can be liquid waste such as toxic chemical sludges used to clean metals; it can even be the radioactive waste of a nuclear power station. Then there is domestic rubbish – plastic bottles, old refrigerators, cars, paper, and so on.

The easy answer is to use it to fill up old quarries or dump it in piles on unused ground. There are plenty of holes dug in Britain for brick and gravel material, for road-making rock, for chalk, for cement, and so on. So dumping is an easy answer, but is it the right one?

1 In Figure 13.12a you see two pits that are to be filled with solid waste. On the surface they look the same, but underground conditions are very different. Explain why no dumping should take place in pit B. Why is A safe?

2 Now consider liquid toxic waste – say spent chemical from a plastics factory that is poisonous in even tiny quantities if drunk. Are any sites now safe?

3 In Figure 13.12b you see two piles of coal waste which have been dumped on a surface which all looked the same. Again underground conditions are very different. Explain why the pile A is safe but the pile B should be removed immediately.

4 Figure 13.12c shows a cross section of the town of Aberfan in Wales, where colliery spoil was tipped on a slope above the town. Why was this an unsatisfactory site? At 9.15 am on Friday 21 October 1966, 140 000 tonnes of spoil flowed down the hill and overwhelmed a village, killing 144 people, 116 of whom were children who had just arrived at school.

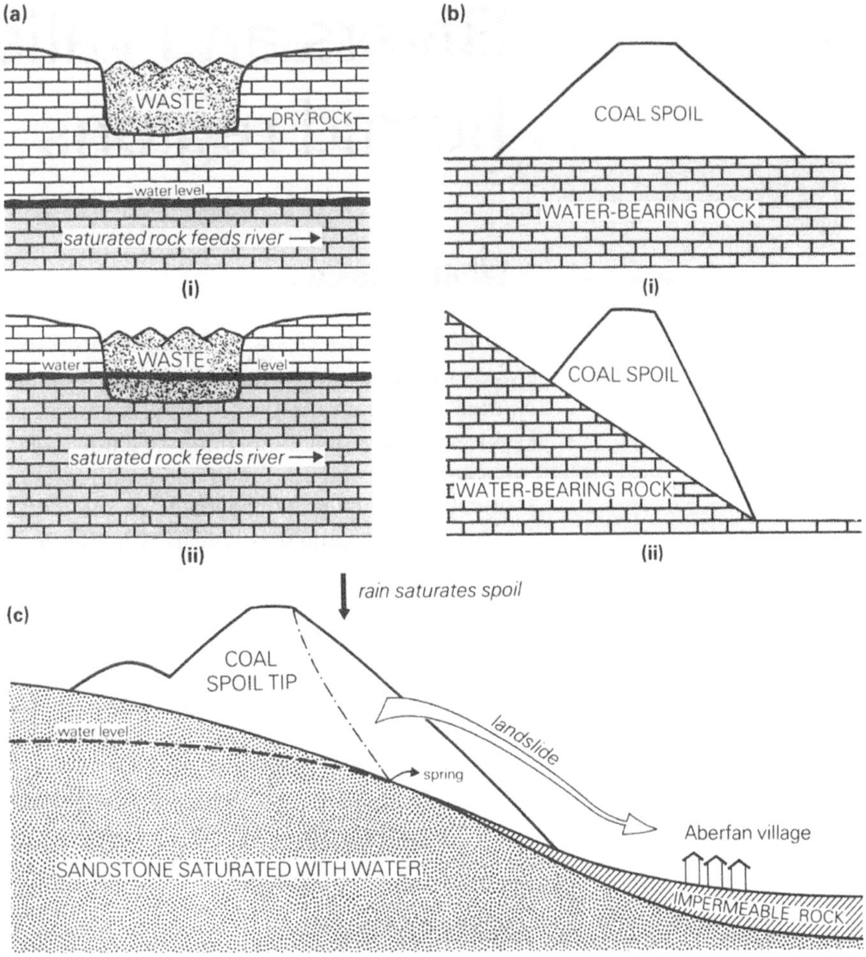

Figure 13.12 Man-made hazards
(a) Waste tips; (b) coal spoil tips; (c) coal spoil tip at Aberfan.

Exercise 13.3 Road building in Nepal

Nepal is a developing country with a very mountainous landscape. Aid programmes are helping to provide better links through the country and with neighbouring India, but there are severe problems created by deep river valleys. Roads connecting towns must have numerous hairpin bends and bridges, all built at a height of more than twice Ben Nevis!

1 List the types of movement you could expect to find in an area of steep slopes.
A preliminary survey of the line of a road was made from maps and air photos, but the dense vegetation hid most evidence of slope movements. A ground survey found slope movements in the places shown on Figure 13.13.
2 Why should slope movements be most common in the region labelled A?
3 Field survey made engineers reconsider their original route line. List the major problems the route would have encountered.
4 Suggest an alternative route for the road and explain how your route is better than that in the original plan.

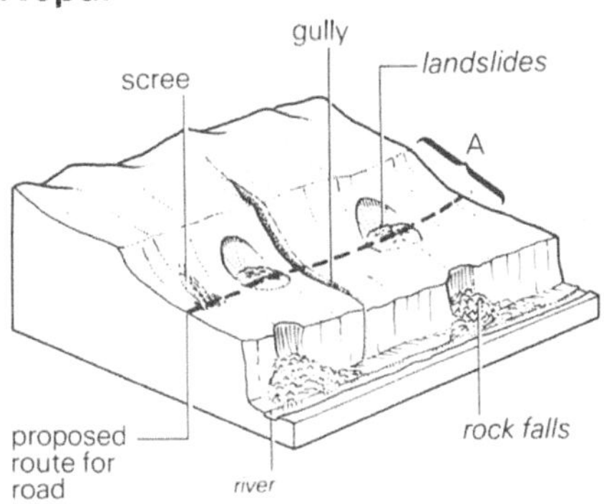

Figure 13.13 Slope movements found from a ground survey in Nepal

14 Rivers and valleys in humid regions

The water cycle and landscape formation

What goes up . . .

Until the first rockets were fired into deep space, one of the most fundamental laws of nature was 'what goes up must come down'. Thus, although water is lifted as vapour from the oceans using the Sun's energy and carried high into the sky where it forms clouds, it must eventually fall back to Earth as precipitation.

Cloud droplets are very small and light but they are still pulled to Earth by gravity. As they grow, they become so heavy that even updrafts of air cannot keep them aloft and they 'fall' as rain. But the pull of gravity does not stop when rain reaches the ground: water is slowly pulled through the soil until it reaches a river bank. Having reached a river, it is finally pulled quickly down to the sea.

If we look at the water cycle in this way – an initial upward movement using heat energy from the Sun followed by a long downward return using gravitational energy – we see that water on land is really pulled through soils and rocks and along river channels. Any winding river channels, small soil pores, and so on, are merely inconvenient obstacles thwarting the return of water to the sea. Anything the water may do to pieces of soil or rock, such as help in weathering and transport, is purely incidental. Erosion is just an accident!

Internal loss of energy

Water at work

Most of the energy available to a river is used up simply in making water flow. This becomes apparent sometimes when water flows quickly in a pipe. In hydroelectric power stations for example, when water is taken from a high-level reservoir to a generating station in a valley bottom, the transfer pipes become warm, as friction between the water and the sides of the pipe is converted into heat. Over 95% of all energy in a stream may be used up in overcoming friction, which doesn't leave much to transport debris or erode bed and banks.

As water tumbles along in its channel much energy is used in pushing against the pieces of rock that form bed and banks. If a piece of rock is pushed hard enough by water flowing past it, then it too will begin to move. Sometimes these pieces are already loose on the river bed and are transported to a new position; in other cases pieces of rock are torn directly from the channel.

Drainage basins

A leaf out of the book of plants

A **drainage basin** is the area of land that supplies water to a river. The number and arrangement of tributary rivers is related to the aim of moving rain as effectively as possible. You can understand how this is achieved by comparing

Figure 14.1 **Drainage patterns**
A leaf (a) has a network of veins which show many similarities with a river drainage basin, (b). Both systems try to perform their functions as efficiently as possible. You can often see **dendritic** drainage patterns eroded in soft bare soil (*right*); the deep channels are called **gullies**; exactly the same pattern would be formed on the slopes of a volcano, a coal spoil tip or any newly exposed land surface.

(a)

(b)

the drainage basin of a river and its tributaries with a leaf (Figure 14.1). The plant has to supply nutrients to the leaf as effectively as possible; the river system has to remove water as effectively as possible. This pattern of water channels (called **dendritic** drainage) or leaf veins is very common in nature and is used to supply blood to your body and to remove oxygen from your lungs: it is a well tried and proven pattern. Sometimes, however, a river system flows over rocks of different resistance to erosion. It is quite common to find easily eroded rocks and more resistant rocks side by side in bands. In the north of England bands of sandstone alternate with bands of shale; in the south, bands of chalk or limestone alternate with bands of clay. In these cases the drainage pattern becomes distorted and more trellis-like. The landscape becomes distinctive too (Figure 14.2). The pattern of rivers is important because each river flows in a valley. Landscapes are made from patterns of valleys and so the whole character of a landscape can be affected by its rivers; and how quickly they remove rock fragments, even if it is by accident!

Exercise 14.1 Drainage basins

1 Select a page from your atlas that shows a large-scale map of your home region.
2 Choose a river that has at least several tributaries and choose one town beside the river.
3 Now trace the plan of the river and tributaries upstream of the chosen town. Is the pattern dendritic or is there some other pattern?
4 Draw a line around this network of channels to separate them from nearby river systems. The area inside this line is the drainage basin for the river upstream of the town.

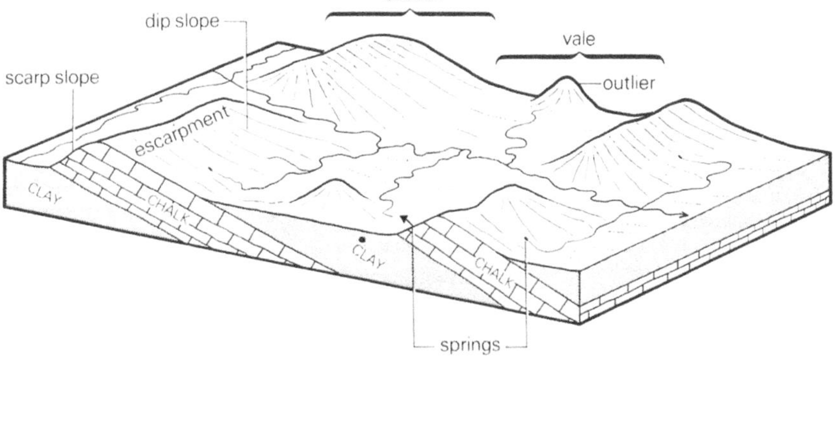

Figure 14.2 **Scarp and vale landscapes**
This region consists of alternating bands of impermeable clay and permeable chalk. Many streams are required to carry away the water from the clay. Fewer streams are needed on the chalk where much water sinks into the rock. The result is a **trellis** drainage pattern. Because the clay is eroded more easily than the chalk, the chalk stands out as ridges called **cuestas** and the clay bands make strips of lowland called **vales.**

A flow for all seasons

We can learn much about river transport and erosion from newspaper reports. They tell us about extreme events when transport and erosion are at their most effective. The three examples below show this clearly. Each quotes an eye-witness account:

Colorado, June 1965

‘The water was running about normal or maybe a bit higher . . . I got curious about the creek . . . , it was raining you know. I got up to look out the front door and there it was coming up as fast as everything. By then it was already around the house. It started coming in under the door and I knew it was going to be bad, but by then it was too late to get out.

I put my dog on the bed and it started jumping around, floating here and there . . . I began to get awful cold and was shivering wet. It smelled awful, that mud and dirt in the water . . . ’

Nevada, 1974

‘I started walking towards the coffee shop [and] I looked up for a second. I became disoriented because I thought the mountain had moved. Then I realised what we were seeing was a wall of water about 20–25 feet [7–8 m] high, stacked with cars, trailers, etc., smash into the coffee shop...and it exploded like there was dynamite inside . . . ’

South Wales, November 1979

‘It had been raining on and off for several days and the ground was thoroughly soaked so that it couldn't hold any more water. Then on Thursday it rained really hard all day. The sky was a dull grey and it was cold, so I stayed inside. I could just see the stream that flows through the village from my window. A pleasant little stream it usually is, bubbling along beside the road and disappearing under the bridge. But for the last couple of days it had become more and more swollen with rainwater and on Thursday it had turned a dull grey brown with mud. The water level was so high I thought it might wash away some of the brickwork from around the bridge. The water got higher and higher and pounded against the brickwork that protected the road. Then during the night I heard a loud crash. As soon as it was light I looked out of my window and saw that the bridge was gone and about 100 yards [metres] of road washed away too . . . ’

Figure 14.3 Water power Water wheels once powered the machinery for industry.

Besides experiencing floods, we often sense the hidden power of running water. We use running water to drive turbines and generate electricity; the first part of our industrial revolution was based on the power obtained from water wheels. The wheel shown in Figure 14.3 is driven by water falling on slats and pushing each one down in turn. The water flowing over the wheel hardly looks up to the job of driving all the machinery of a textile mill or a forge. Nevertheless this is what such wheels did.

Using the evidence of water wheels, perhaps we ought to revise our impression about energy available even in small streams, but we need no imagination to realise the real horror of the effects of floods on homes and people. The three eye-witness descriptions above show that rivers can have many effects. The Colorado is in part of South-West United States which is prone to intense storms. Flow over the soil surface is common and floods are expected. But even here storms can be unusually large as with the 'cloudburst' that caused flooding in a canyon in Nevada. The Welsh story shows flooding due to just the opposite effect – a long period of steady winter rain that saturated the soils so that they could take no more water.

River transport

A hop, skip and bump

Figure 14.4 Sediment transport (a) Material in **solution** is invisible but may make up half the total sediment carried. (b) Clay particles are so small and light they can be kept in **suspension** as long as there is any water movement. (c) Sand particles are too big and heavy to be kept suspended in the water continuously; instead they *bounce* or *hop* along in the lower layers of the water when the flow is fast. (d) Particles larger than sand rarely leave the stream bed and are only **rolled** along when the water is moving very quickly. (e) Together, all of these materials make up the **sediment load** transported by the stream.

Each of the stories above vividly demonstrates the power of a river in flood to break up and carry away most things in its path. In Wales a road was undermined and then carried away in pieces; in Nevada whole cars and trailers were tossed about like toys. But notice too the frequent references to finer debris in statements such as 'It smelled awful all that mud and dirt in the water'. Clearly, a whole range of sizes of material were being carried along.

The descriptions tell us that the finer material was being carried along in **suspension** (Fig. 14.4), but what about the larger? To get some answer to that question we can turn to the mountains of Norway and an ice sheet that straddles the Arctic Circle. Here part of the ice sheet melts each spring and sends torrents of water down steep river beds to the nearby sea. You can stand over a kilometre away from some of these rivers and still hear the roar of the water as it cascades over bare rock steps in its path. And every few minutes the constant sound of the rushing water is broken by a deep thud, like a field gun being fired. These are the sounds of huge boulders, once plucked from solid rock by glaciers but now being rolled over the edge of a rock step. Each thud signifies that several tonnes of boulder has just been pushed a little further by the flowing water.

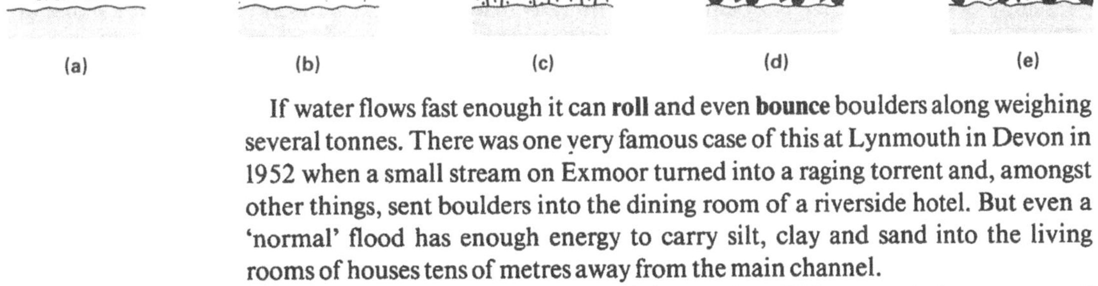

If water flows fast enough it can **roll** and even **bounce** boulders along weighing several tonnes. There was one very famous case of this at Lynmouth in Devon in 1952 when a small stream on Exmoor turned into a raging torrent and, amongst other things, sent boulders into the dining room of a riverside hotel. But even a 'normal' flood has enough energy to carry silt, clay and sand into the living rooms of houses tens of metres away from the main channel.

Where does the transported material come from? The most obvious sources of supply are the banks and beds of the channel carrying the water. Recent investigations of some streams draining through urban areas in smooth, concrete-lined channels give surprising evidence of this. Within a concrete channel streams have little chance of eroding new material. Channels can be eroded only when there is something rough for the water to push against. But while no new material is eroded, the rain from roofs and roads is continually added. As a result the stream gets bigger and acquires more energy, but there is nothing for the energy to be used on. When such streams re-emerge into the countryside downstream of the towns, they are usually allowed to flow back into their natural channels. With such a lot of unused energy the streams then scour their beds and rapidly make their channels deeper.

Scouring is an important effect and it can be dangerous if it happens near man-made structures (Fig. 14.5). It shows how effective a river can be eroding its bed.

Figure 14.5 Scouring and undercutting of a river near a town

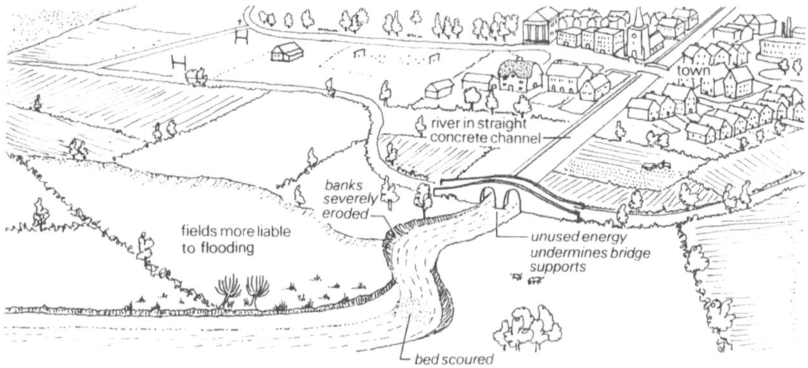

Because much more energy is available in time of flood, the larger material is moved at these times; sometimes with unfortunate effects if man has buildings and roads in the way. But it is all too easy to assume that floods are the only important occasions for erosion. Measurements have shown that often just as much material is moved between floods by suspension and solution. This is because floods may happen only on a couple of days a year. For the rest of the time the water may not carry so much material that it turns brown, but particles are still on the move!

River size

Bank robbery

Although extra energy causes scouring of the beds, it is worth remembering that banks are continually being eroded as well. Think back for a moment to the way hillside material is moved. Part is by solution, but the rest is by slow or occasionally fast movements of soil and rock down towards the river. If the stream did nothing about this, the channel would eventually be squashed narrower and narrower by these hillside movements. Often the river must be kept quite busy eroding its banks just to maintain a width suitable for its flow.

However, even if there is much energy to spare, the river does not become very wide. Measurements show that most river channels have a rectangular cross section: deep channels are always wide and belong with large rivers; narrow channels are less deep and belong to smaller rivers. There is always a balance between width, depth and flow, although streams flowing over gravel are shallower than those over clay, because gravel cannot form steep-sided banks.

SUMMARY: WEATHERING, TRANSPORT AND EROSION

Weathering occurs when a rock breaks down into smaller fragments, or decomposes but remains in place. When it is moved away, then erosion has occurred.

So *erosion = weathering plus transport.*

Rivers erode by breaking pieces of rock from banks or bed and then immediately carrying them away.

The difference between weathering and erosion can be seen in Figure 14.6. A large rock is being weathered into irregularly shaped sheets on its exposed upper surface. Gravity will eventually send the broken fragments sliding into the river, thereby causing the upper surface to be eroded. The underside is being constantly eroded and smoothed by abrasion from particles already in transport.

Rivers use their energy to move material in different ways depending on the size of debris. Small (clay) particles are carried along in **suspension;** larger (sand) particles tend to **bounce** or hop along the bed, provided the water speed is great enough; the largest (gravel and pebble) particles tend to be pushed and **rolled** along only at times of very fast flow. Of course, all the time and at every stage of the river material is also moved in **solution.**

Figure 14.6 Weathering and erosion of a granite boulder

Waterfalls

Mist, spray and thunderous noise

Figure 14.7 Waterfalls Waterfalls such as the Yosemite Falls shown here are major breaks in the course of a river. They will eventually be eroded away.

In upland areas river channels usually have steep gradients and they often cascade over waterfalls or tumble over rapids.

In the Yosemite valley (California) the Yosemite River sweeps round a bend and then plunges nearly 800 m over a huge granite lip before being dashed into seething foam on the rocks below (Fig. 14.7). This spectacular waterfall is impressive, but waterfalls are not rare: they occur throughout the world in upland areas. Indeed they are just as common in Britain as anywhere, even if they do not reach such grand heights. They can be found beside Dartmoor in Lydford Gorge, along the Pennines at such places as High Force (Fig. 1.16) and they abound in Scotland and Wales.

In these upper reaches of their courses, streams appear to have no control over their movements; they are thrown from one cliff to another, forced to create rapids, or plunge through deep gorges. Yet in most lowlands there are neither rapids nor waterfalls. Such a contrast must have an explanation.

Here we are back with nature's fragile balance. A waterfall is an irregularity in the path of a river too great to be formed by the river itself. You can see this by looking at the nature of erosion at a waterfall. There is largely a 'self destruct' mechanism at work. The lip is constantly eroding back in an attempt to give the river a more even gradient. But how far streams succeed in destroying waterfalls depends largely on time and the nature of the rock.

But if waterfalls are not natural results of river action, how are they caused? One clue is that they are common in areas of faulting, where huge pieces of the Earth's crust have been displaced vertically often hundreds of metres above others. In these cases waterfalls can occur on large rivers: the Niagara Falls,

ROGER JONES

(a) Part of the Niagara Falls.

between Lakes Erie and Ontario in North America and the Victoria Falls on the Zambesi in Africa are two of the best known (Fig. 14.8 a & b). Many others are the result of ice action. We shall see later (p. 211) how ice can gouge out steep-sided valleys and give waterfalls to valleys such as the Yosemite.

Many smaller waterfalls show how rivers pick out a resistant band of rock. Horizontal bands of resistant rock are especially difficult for streams to erode. High Force is still trying to erode through a near-horizontal basalt sill. The different types of waterfall are summarised in Figure 14.9.

Figure 14.8 Large waterfalls (above and below)

Rapids and potholes

Nature's slalom

Waterfalls are one feature of upland rivers, but if you drive past the River Wye just to the north of the Builth Wells in central Wales, or near Bala Mill in North Wales, you will see not waterfalls but gaily coloured poles hanging above the

JOHN BURN

(b) The Victoria Falls have both eroded gorges back from their starting points on fault lines. Notice how variations in rock resistance have made the downstream gorge so twisting.

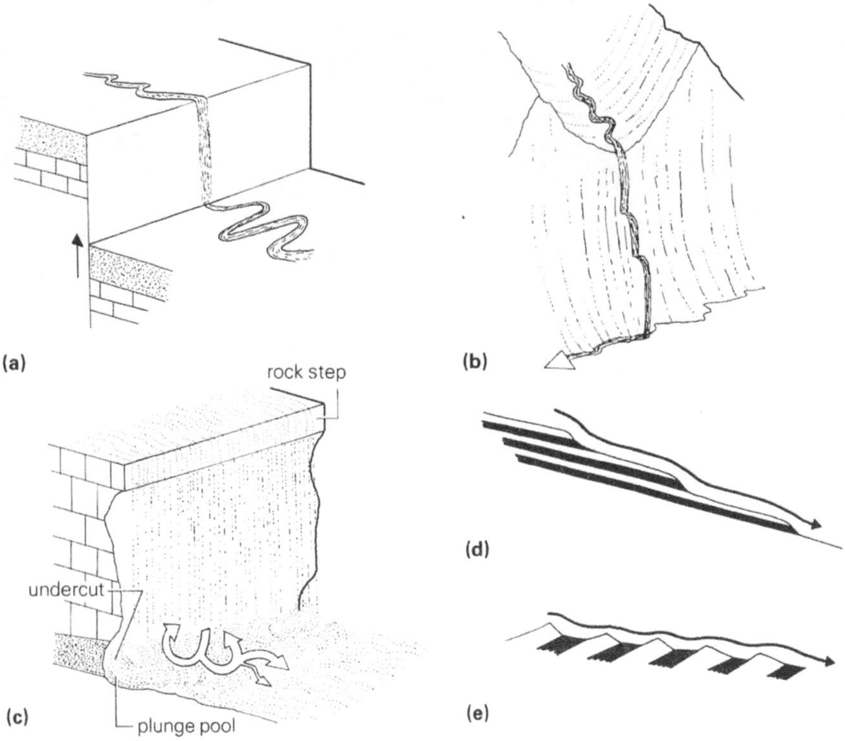

Figure 14.9 Types of waterfall
Waterfalls can occur because of (a) a fault, (b) a hanging valley or (c) a resistant rock band. **Rapids** occur when rock dips (d) down slope or (e) up slope.

river. Stop and look more closely and you may find a canoe go shooting by, the canoeist doing his best to avoid the poles whilst staying upright amongst swirling waters and rapids (Fig. 14.10).

The bed of the Wye makes **rapids** as it flows over dipping rocks. The river has excavated some of the weaker beds and left the more resistant standing out as ridges. As a result the water now cascades over a series of irregular ledges. Most upland stream beds are irregular and usually you can see the rock they are made from. You will also see a variety of pebbles and boulders scattered about in places of slower-flowing water, waiting to be picked up during a storm and

BERNARD PERRETT

Figure 14.10 Rapids
A canoeist shooting the rapids at Bala Mill, Wales.

(a) Potholes form when pebbles are caught in circulating eddies of a turbulent stream. The pebbles are whirled round by the water and they 'drill' holes in the stream bed by abrasion.

(b) Sometimes potholes become so large water has to spill between them as waterfalls.

Figure 14.11 Potholes

hurled against another part of the stream bed. Sometimes such pebbles have produced distinctive circular forms called **pot holes** as they have been caught in permanent water eddies and then used to drill vertical holes in the stream bed (Fig. 14.11 a & b).

We see features such as waterfalls, rapids and pot holes because the turbulent streams in an upland flush out any pieces of eroded rock and carry them away, exposing all the irregularities of bed and banks. So as the weaker rocks are eroded, the stronger rocks stand proud as the ledges of rapids and sometimes as the lips of waterfalls.

Figure 14.13 The line of a river channel from the source to the mouth
It is called a **long profile**.

A deep incision *Gorges*

If there were no hillside weathering and transport a river would eventually cut a deep slot in the landscape (Fig. 14.12a). Sometimes when rivers cut down very rapidly we get close to seeing a slot in formation. It is called a **gorge.**

On the western slopes of Dartmoor the National Trust has acquired a strip of land to be preserved in its natural state for the nation. Lydford Gorge (Fig. 14.12b) is a rather special place. A small stream flowing off the moors suddenly tumbles down the steep slope that leads to the lowlands of the Devon coast. In doing so the River Lyd has eroded quite an impressive 'gorge', some 100 m deep.

(a) A gorge cut by a river where hillslope weathering is very slow. Notice the vehicle gives the scale.

(b) Lyford Gorge, Devon. Notice the shape of the gorge. What does it tell us about the rate of downcutting of the stream bed compared to hillslope weathering? What do the fallen trees tell us about the sort of hillslope movements at work?

Figure 14.12 Gorges

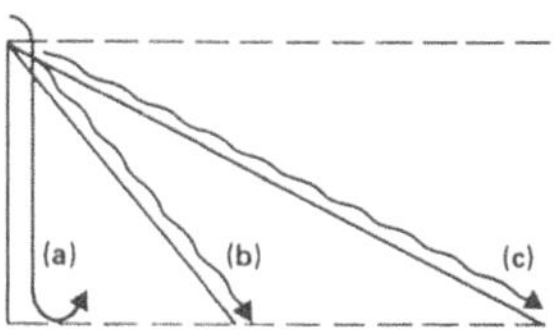

Figure 14.14 Spending stream energy
(a) A stream falling vertically over a waterfall uses most of its energy against the rocks at the bottom, causing undercutting and waterfall retreat. (b) Falling steeply over a short distance, it uses a lot of energy and tends to erode its bed rapidly, forming a steep-sided valley. (c) Falling gently over a long distance, it has to spread its energy out and so can erode its bed only slowly. This gives gentler hillslopes a chance to form.

To cut such a gorge must mean that the stream has been able to lower its bed at a much faster rate than hillslopes have been able to form.

Streams can maintain steep courses only in headwater regions. Remember that the primary task of a river is to move water through the drainage basin as effectively as possible. So the river gradient must always be sufficient to take the water away from the headwaters and out to sea. In practice this means that from source to mouth the river channel has a concave shape (Fig. 14.13). This is called a river **long profile.** It gives steep gradients in headwater regions, but progressively more gentle gradients towards the river mouth. Of course, a gorge or steep-sided valley can be cut only when a river is using up its energy quickly, so these features are confined to upland areas.

Downstream the gradient is more gentle and a different set of valley features dominate the landscape, as we shall see. The point to remember throughout is that energy is used by the streams both in uplands and lowlands (Fig. 14.14). In uplands it is used to cut a channel quickly and leave steeply sloping hillsides; but in lowlands the energy is used in other ways. Some is used to transport sediment eroded from the uplands; some is used to erode more sediment from hillsides and channel bed; and the rest is used by eroding sideways. The results are gently flowing and winding rivers in valleys with gentle slopes.

Topic Features of rivers with gentle gradients

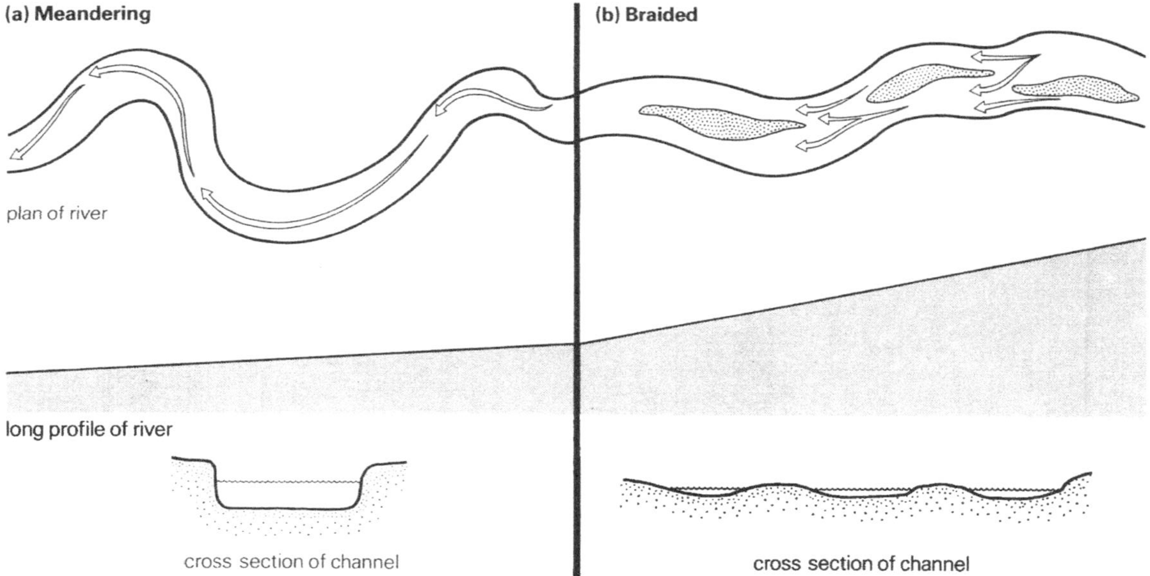

Figure 14.15 **Meandering** rivers have a deep single channel and flow over gentle gradients in clay alluvium.

Braided rivers have a wide shallow channel split into many parts by small islands of sand or gravel. Flowing over relatively steep gradients they are the downstream equivalent of rapids.

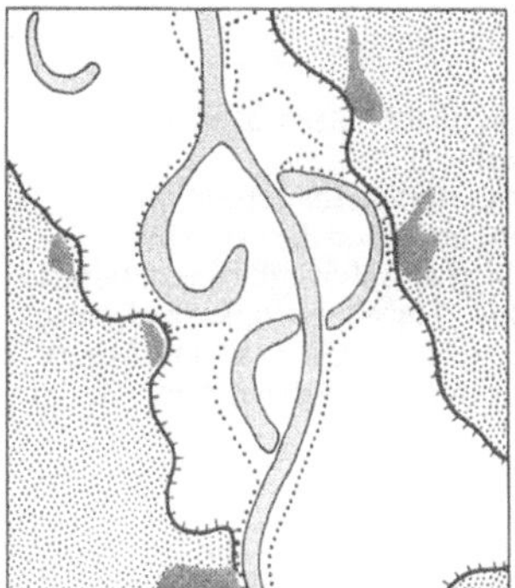

Figure 14.16 A section of the River Rhine in Germany Notice how the river has been straightened for navigation and how it has been dyked to prevent water flooding onto the floodplain.

Swinging rivers *River meanders*

One of the most obvious features of a river flowing over a gentle gradient are its twists and turns. This is an essential part of a river system, as engineers have discovered. But there have been many attempts by engineers to ignore this rule. For example, the **meandering** (twisting) pattern of the Rhine seemed to be quite unnecessary to 19th century engineers who decided to build cuttings through many of the worst meander loops (Fig. 14.16). When they had finished they had a much straighter course, but they discovered that the river now flowed even faster and this made hauling goods upstream even harder, Worse still, it scoured its bed and carried more sediment, but deposited it at the end of the 'improved' reach. The engineers were left with a new problem: how to dredge the sediment away! Today the fastest part of the Rhine is bypassed by canal loops but the events that led up to its construction tell us that rivers flowing on gentle slopes do not meander in a haphazard fashion that can be altered at will.

If we stop to think about the speed at which water flows in a river, we shall see that it largely depends on the slope of the river bed: steepen the slope (e.g. by cutting out meanders) and the speed will increase locally. The story of the river cuttings also demonstrates another point quite clearly. The Rhine has plenty of spare energy to cut down into its bed, but in its natural state it must maintain a balanced gradient to the sea. So the river cannot lose height too quickly. What then does it do with the spare energy? The answer to this seems to be that it meanders. Meanders make a river longer and also give the river something to do with its spare energy. Meandering rivers no longer simply erode their beds; they now have meander bends to erode (Fig. 14.15). This is in contrast to steep straighter rivers that use energy by **braiding** (repeated splitting).

Remember, too, that as with the rest of the river the meandering stretch is on a slope, so the whole meander pattern gradually 'slips' down slope, eroding a flat strip of land called a **flood-plain** as it does so (Fig. 14.17).

A queue at the bank *Floodplains*

Rivers with floodplains and meandering courses often flow in **alluvium.** This word describes any sediment temporarily deposited by a river and waiting in a floodplain or elsewhere to be transported to the sea. During a flood very little

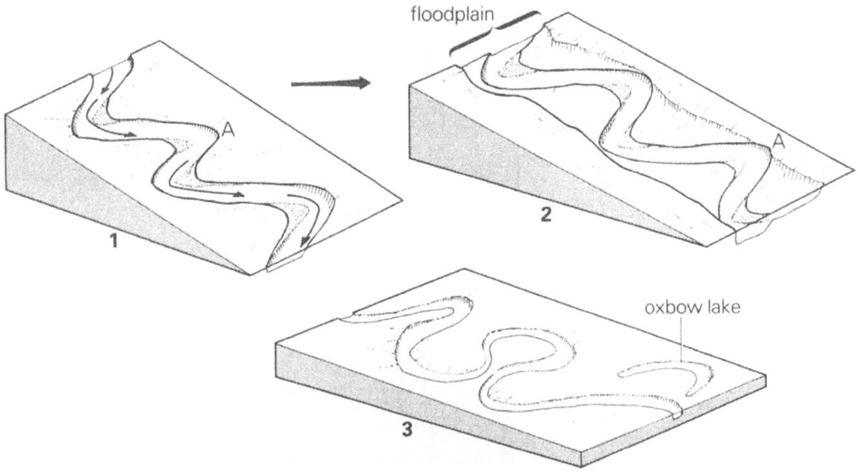

Figure 14.17 Floodplains (1) Gravity pulls the whole meander belt down slope by making water erode the downslope sides of meanders. (2) With time a strip of land is eroded away. This is called a **floodplain.** (3) If the land slope is very gentle the meandering may become exaggerated and the meanders cut themselves off and reform.

Exercise 14.2 River channel erosion

River erosion is complex because three things are happening at the same time: (a) the river is eroding its bed; (b) it is eroding its banks to preserve the channel width; and (c) it is following a winding path.

The winding (meandering) path is a feature of all naturally flowing water. Its effect is to cause uneven bank erosion, as shown in Figure 14.18.

1 What suggests that there is much erosion occurring on the outside bend of the meander?
2 What suggests that the deposition is occurring nearby on the inside bends?
3 Figure 14.18 also shows the line of fastest water flow. Trace the diagram and, using your answers to Questions 1 and 2, mark on places of erosion and deposition.

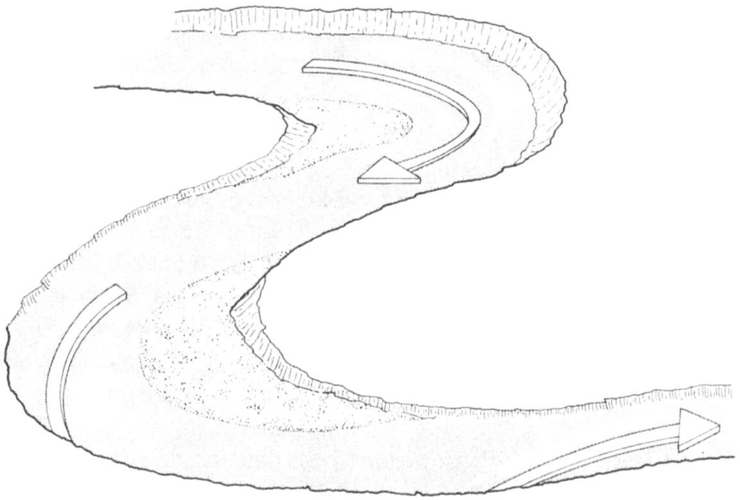

Figure 14.18 Features of a river meander
Notice that the energy in a meandering river is not evenly distributed, so it is quite possible for erosion to occur in one place and for deposition to occur nearby at the same time.

Topic Energy from rivers

Man is increasingly harnessing the energy of even gently sloping rivers. Once this energy drove flour mills or small textile factories. Today it produces very large amounts of electricity for a multitude of users.

Figure 14.19 Man's use of river energy

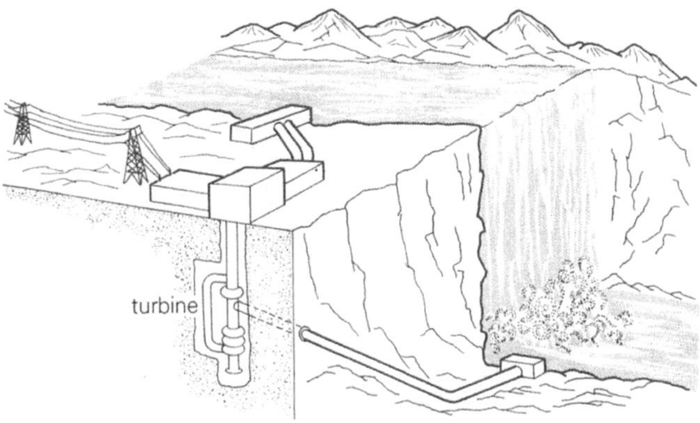

1 Waterfall e.g. Niagara, USA
Schaffhausen on the R. Rhine, Switzerland

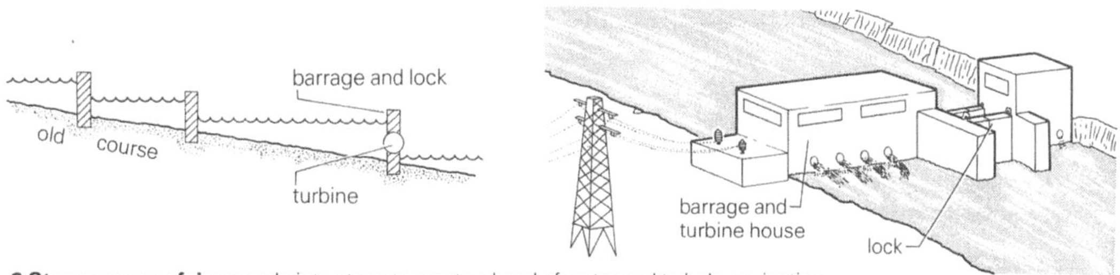

2 Steep course of river made into steps to create a head of water and to help navigation
e.g. Rhine between Basel, Switzerland and Karlsruhe, Germany
Colorado at Hoover and Glen Canyon dams (HEP only)
R. Lulea, Sweden

alluvium is carried right through a river system from where it was eroded to the sea. Most is deposited on the inside bends of meanders or floodplains. A particle of sediment may wait several thousand years before getting from the headwaters to the sea. In this respect the river acts rather like a long road with many sets of traffic lights. Great queues of rock particles develop at each set of lights (inside a meander bend) and have to wait while those at the head of the queue move off. But these particles are mostly caught at the next set of lights. Although several thousand years is a fleeting moment in the life of a river valley, it provides time for the alluvium on floodplains to be reorganised by meandering rivers.

River terraces

Steps of time

Rivers flow only because the land has been uplifted and a slope exists down to the sea. If land can be lifted up once, it can be lifted up many times. When land is lifted, rivers may flow over steeper gradients, have more energy even in lowlands, and so cut into their beds. But this does not stop them from meandering and eventually cutting new floodplains. However, a newer floodplain will be

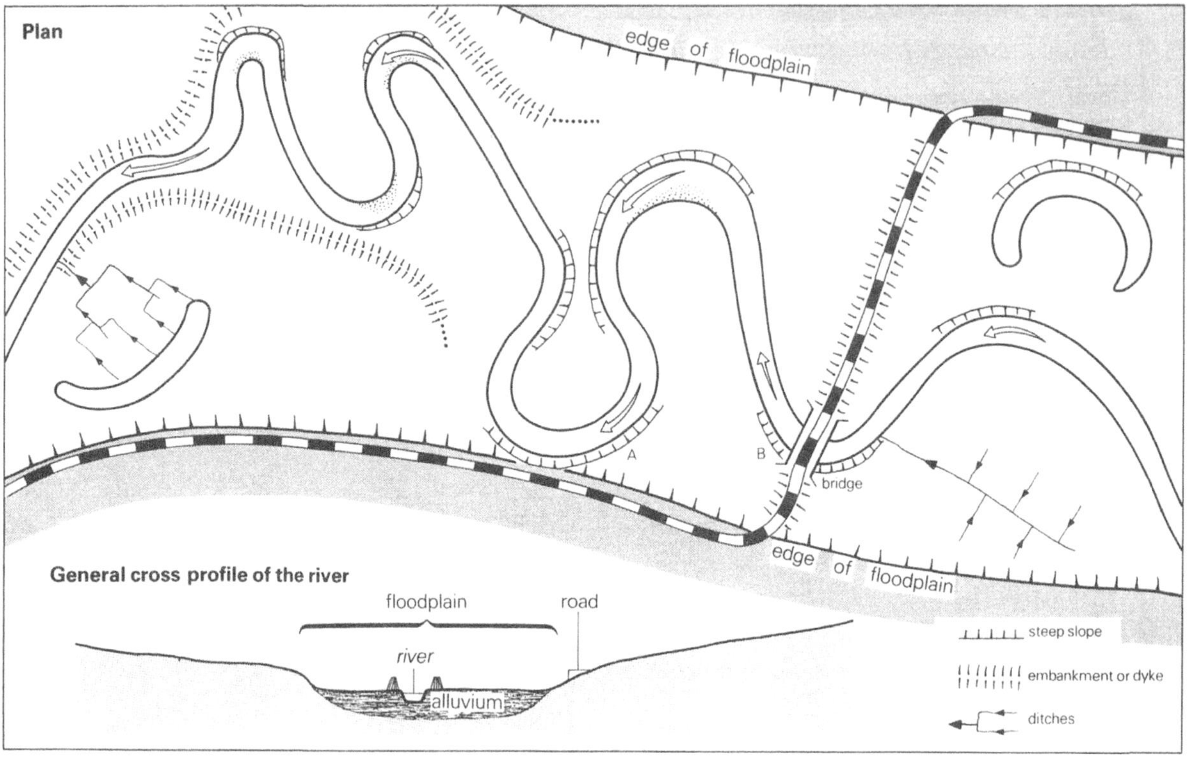

Figure 14.20 The plan of a meandering river

Figure 14.22 The formation of river terraces (1) Floodplain of the river before uplift. (2) Land is uplifted and the river cuts down into its floodplain. (3) A new floodplain is eroded away at the new, lower level, but ledges or terraces remain from the previous floodplain.

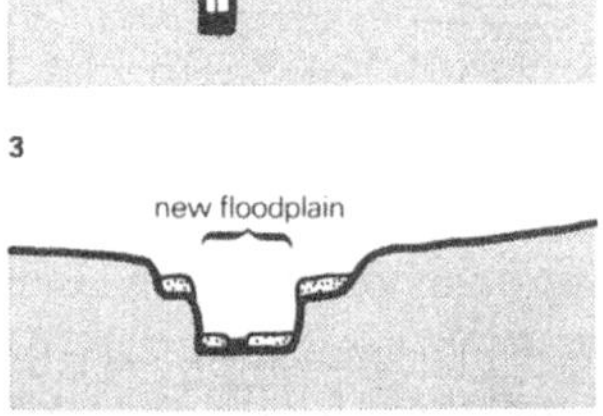

Exercise 14.3 Man, meanders and floodplains

In Figure 14.20 some of man's activities on a floodplain can be seen. Floodplains may provide good grazing land and flat areas on which to build houses, factories and roads. They are also natural reservoirs in times of flood (p. 91). Natural flooding and man do not go together very well, and many attempts have been made to restrict the amount of flooding.

1 What is the purpose of the dykes?
2 Make a copy of the dykes and the river. Now try to complete the pattern of dykes that would be needed to protect the whole area. Remember that they cost a lot of money to build and that a short dyke is therefore preferable unless it isolates much land.
3 Why have people gone to the trouble of building dykes when there are no villages or towns to protect?
4 In some places ditches have been dug across the floodplain. Explain what these are for.
5 Suggest a reason why floodplain ditches can only ever have a limited effect, even in times when there is no flooding.
6 Why would it be unwise to build houses on the floodplain if the dykes were not present?
7 Describe the route followed by the road. Explain why you think this route was chosen.
8 The road crosses the stream by means of a bridge with many arches. Why might this sort of bridge add to the flooding problem upstream?
9 What special precautions would engineers have to take at the places labelled A, B and C?

lower than the one eroded before uplift and it will be formed at the expense of the old. Eventually all of the original floodplain will be eroded away, but until it is, parts remain behind in the valley as 'ledges' which are called **river terraces** (Fig. 14.22). River terraces provide dry flat land very close to rivers. As a result they have been used extensively for the sites of towns and villages. For example two opposing terrace remnants preserved near to a river often make easier the task of river crossing. Similarly, a terrace beside the present river may provide the site for a port.

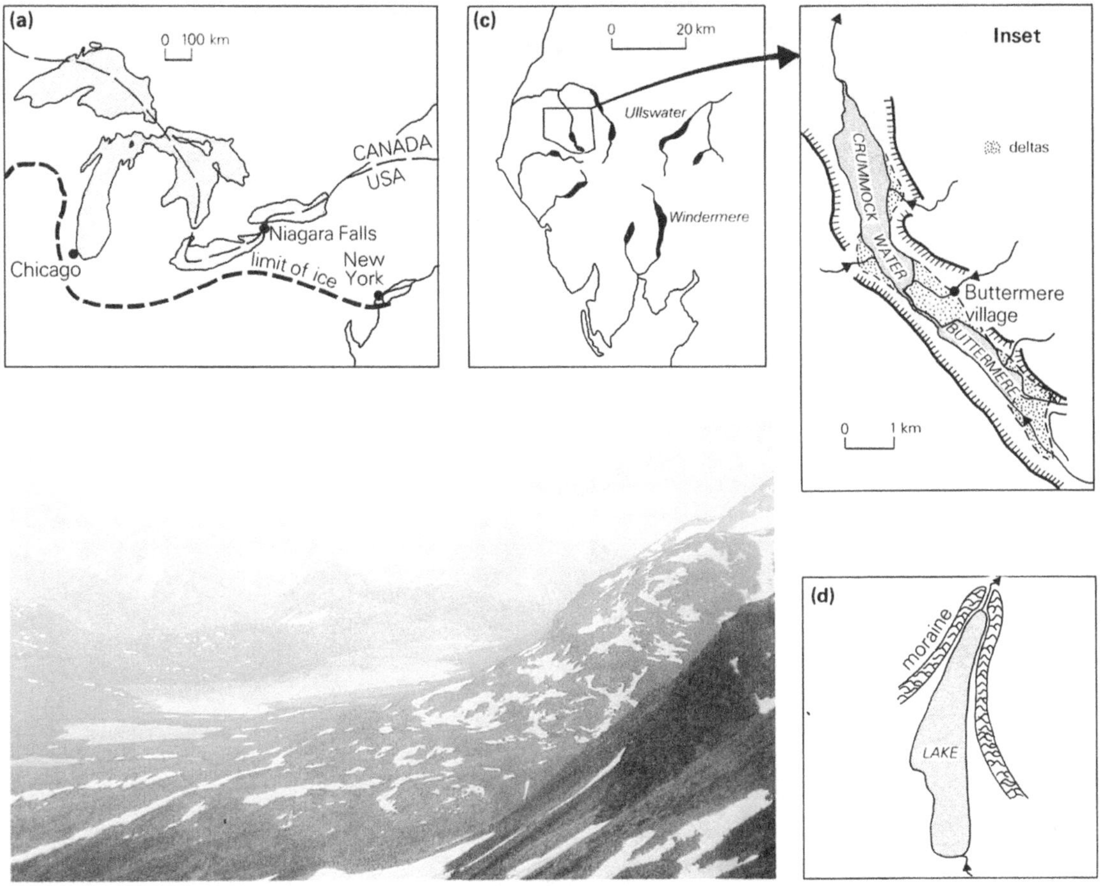

Figure 14.21 Lakes
(a) Filling a depression in the land surface caused by downwarping of the Earth's crust. (b) Ribbon lakes filling over-deepened rock basins in a glaciated valley, Val Bernina, Switzerland. (c) In the Lake District some of these lakes have been partially filled by deltas. (d) Dammed by a terminal moraine, South Germany.

Topiç Lakes

A lake is a natural reservoir of water formed in a depression in the Earth's surface. Water in lakes is almost motionless. To transport sediment, rivers need to maintain a reasonable speed, therefore when they flow into lakes they usually deposit the material they are carrying. As a result rivers fill up lakes; they cannot create them. The biggest lakes are caused by **land sinking:** by folding (see Fig. 3.7) the Great Lakes –Michigan, Superior and others – that form part of the boundary between the United States and Canada represent a downward fold of the Earth's crust so big that the lakes are 240 000 km² (the size of Britain) (Fig. 14.21a). Similarly, **faulted landscapes** often create basins for lakes. Lake Baikal in the USSR, the Dead Sea and the Sea of Gallilee in Israel, Lake Tanganyika in East Africa – all lie within rift valleys.

In Britain most of the **rock basins** filled by lakes have been eroded by glaciers (Fig. 14.21b). The Lake District has a number of long, thin (ribbon) lakes radiating out like the spokes of a wheel. These were formed when glaciers flowed down from the Lake District peaks, eroding and deepening the former valleys well below the levels of the past (Fig. 14.21c). Yet others are impounded behind **dams of glacial debris** (Fig. 14.21d) (e.g. Windermere in the Lake District).

Lakes are temporary parts of a drainage system. They will eventually be filled in by sediment brought from upstream rivers, but while they survive they are important in controlling floods. Lakes store up the flood of a river with perhaps only a few centimetres' rise in level. This surplus water escapes at the downstream end slowly and in a more even flow. We have learned from this, and many of our flood control schemes now use artificial lakes (**reservoirs**, see p. 98).

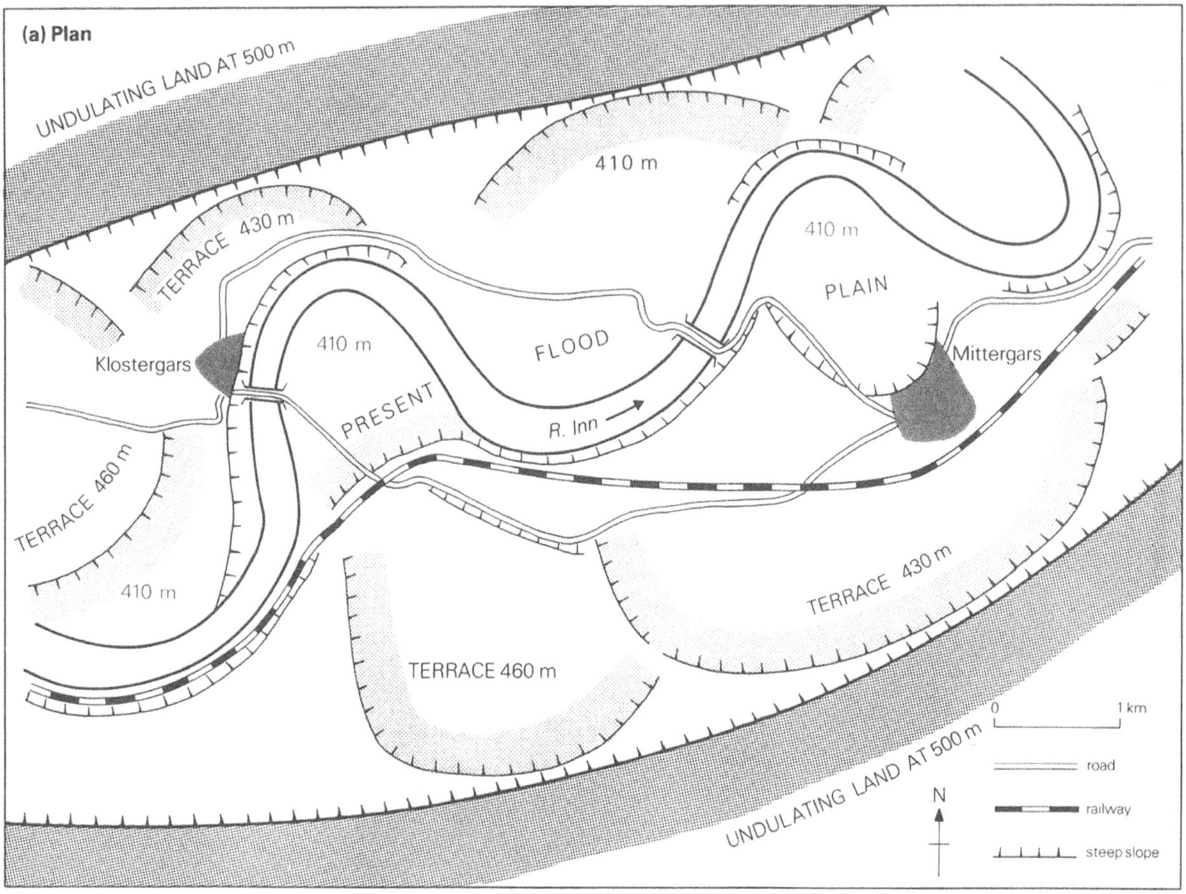

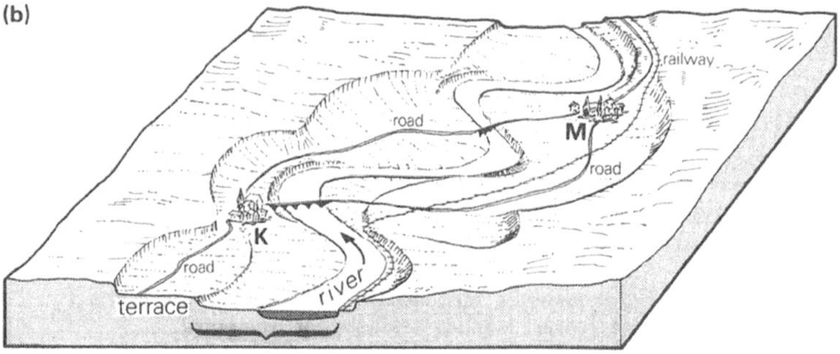

Figure 14.23 River terraces and the floodplain of the River Inn, South Germany

Exercise 14.4 River terraces and man

Figure 14.23 shows part of the valley of the River Inn in Southern Germany. This meandering river has lowered its flood-plain several times recently and many river terrace fragments remain.

1 Copy the river plan. Now draw on the present floodplain which is at a height of about 410 m.
2 Next draw on the two towns, the roads and railway line. Describe the site of the towns, saying how they are affected by the floodplain.
3 Describe the route followed by the two roads, explaining why they follow their particular courses in relation to the river and terraces.
4 Describe the route of the railway line. It is quite different from that of the roads, especially in the east. What special problems of railway construction have forced the railway to run so close to the rivers?

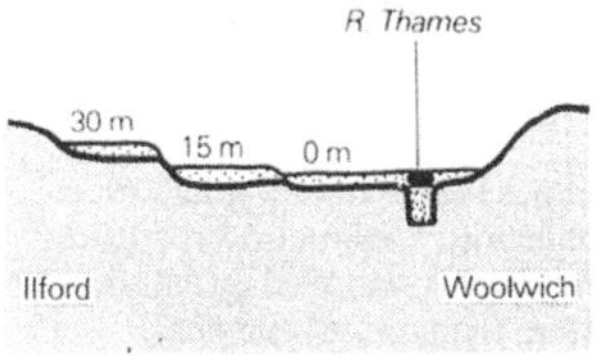

Figure 14.24 The River Thames in London flows on 20 m of alluvium

On aggregate

Uplift is not always the rule. Sometimes the land sinks or the sea rises. This happened quite recently in geological terms (10 000 years ago) when ice from worldwide ice sheets (p. 204) melted and released vast quantitites of water back into the oceans. With a higher sea, rivers had to adjust to a new, more gentle gradient. They did this by depositing alluvium in the lower parts of their valleys. Now, for example, the River Thames in London no longer flows on top of solid rock but on a thickness of 20 m of alluvium (Fig. 14.24). Most rivers are still depositing sediment in their valleys.

Many events have altered the shape of a river valley. Terraces show how the land has risen quite quickly many times. However, the fact that many rivers are now flowing on alluvium and not eroding their beds at all shows that sometimes the erosion of a landscape can temporarily go into reverse!

Exercise 14.5 Levées

Rivers build up **levées** (natural 'dykes' either side of the channel) by depositing sediment brought from up stream (Fig. 14.25). Sea level rises and the river has to raise the level of its bed and banks in step. It is a dangerous condition for people living on floodplains because the river flows *above* the plain, held in only by its levées.

The Mississippi river in the USA is noted for its large levées, and songs have even been written about them. The city of New Orleans has grown up beside the river where no town ought to be. It happened to be the first almost-dry spot found by the early French explorers when they ventured up the river from the sea at the end of the 17th century. Only meant as a temporary settlement, today 'Crescent City' curls around a huge meander of the river, its sole protection being levées, most of which are about 8 m high. The people of New Orleans often say they look up at the water and, before pumps were installed to push out the rainwater, they used to jest that people there were born with webbed feet! It is a very good example of the struggle between man and his environment.

1 If the levées are built of river clay and silt, are they strong and stable or might they be liable to breaching?
2 What are the dangers to man of building a town next to a river flowing within levées?
3 What precautions should the city take in order to avoid the dangers you have identified in your answer to Question 2?
4 New Orleans is built on the outside bank of a meander where there is deep water for a river port. What causes deep water in this position (p. 181)?
5 Explain why a position on an outside bank might be more dangerous than average when a river is held in only by levées.

Figure 14.25 Levées
When a river floods it spills out of its channel and over the floodplain. (1) Water on the floodplain moves slowly, so much sediment is deposited by the sides of the channel. (2) After many floods, sand and silt may have built up ridges of deposited material called levées. (3) Levées are particularly common in valleys where the river cannot readily carry all of its load. (4) Slowly the bed is raised by deposition and is contained in its channel only by the levées.

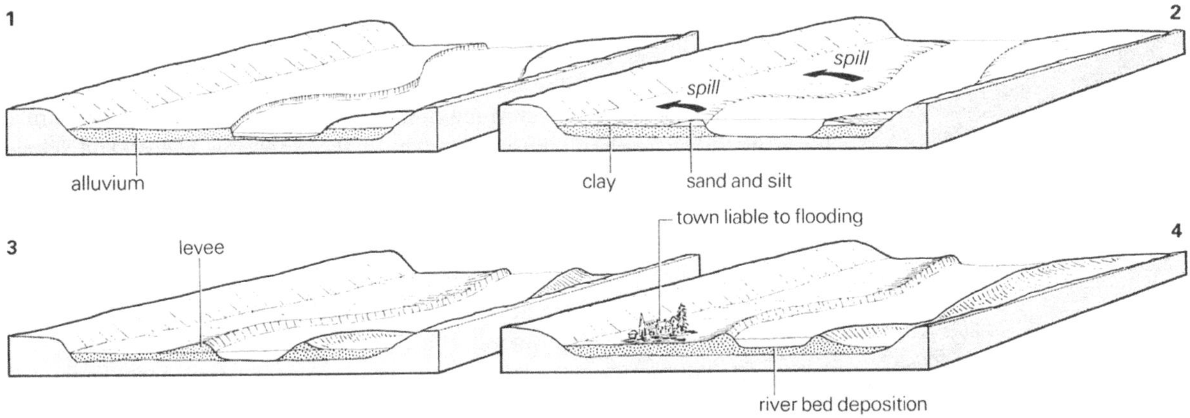

Topic Valley shapes

(1) Rivers normally just erode their beds. Sometimes this can be seen very clearly when the processes that make hillsides do not work very quickly. Some desert areas show these conditions: plateaux cut into by deep gorges. One of the best examples is the Grand Canyon (Arizona, USA) (Fig. 2.1).

(2) In many humid upland areas rain falling on the surrounding drainage basin causes rapid physical and chemical weathering. Loose rock and soil move down to the stream and some of the energy of the stream has to be used to remove it. Nevertheless many upland streams still erode their beds quickly and valleys have steep, straight sides.

(3) With less swiftly flowing streams, weathering produces more debris from the hillslopes beside it, and much more stream energy is used in transport. Channel downcutting into the debris is now slower. Soil is removed by solution, landslides and mudflows rather than rock avalanches.

(4) When streams are near sea level, vertical erosion slows down and floodplains form by lateral erosion. Hillsides continue to deliver soil to the streams until they too have been reduced to gentle slopes.

(5) Rivers in lowland areas cannot erode vertically, and hillslopes only deliver soil to streams slowly. Landscapes change extremely slowly in these conditions. There are many lowland landscapes from Essex and Denmark to the Amazon basin.

Figure 14.26 Valley shapes

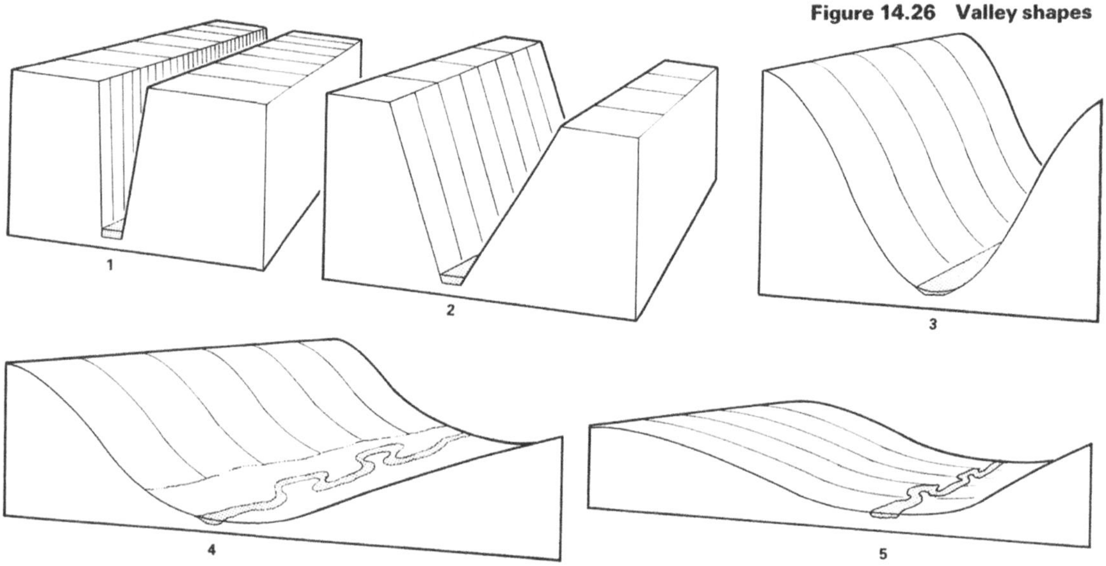

Deposition in valleys

Alpha, beta, gamma . . .

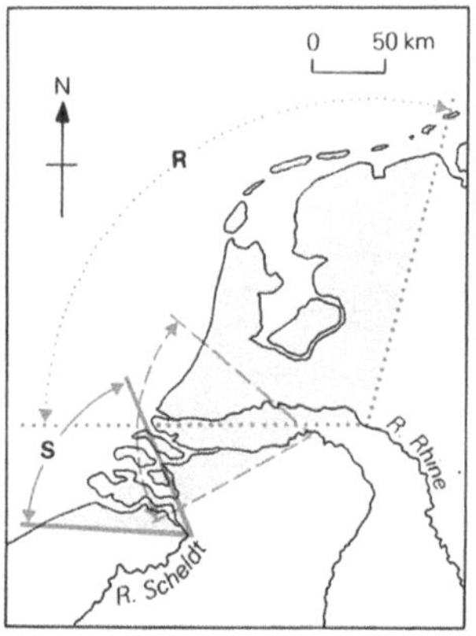

Figure 14.27 The Rhine (R)–Meuse–Scheldt (S) deltas

When the Dutch set about building the dam that closed off the famous Zuider Zee earlier this century, they knew they had to find millions of tonnes of sand. It was no use looking to the land for supplies – they had little enough land already without dumping some in the sea! The only answer was to dredge the sand from off shore. In doing this they scooped back some of the sediment that had been stripped off in the past. In fact much of the Netherlands and part of Belgium are really above sea level only by courtesy of the Rivers Rhine, Meuse and Scheldt. Most of the Netherlands is part of a huge fan-shaped pile of sediment formed as the rocks of the interior of Europe were carried by these rivers and dumped into the North Sea (Fig. 14.27). The regions of sediment accumulation are called **deltas.**

Each year millions of tonnes of sand, silt and clay are pushed into seas and lakes by rivers, usually to build up deltas.

Some deltas are triangular; the Greeks even used the letter (Δ) to describe the triangular shape of the Nile delta in plan; but many can be found that are all sorts of odd shapes. This is because deltas are made from sand, silt and clay which is easily moved about by waves and currents. The final moulding of the delta depends on the balance between deposition of sediment from the river and the energy of the waves and tides to rearrange it.

Delta formation

Down in the dumps

When a river enters a sea or a lake it sends a current of water from the shore containing all of the sediment eroded from its drainage basin (Fig. 14.28). The

Figure 14.28 Growth of a delta

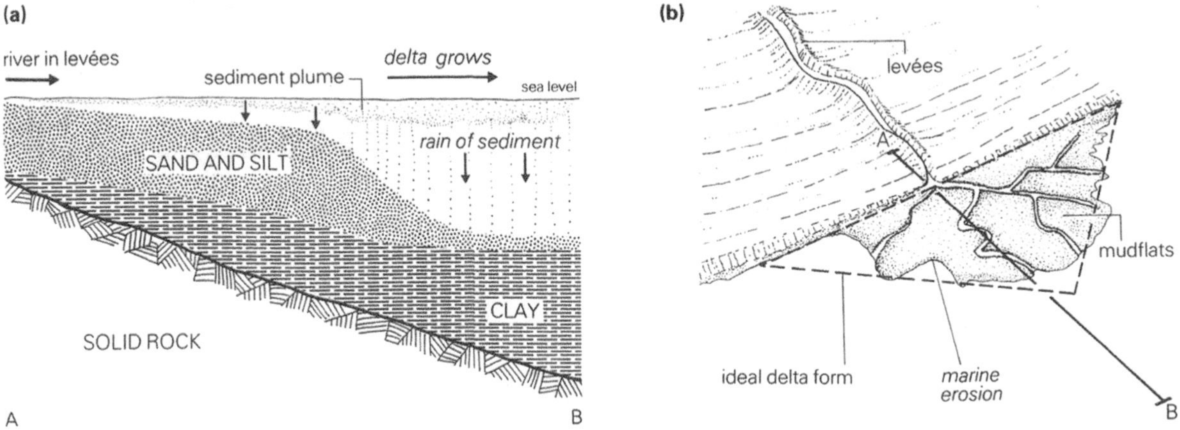

current may extend for several kilometres away from the coast – fresh water from the Mississippi can be detected 50 km out to sea – but all the while the water is slowing down so that the coarser sediment quickly drops out and 'rains down' onto the sea bed. The first material to be dropped is sand, then silt, with clay particles (being so small and light) settling farthest from the shore. Most of a delta is made from silt and clay; in the Mississippi delta only 2% is sand, and even the coarse sediment in rivers flowing quickly from mountains is less than half the total (e.g. the River Po in Italy flows from the Alps but the sand content in the delta is less than 40%).

Small rivers push out a narrow current of water and create deltas that stand out from the coast. Because of this they are rapidly eroded. Large rivers push out a delta which is very broad and has a very gentle slope. Waves cannot erode this as effectively and so it remains as a prominent coastal feature.

A delta grows into a fan shape by depositing material out in the sea and then flowing over this in a channel formed by levées. From time to time these levées are breached (especially on meander bends during floods) and the river finds a new route to the sea. There have, for example, been six major new delta growths in the Mississippi in the past 5000 years alone. In between these channels, mudflats form which are rapidly colonised by either grass (salt marshes) or in tropical regions by mangrove trees (mangrove swamp). When only part of the delta is growing, the rest is eroded by the sea and this gives the delta its irregular shape, as though something had been gnawing at it.

Exercise 14.6 The Mississippi delta

1 Study the plan of the Missippi delta (Fig. 14.29). You will see that, in the past 5000 years it has developed in six major phases. Draw each stage out separately. First draw stage 1. Now on a new diagram draw stages 1 and 2. This was the shape of the delta after two growth phases. Now draw a new diagram with stages 1, 2 and 3 on. Continue with new diagrams for stages 4 and 5.

2 The diagram also shows the rate of erosion of each part of the delta. Can you see a relationship between the rate of erosion at various places along the delta and the fact that the most powerful waves come from the south-east?

3 What are the advantages for using deltas to cultivate rice? (Remember that much of a delta is fed by fresh water from rivers rather than salt water from the sea).

4 In the Netherlands the Rhine delta is used for other purposes. Can you list some of them?

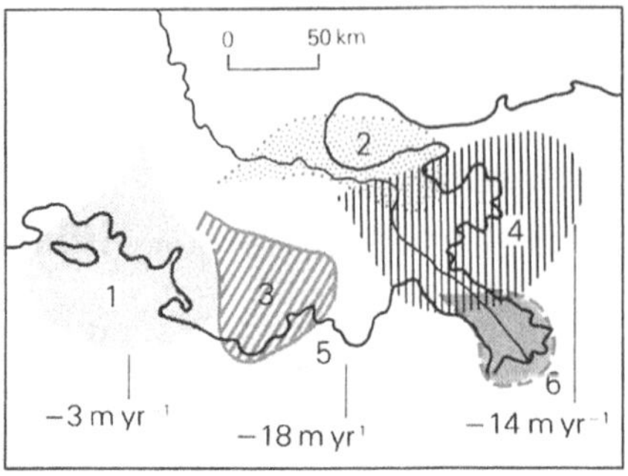

Figure 14.29 Deltas
Deltas grow by producing a series of 'lobes' of sediment. These lobes are then eroded by the sea to give an irregular shape. This diagram shows how the Mississippi delta has grown in the past 5000 years. Numbers are rates of marine erosion in metres per year.

15 Running water and limestones

Underground water flow

(a)

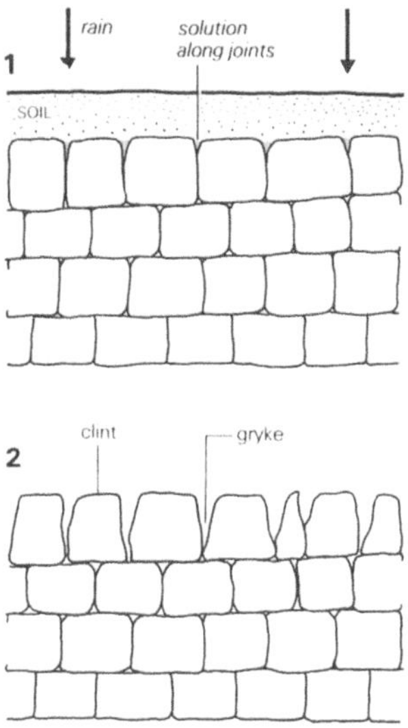

Figure 15.1 Limestone surface features
How a limestone pavement forms: (1) The limestone block surfaces are attacked by acidified rainwater. The limestone rock surface is now very irregular but this is hidden by the soil cover. (2) Glacial erosion strips off the soil cover and exposes the limestone pavement. (3) This one is at Malham, Yorkshire, England.

The inside story

If you go to parts of Britain such as the Mendip Hills (Somerset), travel in the Brecon Beacons of South Wales, or visit the Yorkshire Dales and Derbyshire Peak District, the chances are that you will see small groups of people clad in muddy rubber suits and carrying miner's lamps, helmets, ropes and wire ladders. These are the people popularly called 'pot holers'. Through their investigations we have learned much about the way **limestone** scenery develops.

Limestone is a word used for a whole range of materials from the soft white chalk of southern England to the massive blocks of the 'mountain' limestone of the Pennines. But it is to the 'mountain' limestone that we first turn our attention, because this contains the caves and passages that have made limestones so well known.

Like all limestones, the 'mountain' limestone is mostly made from calcium carbonate, with very little other sediment. Calcium carbonate is particularly vulnerable to reaction with acidified rainwater or streamwater. Indeed it is so reactive that most limestone solution occurs on the surface before the water even gets underground.

Chemical weathering attacks the surfaces of materials and its effectiveness depends on the number of joints in the rock. Mountain limestone may only have joints every metre or so. In a way this concentrates the weathering and causes the few joints to be attacked vigorously (Fig. 15.1). The results of this weathering are sometimes exposed as a **limestone pavement.** Although each joint may start out being just a hair's breadth, it soon widens sufficiently to allow water to run through it quickly (Fig. 15.2).

Figure 15.2 Erosion in limestone
Water rushes through a bedding plain in limestone. Notice the pebbles below the waterfall. These are now the main tools of erosion.

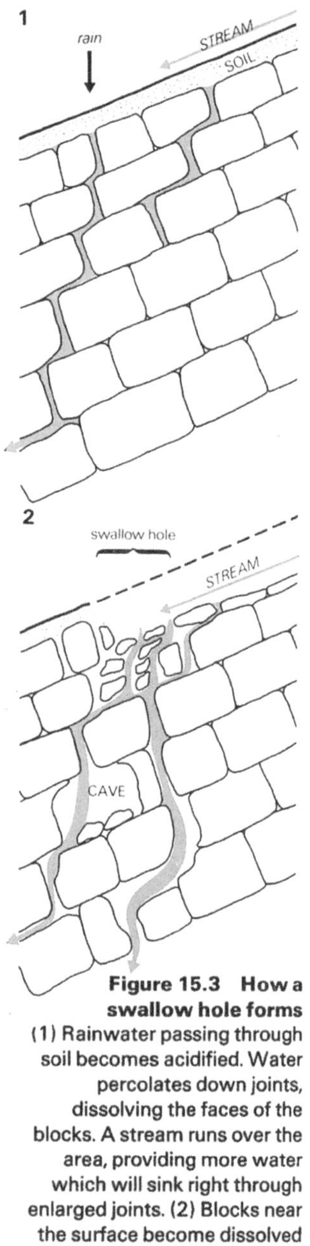

Figure 15.3 How a swallow hole forms
(1) Rainwater passing through soil becomes acidified. Water percolates down joints, dissolving the faces of the blocks. A stream runs over the area, providing more water which will sink right through enlarged joints. (2) Blocks near the surface become dissolved away and they collapse, providing a depression (swallow hole) down which the stream sinks. As soon as the joints between the blocks become wide enough, running stream water carrying sediment causes rapid erosion, producing passageways and caves. Notice how the water can move independently by several routes. This is what makes cave systems into three-dimensional mazes.

Once running water enters a limestone, joints are widened much more rapidly and the effects of solution become much less important. Running water, together with the sediment load it carries, rapidly abrades the joints and makes tunnels – natural drainpipes – through the rock. Occasionally tunnel erosion causes roof falls and a **cave** is born (Fig. 15.3.) Solution continues to be important near the surface, dissolving blocks in such a way that they collapse and give depressions called **swallow holes** (Fig. 15.4). But underground there is very little solution. Indeed much dissolved limestone is precipitated as the **stalactites** and **stalagmites** that make caves so beautiful (Fig. 15.5).

Stalactites and stalagmites are the cone-shaped pinnacles of white calcium carbonate shown in Figure 15.5. Stalactites are formed as drops of water seep through the joints in the upper surface of a passage or cave. The water coming out of the joint is saturated with calcium bicarbonate, but when it comes out into the cave or passage much of the carbon dioxide gas goes into the air. This leaves the water with a reduced ability to keep calcium carbonate dissolved, so some precipitates out as calcium carbonate. Precipitation is very slow and stalactites may take thousands of years to form. Note that perfect cones are unusual. If, for example, the surface on which the drip forms is sloping, the water will run along the rock depositing calcium carbonate in a line. Eventually this builds up a flat sheet of stalactite called a **curtain.**

Water dripping from the roof of a passage or cave will fall onto the floor where further precipitation may occur. In this case a cone of calcium carbonate grows upwards from the floor to produce a stalagmite.

Sometimes water runs over the sides of a cave and then precipitation coats the rock in flowing white **flowstone.** Streams also deposit flowstone, or sometimes create little dams which retain water. The range of forms is endless!

You can form an idea of the nature of an underground limestone landscape by a detailed look at a particular cave system. Swildon's Hole comprises over 8 km of accessible stream passage below the Mendip Hills near the small village of Priddy. The stream sinks into a swallow hole in a field and comes out again at the famous show cave at Wookey Hole, near Wells. The straight-line distance between swallow hole and spring is 3 km and the vertical difference 200 m. However, even the accessible passageway length is 8 km and there must be three or four times as much that people can't squeeze into, so this gives some idea of the extent of the three-dimensional maze that underground streams have eroded in the rock.

This description is intended to give an impression of the nature of underground drainage systems from the viewpoint of the caver. Swildon's Hole cave system can be dangerous to the inexperienced. It is liable to flood in wet weather and there are deep drops. On no account should this or any other cave be entered without proper equipment and an experienced guide. If you want to see a cave, you should visit one of the many safe show caves that are open to the public. Any other form of caving should only be done through a recognised caving club.

'Swildon's Hole is entered feet first between a pile of large boulders (the swallow hole) in the bed of a small stream. An intricate network of meandering stream passages lie ahead. The stream first falls over a 3 m drop in a passage about 60 cm wide and 3 m high. This stream tumbles steeply down towards the

first cave, but the passage height quickly lowers so that a couple of crawls in water are needed. Eventually the streamway opens out into a cave 10 m wide, 20 m long and 8 m high, with some stalactite formations but many fallen boulders. The cave continues through a passage some 2 m wide in places and 8 m high, which leads on after 100 m to a wet stoop and then an awkward 4 m climb down the side of a pot hole. Following the stream past much flowstone decoration soon leads to the head of a 7 m pot hole which needs wire ladders and safety lines for descent. After climbing down the ladder and through a waterfall it is noticeable that the pot hole has rounded boulders in the bottom. A little further on and there are two more pot holes – a fine pair of water-worn circular features about 5 m in diameter . . .'

This description is a small part of the cave system that contains a stream, but there are kilometres of passageways that are now dry and abandoned by the stream. These are parts of the underground networks that were formed at an earlier stage when the outlet for the stream was higher.

Figure 15.4 The entrance to Gaping Ghyll cave, Yorkshire, England

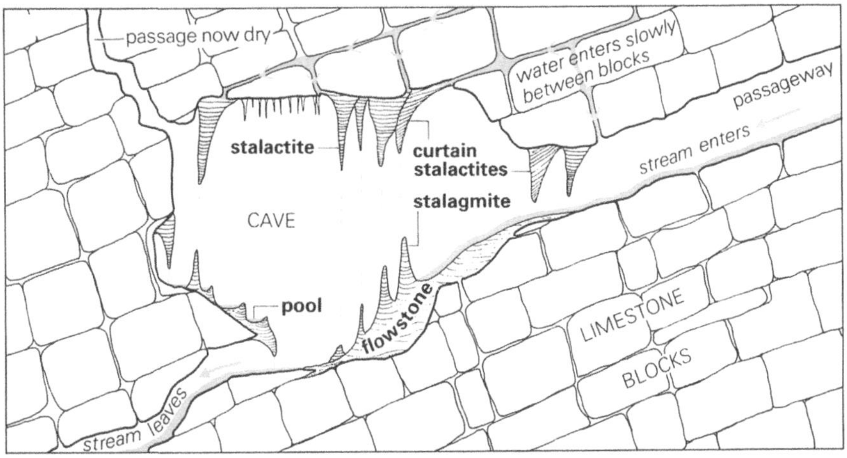

Figure 15.5 Some of the features of a well decorated cave in limestone
Below. Flowstone and curtain stalactite, the most common type of precipitate. *Below right.* The less common cylindrical stalactites and stalagmites. *Right.* Cross section through a cave.

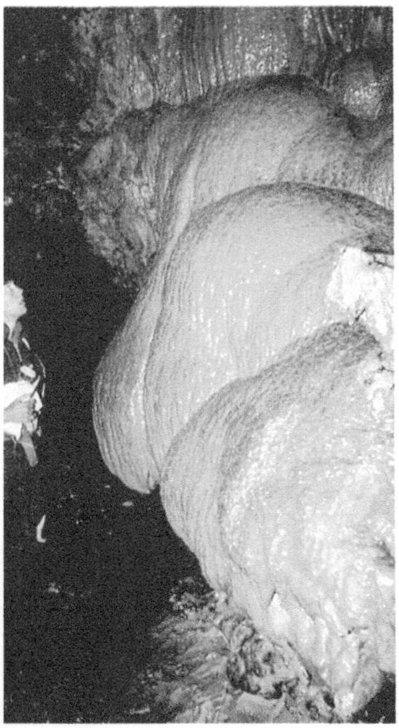

Spot the difference

You might think that there ought to be much similarity between chalk and mountain limestone landscapes. The rocks are, after all, both made from calcium carbonate, both are highly reactive with rainwater, and neither have any surface drainage, all the water flowing underground leaving abandoned dry valleys or gorges. Yet the landscapes are very different – chalk landscapes have smoothly rounded hills and dry valleys with gentle slopes; mountain limestones have sharp angular outlines with vertical-sided gorges. The difference is in the nature of the rock. Mountain limestone is a physically hard rock that will not let water pass through; its permeability is all due to the interconnection of cracks which divide the rock into the large blocks. In fact, the cracks are so regular that you can think of limestone as a pile of bricks. Stacked one upon the other these blocks form stable vertical sides to gorges such as Cheddar Gorge in Somerset or Gordale Valley in the Yorkshire Dales. Similarly there is plenty of support for stable passageways and caves underground. By contrast chalk is a physically soft permeable rock, without well defined joints; instead it is crossed through and through by multitudes of hairline cracks which divide it into many small blocks. Without the brick-like structure chalk will not support caves or permit steep slopes to remain for long in valley sides. Chalk cliffs along the coast are steep only because they are being eroded rapidly by the sea.

16 Deserts

Flash floods

Lucy in the sky with diamonds

One night Lucy was sleeping on a dried-up river bed in semi-arid Ethiopia when a rainstorm broke over the nearby mountains. Lucy neither heard or saw the storm, but as the rain gathered into torrents and flooded out of the mountains she was already doomed. Soon this sudden flood of water **(flash flood)** was racing down the dry river bed to engulf and bury Lucy and her companions in a mud and water mixture like warm cement. That was three million years ago. Lucy is the oldest known 'human' skeleton found on Earth. Named after a Beatles' song, she must have been one of the first to die in a sudden desert flood (Fig. 16.1), but many people suffer the same death today.

Deserts (arid lands) are areas where so little rain falls that vegetation survives only in specialised forms and it covers a very small fraction of the land surface. Along the margins of deserts are the semi-arid lands where rainfall is greater but still not sufficiently plentiful to support an unbroken vegetation cover. There is no easy way to define an arid or semi-arid area. It is best attempted by using vegetation, although some people have defined a desert area as having less than 250 mm of rain a year.

Arid and semi-arid lands cover 30 % of the earth's surface and are inhabited by 600 million people, so although they may be unfamiliar to us, they are important and should be understood.

Alluvial fans

Fans of the mountain front

About 8 km from Suez town in Egypt a gently sloping gravel plateau **(hamada)** ends abruptly at the rising front of a mountain range (Fig. 16.2). This sharp

BERNARD SMITH

Figure 16.1 A rare photograph of a flash flood moving down a valley in a desert
In the foreground the valley flow is totally dry and there are no storm clouds in the sky. The flood has been caused by a local storm somewhere in the mountains.

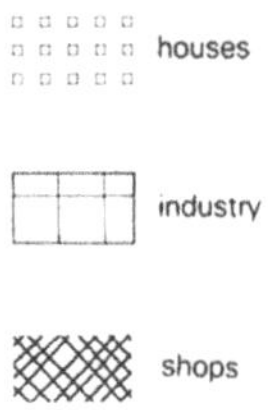

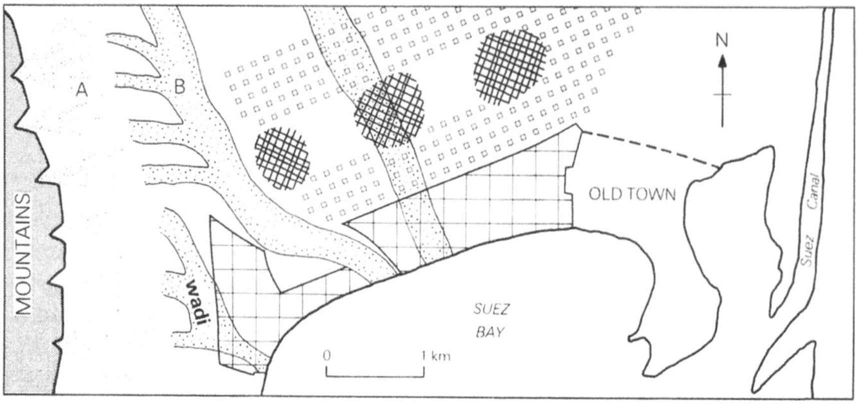

Figure 16.2 The area near Suez city, Egypt

contrast in landscapes is typical of deserts. The mountains are there because sometime in the past a huge faulted block has been tilted and thrust up 800 m above the surrounding plains. Now these mountains tower over the Gulf of Suez, their stark edges bitten into by the results of torrential rain over thousands of years.

Storms are rare, perhaps occurring only a couple of times a year, but each storm causes water to cascade over the steep slopes and carry away broken rock debris that litters the slopes. On the floors of the valleys (called **wadis**; Fig. 16.1) water and debris combine to cause torrents that surge and boil with clay, sand and pebbles. Streamflow in a wadi is very like warm, runny, liquid cement with lumps in. Yet the rain is so short lived and the ground so parched that the water barely has time to pick up the debris, let alone move it very far. On steep hill slopes water has enough energy to carry even coarse debris, but valley floors are much less steep and the plains beyond the mountains even more gentle. As a result much of the coarser material is dumped at the foot of the mountains as huge **fans,** leaving only the finer debris to be carried on across the plain. These fans are enormous: many are more than 300 m high and stretch out several

Figure 16.3 Alluvial fans stretching out from a mountain front
There is a road in the bottom left-hand corner of this photograph. The distance from road to top of fan is about 3 km.

kilometres from the mountain front (Fig. 16.3). Very often they spread into one another to form a continuous 'fringe' of debris around a mountain range.

Storms in mountains are dangerous because they arrive with little warning. People have been drowned when venturing into a dry and apparently safe wadi, only to see a wall of water approaching them! Roads are destroyed by flash floods, and damage can be very costly. Both Israel and Jordan have suffered from roads being washed away in the past few years. The road in Figure 16.3 is vulnerable where it skirts the base of the fan.

Wadi floors

Blooming pebbles

At first light every morning women from the rural villages near Erfoud in Morocco make their way out across the nearby barren and pebble-strewn plateau. Slowly they guide their herds of goats to places where a few hard and dry plant shoots might still survive (Fig. 16.4). All day long they wander with the goats under the blazing sun, but they are careful to keep to the barely discernable broad and shallow channels in the pebbly desert.

The importance of these channels becomes very clear only a few times a year. At these times heavy convectional storms sweep over the area and deluge the plateau and nearby mountains. For a few hours water cascades from the mountains and floods out over the plateau, using the shallow channels on its way to the sea. Almost immediately after the rain starts, the tiny hard and shrivelled buds that have been hiding between the pebbles swell and burst into life. Now the landscape takes on a green haze as millions of desert plants come into leaf and flower.

Desert flowers are typical of the way arid regions work. For months, perhaps years, there is no water, no erosion, no plant growth. Then an occasional violent storm causes the whole landscape to burst into life – rivers flow, rock debris is washed from slopes and swept along river channels, and plants spring up and bloom. Then the storm waters have gone, absorbed by the pebbly alluvium, evaporated by the blazing sun, and within a few hours the landscape is once more parched and dry, the flowers fading away to await the next deluge.

After a few days the brief period of flowering is over, the seed scattered, and the desert lifeless once more. But because more water sinks into the channels than elsewhere, this is where most plants survive most easily. Consequently this is where the herds of goats wander in the search for the scant foodstuffs on which their lives depend.

Figure 16.4 The gravel-strewn plateau over which flash floods sometimes flow
The goatherd is to be seen in the middle distance collecting water from a well.

Figure 16.5 The shallow channel that signifies the track of a flash flood
Notice the road is built straight across this route, which gives some indication that flooding is very rare.

Exercise 16.1 Water erosion in the desert

1. Figure 16.1 shows a wadi with a rare but well known desert feature. It is a flash flood. Explain why flash floods occur. How can you tell that this is a photograph of one.
2. When a flash flood flows down a wadi it moves over sand and pebbles such as are shown in Figure 16.5. What do you think happens? What would happen to the car if it were not moved clear?
3. Why do alluvial fans occur at the edge of the mountains?
4. Flash floods travel only a short distance over plains because the supply of rainwater soon ends. What happens to the floodwater and the sediment it is carrying?

Topic Inland drainage basins

Because storms are so rare and restricted in area, much stream water never reaches the sea. In many areas, especially those where large-scale faulting has produced enclosed basins, storm water builds up into temporary lakes, called **playa lakes** (Fig. 16.6). These lakes quickly evaporate, leaving behind dissolved salts. Thousands of years of evaporation has formed 'salt lakes' which may have salt hundreds of metres thick accumulated in them. The Great Salt Lake, and Bonneville Salt Flats, Utah (USA) are perhaps the most famous of these.

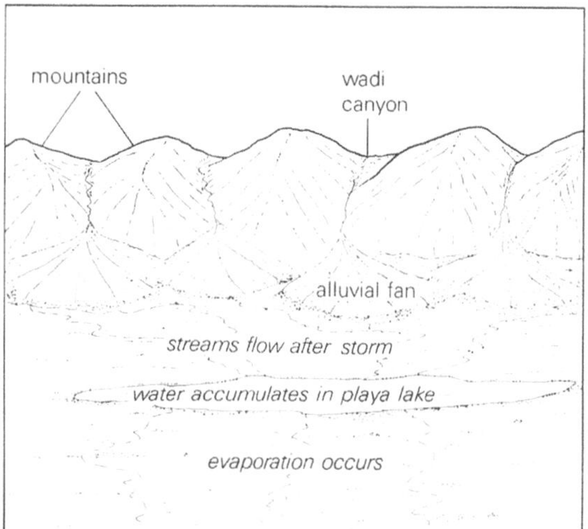

Figure 16.6 Salt lakes
Playa lakes are salt lakes formed when local drainage basins dry up. This one is in Death Valley, California.

Exercise 16.2 Urban development in the desert

At the southern end of the Suez Canal lies the town of Suez. It is on a flat featureless strip of ground about 5 km wide, covered by sand and pebbles (Fig. 16.2).

The original site of the town dates back into ancient history but the plan to develop a new city will cause considerable development. Houses and industry will be built on the previously unoccupied sand and pebbles. Civil engineers needed to know whether the new city would be built over any problem areas.

1 Look at the map that was prepared for part of the area. At A there are some spreading channel features. What do you think they are? The material in which they are cut is sand and pebbles, so of what do you think the area at A forms a part?
2 Water leaving an alluvial fan is very unpredictable, but there are some areas where it is more likely to flow than others. They are mostly marked by broad shallow channels such as that shown at B. What are these channels called?
3 Compare the plan for the expansion of the town with the position of the most likely route for water from flash floods. Describe what you notice. Can you suggest modifications to the plan that would help prevent houses and factories being built in the path of a flash flood?

SUMMARY

SUMMARY: FEATURES PRODUCED BY RUNNING WATER IN DESERTS

The main features of an arid area are that (a) weathering is due to exfoliation (see Ch. 9); (b) rainfall occurs in occasional, intense but short-lived storms; and (c) there is no soil to store water. Water cascades over the bare valley sides and reaches channels quickly as **flash floods.**

Without constant moisture and a vegetation cover rock fragments cannot easily be weathered down to clay size. Weathered debris littering slopes is therefore coarse – mostly sand, pebble and boulder size. Valleys in desert or semi-arid areas are called **wadis** in Africa and **canyons** in America. Where wadis meet lowland areas the river gradient becomes less steep and the rivers are no longer confined to narrow channels. As a result, the river water spreads out and is able to carry less debris. At junctions of upland and lowland it is usual, therefore, to find large **alluvial fans.** If the wadis spill water into enclosed basins, **playa lakes** form (Fig. 16.6).

Sand dunes

Frozen waves

Rocky mountains and pebble-covered plains provide one type of desert landscape. But there is another type of desert landscape where wind-blown sand collects. These are the **sand seas.**

Sand dunes are probably the most famous of desert features even though they occupy less than a quarter of desert lands. They are really only piles of wind-blown rock fragments, yet to many they are beautiful and intricate patterns of smooth flowing shapes, like frozen waves in a golden sea (Fig. 16.7). Locally, such as in Libya, sand seas cover vast areas and provide the

Figure 16.7 Part of a sand sea in the Sahara desert of North Africa

world's most impenetrable and hostile land. These are the lands of the Sahara – the Grand Erg Occidental and the Grand Erg Oriental with their camel trains, the Bedouin tents and abandoned forts of the French Foreign Legion. But elsewhere they tend to occupy very small areas.

Sand transport

Borne on the wind

People with much experience of desert life can sometimes tell when a dust storm or sand storm is coming, but for most people they arrive without warning. Often the first sign is a change in the direction of the breeze. Next, dust will start to lift off the ground in an almost ghostly way, just as though someone had disturbed the floor of a haunted house. After a slight but noticeable rise in the wind speed sand begins to move. At first sand blows in wispy trails over feet and around ankles. Then a brilliantly clear and hot day suddenly becomes obscured by a dark cloud stretching from the ground high up into the sky and getting nearer every second. The dark sky is not a sign of rain, but of parched hot air, dust and sand. The dust (silt and clay particles) gets into the lungs, hair, eyes and ears; the sand beats against any exposed flesh with a force that can be painful. The wind has become stronger now causing violent swirling around any obstruction and taking sand and dust into every nook and cranny and piling it up against rock, wall and fence alike.

A mass of air blowing across a landscape behaves in a very similar way to a river. In the desert air moves quickly over an irregular bed of sand and pebbles. Any loose material is picked up and moved in suspension or by bouncing or rolling. The faster the wind blows, the more energy it has to move bigger pieces of material and the greater the total amount of debris transported (Fig. 16.8).

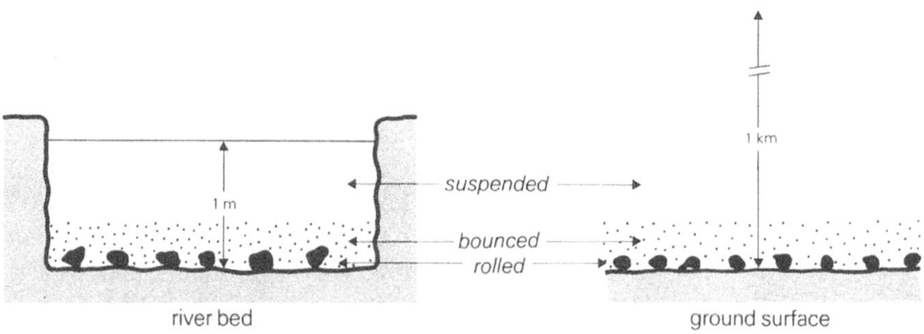

Figure 16.8 The movement of material Material is moved by wind and water in exactly the same way. Only the height to which suspension and bouncing can occur varies.

The similarities between wind and water do not end with transport. Water carries away weathered debris or abrades its channel by continually rolling and bouncing stones across it; wind abrades by driving sand across any exposed rock. There can be no doubt about the abrasive efficiency of sand-filled air – the technique is often used to clean buildings. Nevertheless, a jet of air and sand from a compressed air line is vastly more powerful than the wind and we must understand that, in natural circumstances, sand abrasion works very slowly. By contrast, dust does not abrade and so natural sandblasting is confined to the height that sand can be lifted by the wind – about a couple of metres. Of course, the larger pebbles are rarely moved, so that eventually differential removal of sand, silt and clay particles from the surface of a desert may lead to the development of a stone pavement (Fig. 16.9).

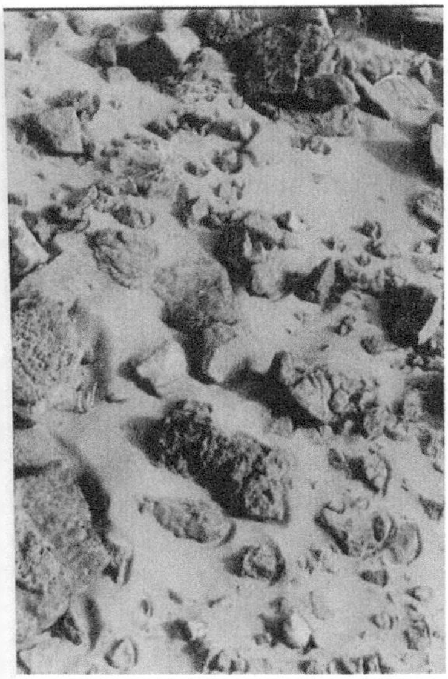
HARRY WALKLAND

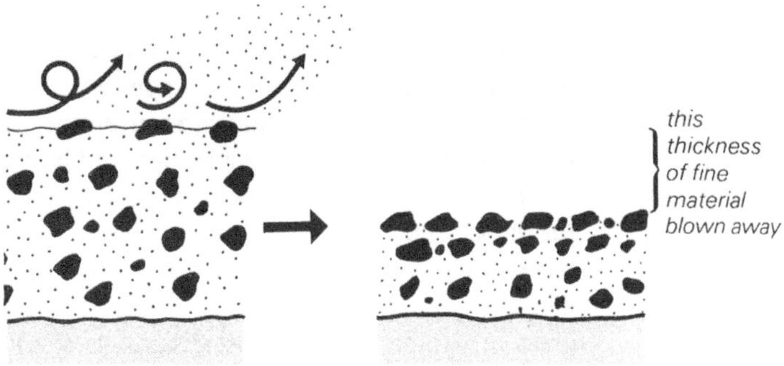

Figure 16.9 Stone pavements
Stone pavements may develop when wind blows over a soil surface which is dry and has no surface vegetation cover. The fine material is blown from between the pebbles so that they eventually form a complete surface cover.

Exercise 16.3 Wind transport and deposition in the desert

1 There is no doubt that without a protective vegetation cover soil is easily moved by both wind and water. Which size of sediment could be moved by wind action: boulders, pebbles, gravel, sand, silt, clay?

2 There are, in fact, several grades of material moved by wind. Particles of the largest moveable sizes are made into the features shown in Figure 16.7 and they form a sand sea. What are the irregular mounds of a sand sea usually called?

3 Silts are rarely found in a desert because they are blown away. However, in areas marginal to deserts they may settle out and accumulate to depths of over a metre. Silt deposits that have formed in this way are called loess. Sometimes when the wind blows from the south-west, summer rainstorms in Britain wash a red dust – from the Sahara desert – out of the air. It is deposited as a fine layer over cars, roads and houses. What size of material is this and how did it manage to travel so far from the Sahara?

4 The shape of sand dunes usually reflects the direction of the dominant wind. Therefore, if they are regular in shape like **barchans** (Fig. 16.10), the wind must come predominantly from one direction. Describe the shape of a barchan, make a sketch of it and try to explain why the 'horns' of the barchan move ahead of the main dune.

5 Barchans are rare. Similarly, long straight sand dunes, called **seif** dunes, are found only in the places where wind directions are constant. So what sort of wind pattern might have caused the irregular surface of the sand sea in Figure 16.7?

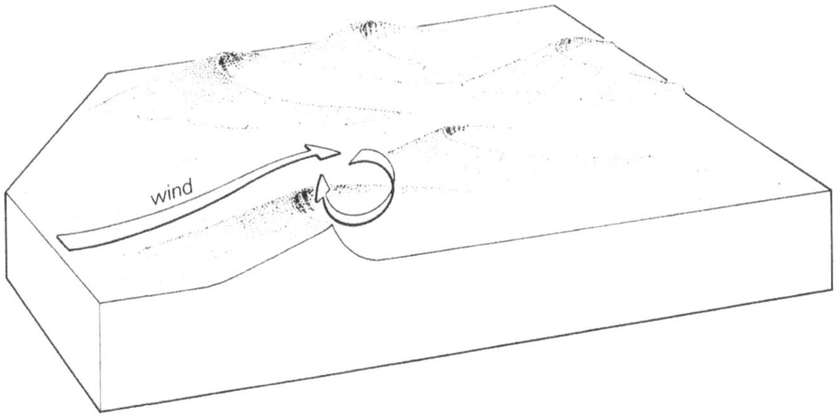

Figure 16.10 Barchans
Barchan sand dunes are driven forward by the wind. The sides of the dunes are called 'horns'.

So, except in areas near to the ground, wind is more of a transporting than an eroding agent. *Very few landscape features have been produced by wind alone.* You must remember that both wind and water are at work in a desert. Water aids weathering and breaks the rock up into fragments which can then be carried away by the wind.

Wind transport is not confined to deserts. Wind will blow around in any part of the Earth's surface where vegetation is sparse. Sand dunes and beaches around Britain's coast are obvious places where wind action can be experienced, as anyone who has eaten sandwiches on a beach knows. Near to glaciers the land is also bare and fine material is easily blown away. Indeed the fine material from both hot deserts and glacier margins has, in the past, been blown over and settled onto much of Northern Europe, America and China. Today it provides some of the world's most fertile soils (called **loess**).

Recently wind erosion has been given the chance to work in new areas. Man has been busily removing vegetation and ploughing land for thousands of year. Over the past hundred years he has been very clever at preventing the return of the unwanted natural vegetation (weeds) with herbicides and has left large barren fields which dry out in spring before plants can grow and protect the soil. These areas are ripe for wind erosion (Ch. 12). Soil is blown away from many areas in winter, but the most disastrous results occur when seedlings emerge, because wind erosion will then blow millions of sand-size soil particles across the fields and slowly shred seedling leaves by abrasion. Not surprisingly, the process is called **desertification.**

Islands of water, seas of sand *Oases*

Despite its widespread effect, wind action rarely creates distinctive landforms. Abrasion mainly results in smoothing the base of cliffs or upstanding rocks. However, one feature in a desert might not exist but for the wind. In areas where water is near the surface, it can rot the rock and make it easy for wind to blow away. This forms large, shallow depressions which eventually expose the water source. Shallow water-filled depressions in deserts are called **oases** (Fig. 16.11).

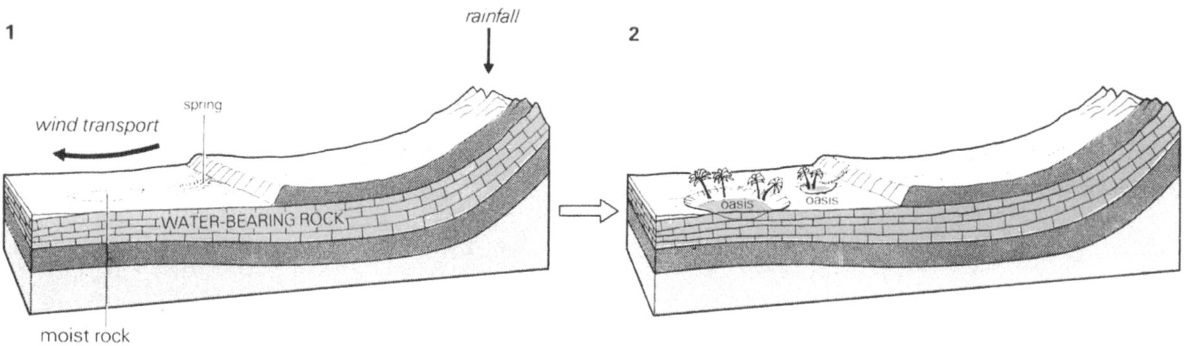

Figure 16.11 Oases (1) Water-bearing rock near the surface allows chemical weathering and the break up of surface rock. (2) Wind removes loose, rotted rock and exposes water-bearing rock. The permanent pool thus provided allows vegetation and human occupation. Some oases stretch over many square kilometres.

17 Glacial landforms

Ice advance

A chilling tale

Chamonix, in the French Alps, developed as a small thriving market town during the Middle Ages. In the nearby fields the farmers grew barley and other crops. High above the valley in which Chamonix lay stood the majestic snow-capped peaks of the Alpes Maritimes (Fig. 17.1). It had always been like that as far back as anyone could remember. But from the year 1600 terror and death came to the townspeople of Chamonix as, season by season, the weather deteriorated until what had once been a green and fertile valley was invaded by a huge tongue of ice.

By the 1640s government officials, who had been sent to the area to investigate the low payment of taxes, reported:

‘The glacier "Des Bois" advances by a musket shot every day, even in the month of August, towards the said land of Chamonix. We have also heard it said that there are evil spells at work among the said glaciers and that the people implored God's help to preserve them against the said peril. The people sow only oats and a little barley which, throughout most of the seasons of the year is under snow, so that they do not get a full harvest in three years, and then the grain rots soon after. The people there are so badly fed ’
they are dark and wretched and seem only half alive.

Soon this advance of ice into the once-fertile valleys of the Alps became 'normal' and the area was abandoned. However, eventually the climate improved again, the glacier retreated, and today Chamonix is once again an agricultural market town. And the ice? It is back up in the mountains and once again just part of the backcloth to be admired by the visiting tourists.

Figure 17.1 The French Alps near the town of Chamonix

SIMON BURN

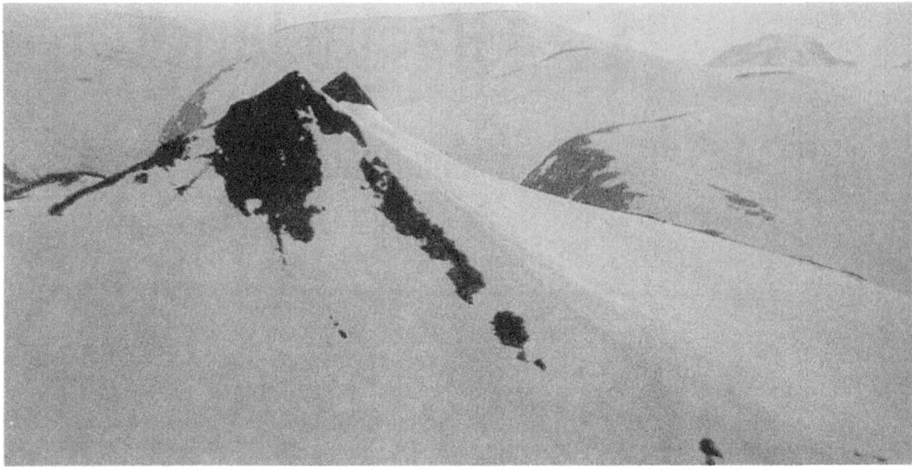

Figure 17.2 Arctic Scotland?
With just 1°C drop in average temperature this is what the mountains of Scotland might look like in summer!

Glacial advances or surges, like the one that threatened Chamonix, can happen quite suddenly, within a few years of a small downward fluctuation in average temperature. Not even Britain is secure: a decrease in average temperature of just 1°C would bring glaciers back to the mountains of Scotland (Fig. 17.2). And if the decrease in temperature falls a mere 6°C on average, then we will back in conditions like the 'Ice Age', with ice sheets spreading down over Aberdeen, Edinburgh, Manchester, Newcastle, Leeds . . . Indeed ice in Britain began to retreat only about 12 000 years ago (Fig. 17.3). So the fortunes of Britain, and for that matter, North America, Northern Europe and Northern Asia hang in a very fragile balance.

Clearly the changes required to go from one climatic regime into another are small. In the past half a million years (the period often referred to as the 'Ice Age') there have been several advances and retreats of ice sheets. Between each **cold period** there have been warm periods (called **interglacials**) during which the temperatures were often higher than those we experience today. Some

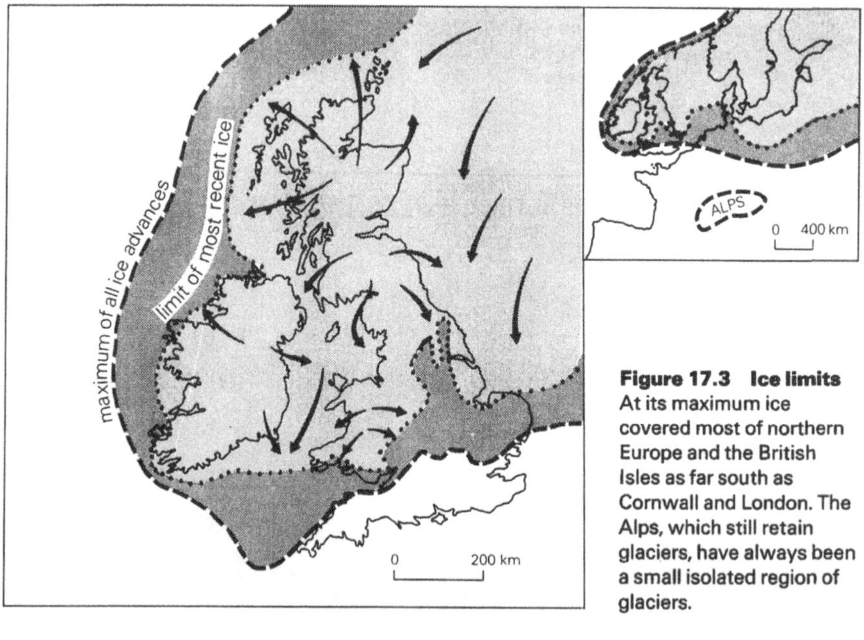

Figure 17.3 Ice limits
At its maximum ice covered most of northern Europe and the British Isles as far south as Cornwall and London. The Alps, which still retain glaciers, have always been a small isolated region of glaciers.

scientists think we are just in another interglacial period today. Records show that interglacials usually last between 8000 and 12 000 years, and it has been well over 10 000 years since the present interglacial began! It seems we can confidently expect colder times ahead, although the change is not likely during our lifetimes. Nevertheless, this 'Little Ice Age' that sprung so suddenly upon the 17th century, shows how quickly conditions can change.

Permanent snow

In search of snow

One of the newest of Europe's ski resorts – and one of the most successful – is not in the Swiss Alps, but in Scotland. Here, among the grandeur of the Cairngorm mountains the A9 trunk road follows the River Spey and the railway north towards Inverness. Along this road the town of Aviemore has developed as a ski resort because it can rely on a long period of snow each year.

The ski slopes of nearby Cairngorm are reached by road along the shores of a forest-clad loch, and then a spectacular trip by chair lift takes the skier to the summit of Cairngorm itself. From this summit thousands of skiers make the descent of the northern slopes where the Sun rarely shines and the snow lasts longest.

Cairngorm has a long snow period because it is over 1400 m high and 57°N. Nevertheless the snow does not last all year except in a few small sheltered hollows on north facing slopes. But in the Alps 1400 m is only the height of some valley floors. The permanent snowline at places such as Diavolleza near San Moritz in Switzerland starts 2000 m higher because the area is over 10° closer to the Equator. By contrast, the Svartisen area of Norway lies astride the Arctic Circle. Here, 10° north of Cairngorm, snow lies near to sea level for the whole Alpine skiing season, and even in August it lies at 1500 m.

These examples show that snow survives because of a combination of *altitude, aspect* and *latitude,* with the permanent snowline being lower towards the Pole. So when a cold period begins and snow and ice stretch out beyond their present limits, the areas most quickly affected are those in northerly latitudes. Even at the maximum spread of the ice during the Ice Age, Alpine glaciers could never spread very far from their mountains. They were simply too near the Equator. When the ice spread over Europe, America and Asia, it came from the north, overwhelming everything in its path.

Exercise 17.1 The snowline

1 Draw a graph of altitude (vertical scale) against latitude (horizontal scale) and mark on the positions of the permanent snowline of the following mountains:

Mountain	*Snowline height (m)*	*Latitude*
Mt Kenya	5500	0
Diavolleza	3200	46
Svartisen	1500	67
Arctic pack ice	0	80

Join the plotted points with a smooth curve. What does the graph show?

2 Cairngorm in Scotland has a height of 1400 m and is at latitude 57°N. Plot this point on the graph. Would you expect Cairngorm to have a permanent snow cap?

The freeze

Figure 17.4 Snowflakes Snowflakes form slim, star-shaped crystals which add complexity to their shape, size and design by attracting more water molecules. It is the intricate shapes that allow snowflakes to lock together and hold snow at steep angles on mountain sides.

Snowflakes are amongst nature's most delicate objects (Fig. 17.4). When snow falls, millions of tiny flakes drift silently down to the ground and settle into a light, easily blown mass. It can take a metre of newly fallen snow to make the equivalent of only 120 mm of rain.

When snow falls it clothes trees in a delicate cloak of sparkling white. But it also falls onto roads where it is flung onto pavements by the passing traffic, the delicate flakes torn and crushed within a fraction of a second. In this form snow becomes brittle and hard. Crush snow into a snowball and the same effect is achieved. Out on the mountainsides the same process is at work but it takes months to crush snowflakes by the sheer weight of banked-up snow. Nevertheless, the snow eventually loses its delicate form, being transformed into small round crystals that are fixed together. Further compaction can take place quite quickly on a warm spring day. Warm air blows over the snow and transfers enough heat to melt the surface. The melted snow sinks into the gaps between the crushed flakes and then refreezes, helping to make the snow more and more like a solid mass of ice.

These two processes of crushing and refreezing gradually turn a hillside of snow into a brittle crystalline material that can rip open the hand of a falling skier. It is called **névé** by the French and **firn** by the Germans. But it still doesn't make real ice; many years are required over which crushing and refreezing work while the snow gets thicker and thicker.

Ice behaves quite differently from snow. Snow is made from tiny independent flakes which have little combined strength; ice is a solid mass that has a large strength both to support a weight and to tear at a rock (Fig. 17.5). When snow turns to ice then gentle hillside hollows can be eroded into deep bowl-shaped features, boulders ground to rock flour, and blocks torn from the solid rock.

Figure 17.5 Glacial erosion Ice can tear at rocks and grind them down. The large boulder (an **erratic**) has survived, but many others surrounding it are much smaller. Together these particles are called **moraine**, a jumble of glacial debris from boulders down to 'rock flour' in size.

Figure 17.6 Under the ice *Above.* Ice is a solid mass that has strength enough to support a cave, but it also flows slowly to give the beautiful arch shown here. *Below.* As ice moves from the rock and arches out to form a cave, it 'sticks' to the rock and pulls small pieces away. In effect the ice is **quarrying** the back of the rock step. This material is carried across the cave roof and shows as the dark smear moving from right to left. Pieces of rock quarried from the cave back wall are partially frozen into the bottom surface of the ice. *Below right.* When they reach the lower end of the cave, they are pressed into the ice and form part of the glacier's debris load called **ground moraine**. At the same time they abrade the rock to form grooves called **striations**.

Nature's stonemasons *Glacial erosion*

Some of the most intricate shapes created by man are in the stonework of our cathedrals. They are the work of generations of patient effort by stonemasons. The tools of the stonemason are the mallet, chisel and file. The mallet and chisel chip off pieces of the rock and leave the surface rough and uneven. The file, on the other hand, scratches the surface with many fine teeth and leaves a much smoother finish.

Glacier ice is nature's most effective stonemason. Ice is strong and it can carry blocks of rock on its undersurface across other rocks (Fig. 17.6). This is the glacial file. The process is called glacial **abrasion.** Ice can also pluck at rock and rip it away from hillslope or valley floor. This is the mallet and chisel. The process is called **quarrying.** You can get an impression of the pulling force of moving ice because food packages often become stuck to the inner surface of the freezing compartment of a fridge, but it is best seen in an ice cave (Fig. 17.6).

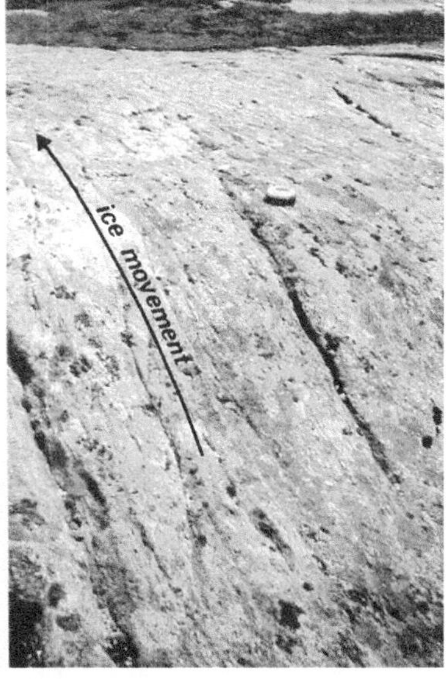

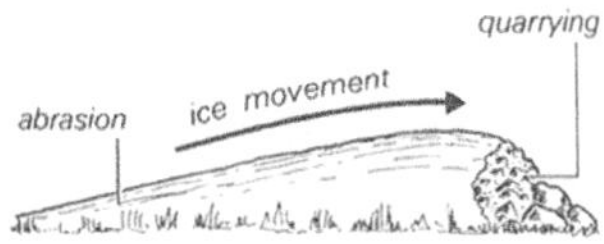

Figure 17.7 A roche moutonnée

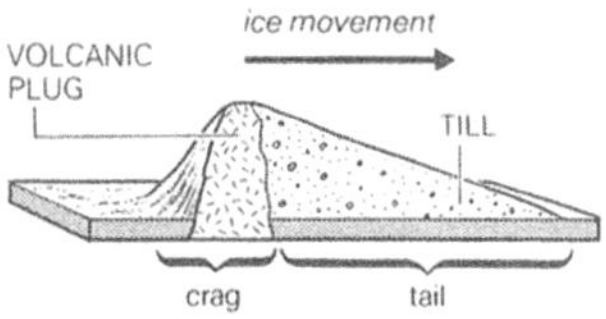

Figure 17.8 A crag and tail

Ice moves because of gravity. The weight of a hollow or valley full of ice will cause it to slide very slowly. All the time the ice is sticking to its floor, quarrying loose blocks away and then pressing them down hard and scraping them across the rock that remains in place. The results are very distinctive. Any projecting rock on the valley floor becomes ground smooth and covered with scratches on the side facing the ice movement. The scratches are called **striations** (Fig. 17.6c). On the downslope side the rock is quarried and becomes jagged. The final sculpted rock is sometimes called a **roche moutonnée** (Fig. 17.7). It is not usually very large, being a few tens of metres long, and is just an irregularity in the bed-rock. Larger obstructions effectively protect their downslope sides and may even be a place of deposition for glacial debris – as though the glacier had swept its chippings out of sight behind a rock (Fig. 17.8). This is called a **crag and tail.**

However, in the mountains erosion is dominant and glacier ice uses the abrading and quarrying tools to create the very special type of landscape.

Corrie formation

Giants' armchairs

The first places snow turns into ice at the start of a cold period are the headwater hollows that once fed rivers. These headwaters are high and so will be in the first areas to experience freezing temperatures. Together, low temperatures and

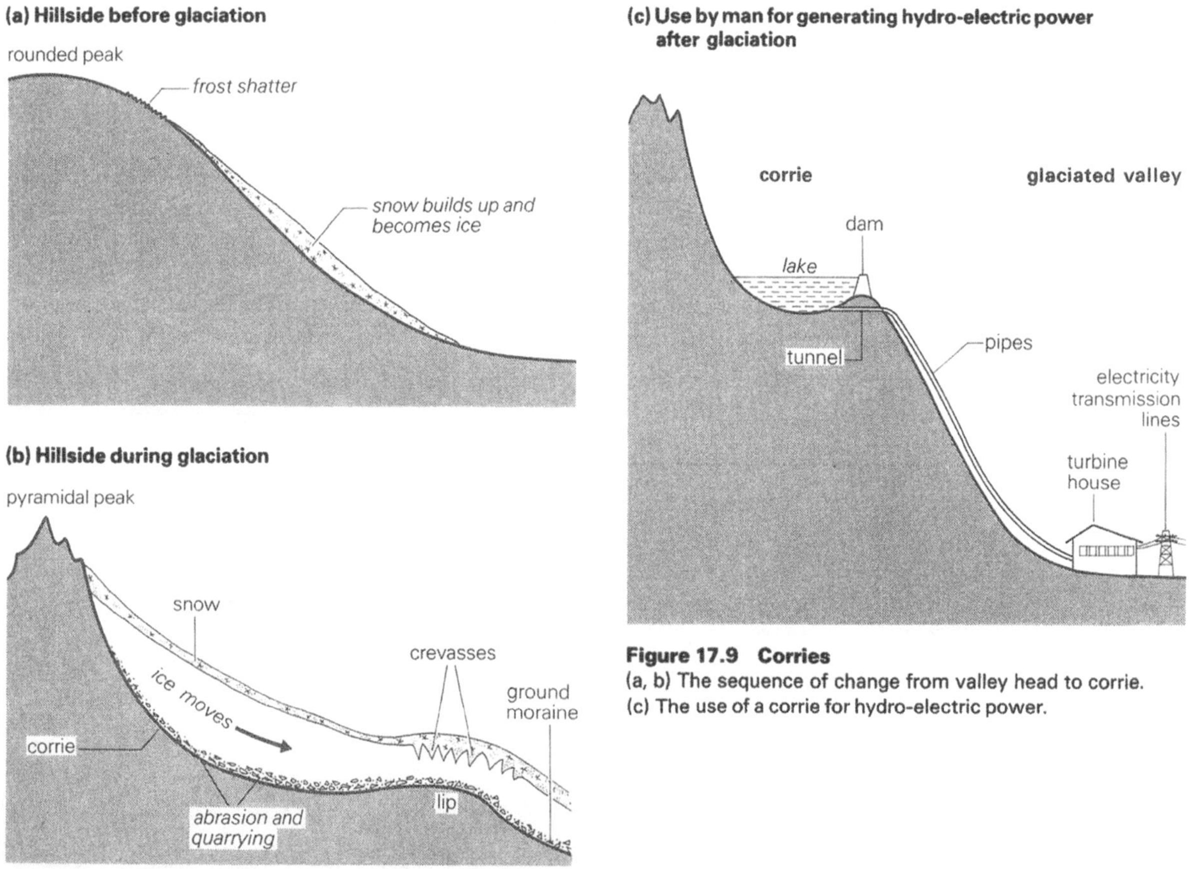

Figure 17.9 Corries
(a, b) The sequence of change from valley head to corrie.
(c) The use of a corrie for hydro-electric power.

Figure 17.10 Ice in a corrie A corrie with ice still eroding it, *above* from straight on and *below* from the side. Notice how the ice appears to have slumped down from the back. Crevasses show where the stresses occur in the ice except near the top where it is covered by snow. The photographs were taken in late summer when snow cover is at a minimum.

sheltered hollows allow snow to build up and be crushed into ice. Slowly the ice slips down the hollow and down the hillslope towards the valley. As it does so the ice mass rotates and scrapes at the rock so that, in time, the hollow changes shape to allow the more efficient movement of ice (Fig. 17.9). All the time the ice is moving out of the hollow, more snow falls and builds up into new ice. After thousands of years the result is to produce a bowl-shaped feature like a giant's armchair set in the mountain side. Ice-eroded hollows are very distinctive. They have been called **corries** by the Scots, while the Welsh call them **cwms** and the French **cirques.** They come in many different sizes and are hewn from all types of rock, but through all this their distinctive shape remains (Fig. 17.10). Indeed, they are probably the clearest evidence of the differences between erosion

Figure 17.11 Ice in the landscape
(a) Before glaciation, (b) during glaciation, (c) after glaciers have melted.

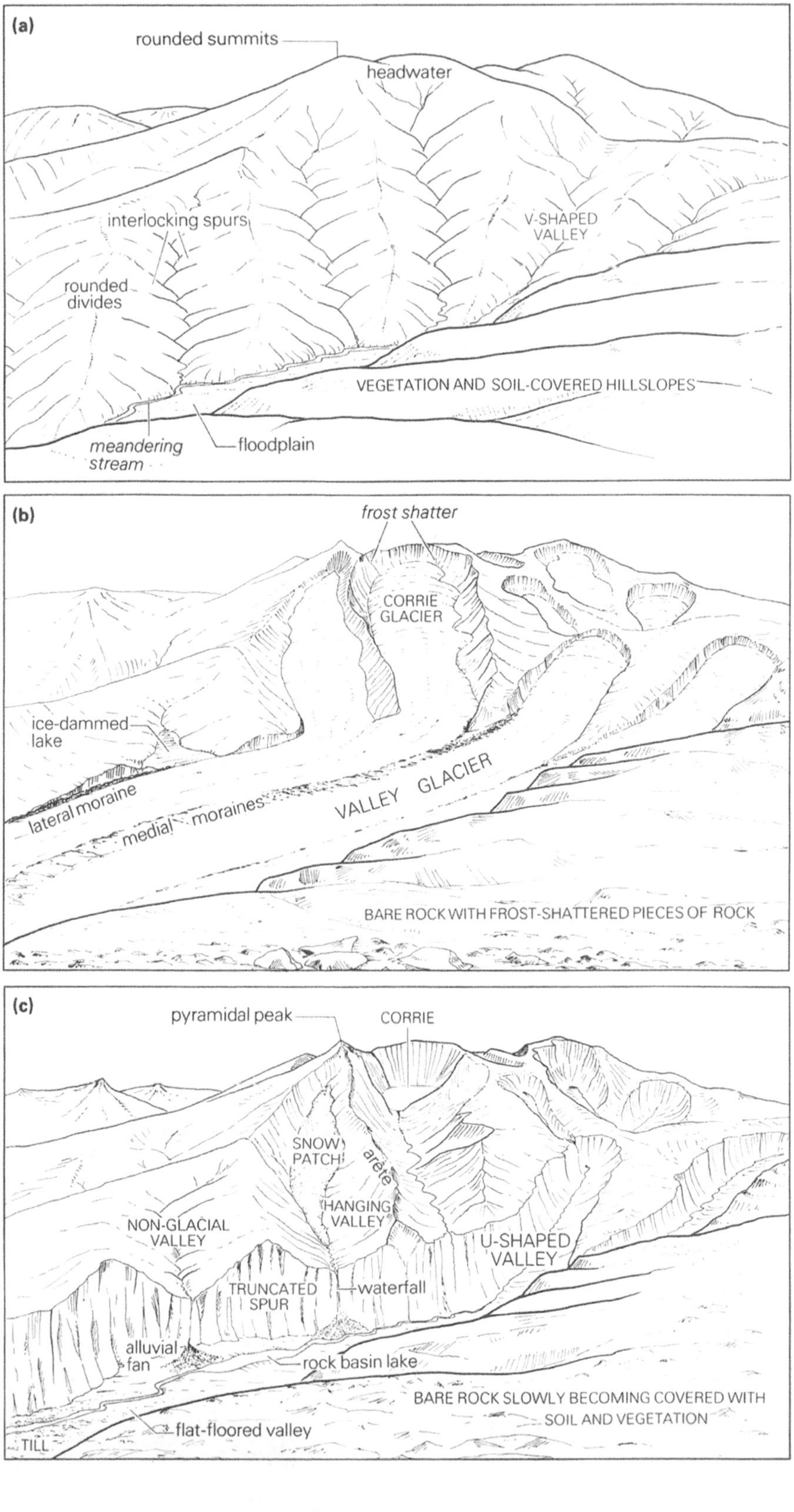

by water and by ice. Bowl shapes, where the outlet **lip** is higher than the centre of the bowl, would not be possible with water erosion. This is why many corries fill with water and form the high lakes (**tarns**) that make ideal sites for HEP stations.

Before glacial conditions arrived, a mountain summit may well have formed the starting point for several streams and it will, therefore, be the site of several headwater hollows (Fig. 17.11a). Gradually, ice develops in each hollow and, as described above, enlarges them into corries. At the same time the mountain peak will become more jagged. Indeed, corries eventually compete for space around the peak, enlarging themselves until only knife-edged ridges called **arêtes** remain to divide one corrie from another. The Matterhorn in Switzerland is perhaps the most famous example of a glaciated mountain where four corries have reduced the peak to a jagged pyramid. However, there are many other areas with **pyramidal peaks,** including those in Figure 17.13 and Mt Lyell in California.

Glaciated valleys

Battle in the valleys

Figure 17.12 Crevasses Ice breaks and produces deep crevasses when it is forced to change gradient quickly such as at an ice fall.

As ice slowly squeezes out over the lip of a corrie and down the hillside it is already carrying a vast supply of debris of all sizes. Now these weapons are used to do battle with the valley rock below.

When ice first enters a valley it moves into a landscape created by flowing water (Fig. 17.11a). It moves across hillslopes that were once covered with soil, grass and trees. But in the cold climate that allows ice advance the trees and grass have long since died and the exposed soil has been stripped off by avalanches and other mass movement processes. Now, only bare rock remains strewn with frost shattered debris. Nevertheless, the valley shape has to be changed until it is more effective at allowing ice movement. Water can meander and flow quickly in a narrow channel, but ice is slow moving and does not easily change shape to flow around bends. So whereas water once tumbled along the bottom of the valley, it now becomes locked up in a dirty grey creeping mass of ice hundreds of metres thick. (Fig. 17.11b).

Valley and corrie glaciers erode in similar ways, but in a valley the volume of ice is greater and the area of abrasion much larger. Here the task is to create a new valley shape which is more suitable for ice flow. It must be broad and deep and have no obstacles or tight bends. Abrasion and quarrying work fastest where such constrictions and obstacles exist, because in these places ice must flow faster and so scrape more debris over the valley floor. Erosion continues day and night, year in year out until *a new valley shape is formed with concave sides, a wide floor and reasonably straight plan.* No longer do hill spurs project across the valley. They have all been **truncated** (cut off) by ice erosion. The small streams of ice that flow in tributary valleys are not able to keep pace with the downward erosion of the glaciers in the main valleys and soon they are left as **hanging valleys** with icefalls cascading from them (Fig. 17.12). To achieve this, millions of tonnes of rock have been removed and carried away in the lower layers of the ice. Erosion by valley glaciers continues so long as new ice forms in corries. But although the valley shapes have been modified, gravity still ensures that glaciers follow the old paths of the rivers to the sea.

Sometimes valley glaciers erode some parts of their valley floors deeper than elsewhere. This may be because the bedrock is less resistant, or the glacier might be more powerful after receiving ice from a tributary. Overdeepened sections of valleys may fill with water and become **rock basin (ribbon) lakes** when glaciers melt away (Fig. 14.21b).

Unless ice overwhelms a valley, slopes above the valley glacier will remain bare and vulnerable to frost shatter. As blocks are shattered from the slopes, they fall and bounce onto the edge of the ice. Accumulations of frost-shattered debris at the sides of a glacier are called **lateral moraines.** Where two glaciers join the common lateral moraine will be carried along some distance from the combined ice margin. Strips of surface moraine carried away from the margin of the valley are called **medial moraines.** It is rare for any surface moraines to remain as distinct ridges after the valley glacier has melted.

The main supply of debris is carried in the lowest layers of the ice and is obtained by abrasion and quarrying. It is called **ground moraine** or **till.** When a glacier melts away, till covers the valley bottom and may well give the valley a flat floor. If the glacier has remained stationary in a valley for some time, a **terminal moraine** may form. When the ice melts, this may hold back some of the **melt-water** and create a lake (Fig. 14.21d).

Exercise 17.2 Alpine-type glaciation

In some isolated high mountain regions such as the Alps, Himalayas and Rockies, glaciers did not fill their valleys, even at the height of the cold period. As a result mountain summits continued to be eroded by frost shatter over long periods and they became sharp and jagged.

1 Study Figure 17.13, which shows part of the Oberaar valley in the Swiss Alps. Notice that ice occupies only a small part of the valley today.
2 Match up the features labelled A, B, C, D, E, F, G, H with the following descriptions: glaciated valley, arête, frost-shattered peak, pyramidal peak, corrie, truncated spur, ice fall, and hanging valley.
3 Explain how truncated spurs and hanging valleys form.

Figure 17.13 The Oberaar glacier, Switzerland

Figure 17.14 Bogbreen glacier, Svartisen, Norway
In the foreground is a rock basin lake.

Exercise 17.3 British-type glaciation

Most upland in Northern Europe has been glaciated and yet it lacks many of the features of Alpine glacial landscapes. This is because ice completely overwhelmed many of our mountains and was thereby able to erode the tops of the hills and mountains as well as the valleys.

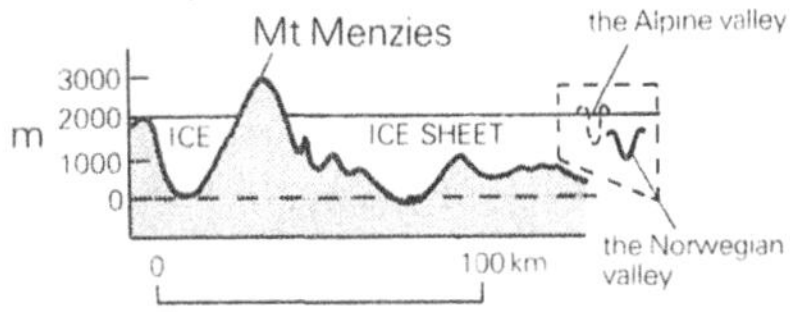

Figue 17.15 Antarctic ice
A cross profile of a small part of Antarctica showing peaks standing above the ice and large numbers of valleys engulfed by it. (The Alpine valley, Fig. 17.13, and the Norwegian valley, Fig. 17.14, have been drawn into this diagram to scale and in their correct relationship to the ice level.)

1. Study Figure 17.14, which shows the Bogbreen glacier in northern Norway. Of the features in Exercise 17.2, question 2, list those that cannot be found in Bogbreen (and are therefore uncommon in Britain).
2. In front of the Bogbreen glacier is a rock basin (ribbon) lake. Explain how ribbon lakes are formed.
3. Figure 17.15 shows part of Antarctica. Notice that most of the continent is under 2000 m of ice even today. If all the ice were to melt away, what features would you expect to find on Mt Menzies that would not be found on other mountains in the cross section?

SUMMARY: FEATURES OF MOUNTAIN GLACIATION

Snow first builds up in the sheltered north-facing hollows that were once the headwaters of streams. As snow accumulates it is crushed into **nevé** and finally into **ice.** Ice within a hollow moves under its own weight, **quarrying** rock from the hollow and using it to **abrade** yet more rock. Eventually a **corrie** (cwm, cirque) forms with its characteristic bowl shape. Ice within a corrie is called a corrie glacier.

If corries form side by side, the rock separating them may be eroded back until it is a sharp ridge. This is called an **arête.** Sometimes several corries form around a mountain and erode it into a sharp pointed **pyramidal peak.** These peaks have arêtes radiating from their summits.

As ice from several corries flows into a valley, the ice makes up a **valley glacier.** Valley glaciers abrade and quarry in just the same way as corrie glaciers do. The telltale scratch marks of glacial abrasion are long thin grooves in the rock called **striations.** Quarrying leaves rough jagged rock surfaces. A small part of valley floor that has been abraded on one side and quarried on the other is called a **roche moutonnée.**

Prolonged quarrying and abrasion straightens out river valleys and produces a **U-shape valley** with **truncated spurs.** Because main valley glaciers are more effective at erosion than their smaller tributaries, glaciers in tributary valleys are often left as hanging glaciers in **hanging valleys.** When the ice melts, the hanging glaciers are replaced by waterfalls. Uneven erosion leaves many over-deepened sections to the valley floor. After glaciation these will be filled with water and form **ribbon lakes** (Fig. 14.21).

Some of the eroded debris will be left behind as **ground moraine** or **till. Lateral** and **medial moraines** that formed on the ice surface may remain as ridges on the till, but they are not very commonly seen.

A change of pace

Sooner or later glaciers reach the more gentle gradients of lowlands; and as they move onto lowlands their paths are no longer guided by valleys and they can spread out and coalesce to form large continuous areas called **ice sheets.**

Lowlands have gentler gradients and ice therefore moves more slowly, often pushed on only by the mass of ice accumulating in upland valleys. Even so it can still perform much erosion, and evidence of this is widespread. For example the chalk hills of Cambridgeshire have been eroded back several kilometres by an advancing ice sheet that overrode them. Later, when the ice stopped moving and melted away, the debris was deposited all over Essex (Fig. 17.16). The rolling hills of Essex are made of material on 'permanent loan' from Cambridgeshire. Deposits of ice-eroded debris (till) also occur at Whitby, on the coast of North Yorkshire, but here material has been eroded from the floor of the North Sea by an ice sheet approaching from the north-east. Some pieces of rock have been transported all the way from Norway and now lie embedded in the Whitby cliffs. They are called **erratics.** Similarly, former parts of the Lake District are now turned over by the ploughs of Shropshire, and farmers on the North German Plain cultivate land brought from Sweden's coastlands.

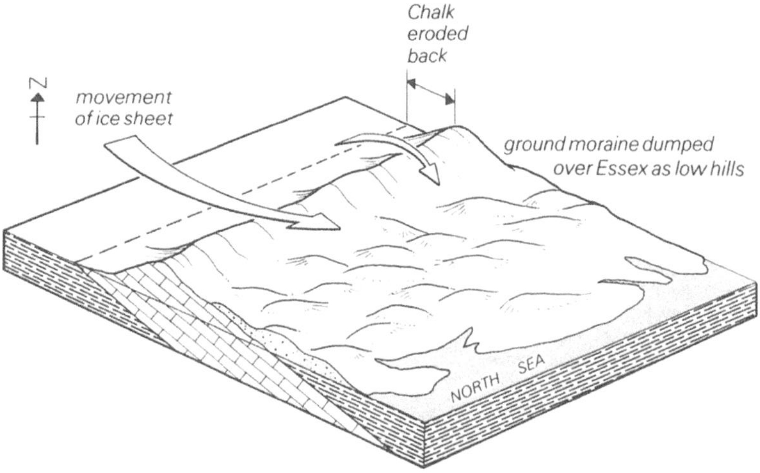

Figure 17.16 Former ice sheets
An ice sheet, moving from the north, swept over East Anglia eroding the chalk cuesta. This, together with other ground moraine, was eventually deposited over the lowlands of Essex, creating gently undulating land and fertile soil.

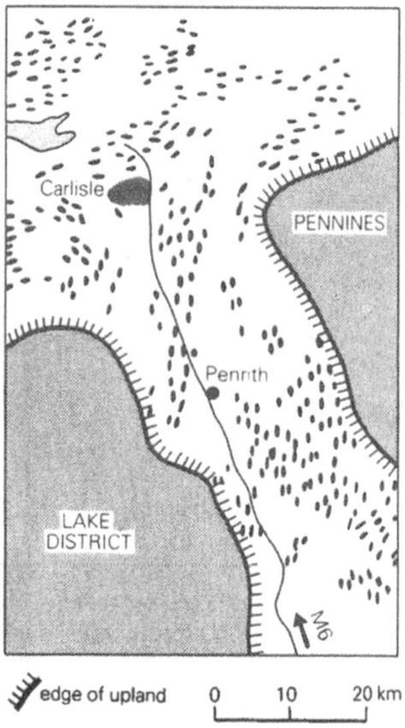

Figure 17.17 Drumlins
Drumlins are piles of ground moraine (**till**) that an ice sheet has deposited and then smoothed into 'half egg' shapes. There are many shown on this map of N.W. England and Southern Scotland.

Although ice on lowlands can perform much erosion, the more it erodes, the more debris it has to carry. Carrying millions of tonnes of debris is an energy-consuming business, especially on a lowland where the supply of energy is low. Finally a stage comes when the ice runs out of energy and erosion ceases. It is like a train full of passengers – it can move onward but it cannot pick up any more until it has dropped some. Sometimes dropping zones are very localised and piles of till are deposited as hummocks called **drumlins** (Fig. 17.17), but most deposition has to wait for a climatic warming and ice melt. In the meantime ice continues to spread until it is no longer pushed forward from behind.

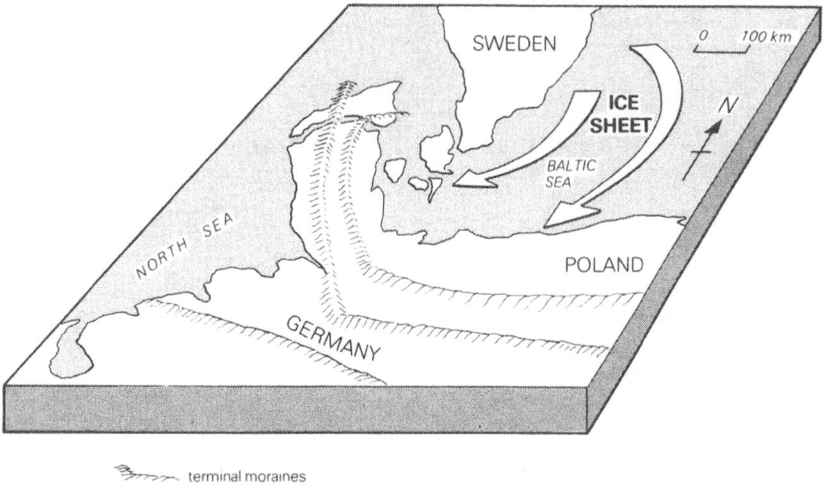

Figure 17.18 Deposits of an ice sheet

Exercise 17.4 Features of ice deposition

Till deposition has clearly been important to Denmark, for without it there would be little left of the country! This is also the case with northern Germany, the Netherlands, parts of eastern England and many other places. However, besides till there are some other depositional features. *All of these other features are of little importance compared with till.* Nevertheless, they are sometimes significant locally and they tell us that deposition from ice has occurred.

1 What is the name for oval-shape mounds of debris that have been deposited by an ice sheet?
2 What is a terminal moraine? Explain the conditions under which it might form.
3 The western half of the Jutland peninsula in Denmark is covered with sands and gravels. Find a map of Denmark in your atlas and identify the Jutland peninsula. The area to the east of this is made of many islands that are the tops of a till plain left sticking out of the sea. Compare the map with Figure 17.18. Why should there be till in the east and outwash sands in the west? What do you think makes up the backbone of Jutland and separates the till from the outwash sands?
4 Britain has many examples of landscapes masked by till. Some of the results are quite surprising, as shown in Figure 17.19 for the Cheshire basin. Now turn back to Figure 17.3. Using this map describe what you think the site of your school must have been like during the maximum of glaciation.

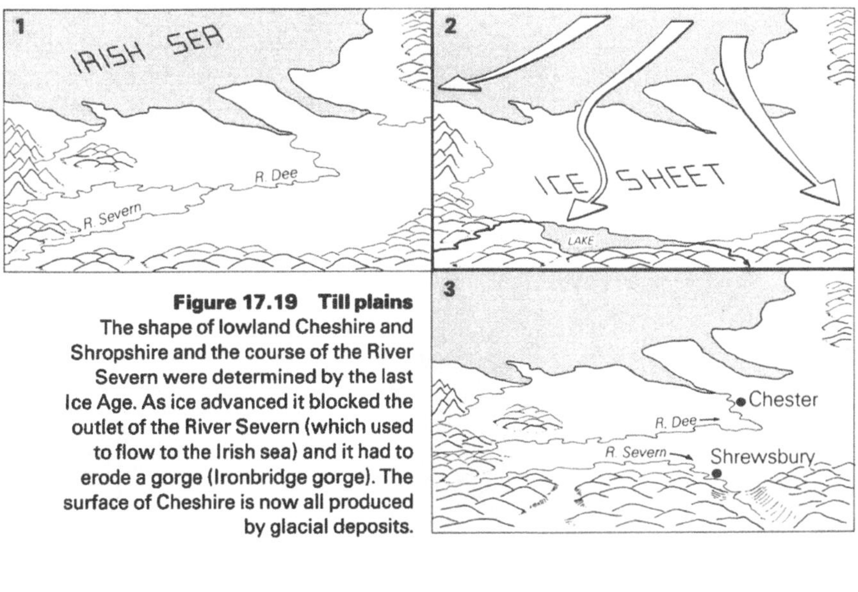

Figure 17.19 Till plains
The shape of lowland Cheshire and Shropshire and the course of the River Severn were determined by the last Ice Age. As ice advanced it blocked the outlet of the River Severn (which used to flow to the Irish sea) and it had to erode a gorge (Ironbridge gorge). The surface of Cheshire is now all produced by glacial deposits.

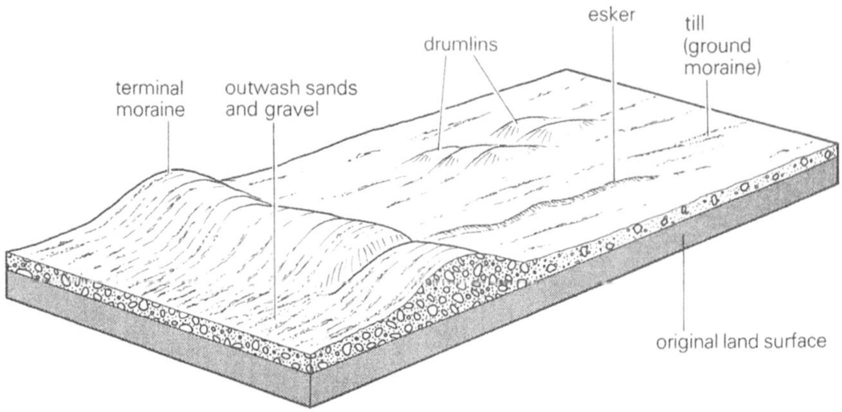

Figure 17.20 Features of glacial deposition in Denmark and north Germany

Figure 17.21 Eskers
This winding **esker** ridge is used by a road to keep it away from the less drained till plain. Trim, Co. Neath, Ireland.

The melt

Eventually the leading edge of the ice spreads so far from its source regions that it melts as fast as it is replaced by new ice from behind. When this happens the ice front becomes stationary, but it is far from inactive. Ice and rocks are still delivered to the melting front. Here ice melts, shedding its debris load which slowly accumulates as a ridge. This is a **terminal moraine** (Fig. 17. 18). Some terminal moraines are major landscape features. Ridges nearly 100 m high, several kilometres wide and hundreds of kilometres long occur in North Germany (Fig. 17.20). The 'backbone' of the Jutland peninsula of Denmark is a terminal moraine. Terminal moraines are rather scarce in Britain, which suggests that the ice sheets were not stationary but were melting back for long periods.

These **meltwater streams** begin not at the ice margin but often tens of kilometres back within the ice. This reminds us that warmer air causes ice to melt over a wide zone near the ice margin, not just near the ice front. As a result of melting, streams form on the ice surface but quickly disappear down crevasses until they are flowing between ice and bedrock. As they flow they erode tunnels in the ice and also pick up much till and carry it towards the ice margin. Sometimes the beds of such streams can still be recognised as winding ridges of forested gravel. These ridges are called **eskers.** Although few eskers have been preserved in Britain, they are still well preserved in Ireland (Fig. 17.21), and in Finland they are very common and have dammed up water to form the famous lakeland region.

Meltwater streams bring much debris to the ice margin and then carry it away. No longer confined to ice tunnels with efficient channels, they spread out into braided channels in front of the ice where they deposit the pebbles and gravel in a wide zone called an **outwash plain.** Like till, outwash plains are very extensive, but they do not make distinctive landscape features. The gravels have since weathered to produce only infertile soils and are mostly hidden under forests.

SUMMARY: DEPOSITS OF GLACIERS AND ICE SHEETS

Glacial deposits are of two types: (a) angular scree that falls on the ice surface from nearby frost-shattered slopes and (b) rounded material that has been quarried from beneath the ice.

Screes carried along by a glacier on its surface are called **lateral moraines.** The inner stripes are made from the same material and they also rest on the glacier surface. They are called **medial moraines** (*medial* means middle). Medial moraines are formed by two lateral moraines when glaciers meet (Fig. 17.11).

The majority of the material carried by a glacier is in the bottom ice layers and is called **ground moraine** or **till.** On lowlands till often masks the preglacial landscape to a depth of tens of metres. Sometimes it forms streamlined mounds called **drumlins** (Fig. 17.18). Most often till is deposited haphazardly over the landscape, filling in some valleys and making plains into areas of gently rolling hills.

At the limit of ice advance a **terminal moraine** sometimes accumulates. This is a place where ice melts as fast as it arrives. Melting water carries the debris away in braided streams, gradually sorting fine from coarse over an **outwash plain.** Meltwater from behind the ice margin flows down crevasses and then runs in tunnels along the base of the ice, carrying washed-out till with it. The winding ridges deposited by these streams are called **eskers** (Fig. 17.21).

SUMMARY

Glacial melting

The end?

As a cold period comes to an end and the climate warms, ice sheets seem to melt quite rapidly. Britain's ice sheets disappeared from the lowlands in a couple of thousand years. They vanished from North America within about eight thousand years. But their retreat into the mountains was slower. Many stopped in their tracks several times and deposited terminal moraines. Some still block valleys in Scottish mountains, but most have been carried away by rivers. In a corrie lake in Snowdonia called Cwm Idwal, the narrow ridge of a terminal moraine juts out into the waters of the lake – a last reminder of the corrie glacier in its death throws (Fig. 11.22). But the ice does not give up easily. It has made a comeback three times in the past half a million years and it shows every sign of trying to do the same in the near future. In the meantime we can admire the shrunken remnants of ice sheets that we now call spectacular Alpine glaciers and wonder about the time that ice spread all over Britain to a depth of 200 m. Ice left Britain and other northern continents less than 10 000 years ago; when it returns these areas will look like Antarctica today (Fig. 17.15). But where shall the people live?

Past glaciations and man

Glacial gifts

Glaciers are natural features that do not have much effect on man's day-to-day life. Most glaciers are up in mountains, and any changes in their size affects only the small numbers of people who live near them. Today people go towards glaciers; they do not run from them. Glaciers are tourist attractions, and each year thousands of people are taken for guided tours over them.

Ice sheets are rare. Small sheets of ice occur in Norway, and somewhat larger sheets in Iceland. The only two areas where really large ice sheets exist on land are Greenland and Antarctica – both places almost uninhabited.

But despite the remoteness of ice sheets today, the impact of past cold periods has been substantial. For example, the main build-up of ice in Scotland was over the Cairngorm mountains in the east. As glaciers spread out from this source their main route was westward, plunging over high mountain ridges and down towards the Atlantic. In time the ridges were lowered to produce the passes that are used today (e.g. the famous pass of Glencoe). Routeways resulting from ice erosion *across* high mountains are important in the Alps and North America as well. At the same time deep glacially eroded valleys with tributaries left hanging above them have also provided sites for hydro-electric power stations.

Probably the widest impact of glaciation is the least spectacular and least well known. Glacial deposition formed whole countries such as Denmark, and substantial parts of many others. North of a line joining London to Bristol, most of the farmland of Britain, is formed on glacial till, yet most of us hardly give it a thought.

18 Coastal landforms

To catch a wave *Wave formation*

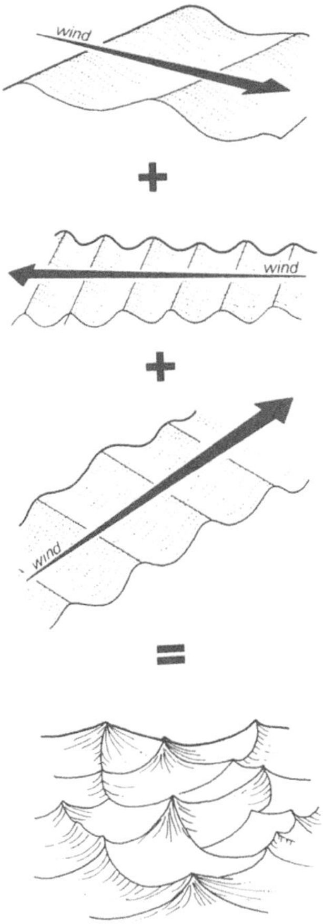

Figure 18.1 Ocean waves Waves in the ocean appear random and without pattern, but the peaks and troughs are the result of quite simple wave patterns that have been formed by winds far out to sea. Each pattern becomes superimposed on the others to produce an apparently inexplicable complexity.

When storm waves hit a cliff the pounding it receives is comparable to that of a huge hammer being struck again and again, ten times a minute, 24 hours a day. If they are to resist such forces most cliffs have to be very strong. Man has often underestimated the power of the waves. At Wick, in Scotland, the end of a breakwater wall was built using an 800 tonne block of concrete secured to the wall with iron rods the thickness of a man's arm. Yet soon after it was finished a storm wave threw the block and half the wall back into the harbour it was designed to protect. Together, wall and blocked weighed 1350 tonnes. So the engineers tried again, this time with a block weighing 2600 tonnes. Imagine their surprise when, a few storms later, this too was thrown into the water. In America there is a lighthouse beacon 45 m above sea level which has had so many rocks thrown through its windows by storm waves that a steel grating had to be built to protect it. And then there are storms lashed on by hurricanes. There are many reports of boats being carried several kilometres inland by waves tens of metres high. These are examples of **storm waves.** They are quite different from the **swell** waves that lap the beaches on a sunny summer's afternoon. But any wave can soon be whipped up by gale-force winds. A few hours of a strong wind blowing over an ocean surface can turn waves a few centimetres high into gale-lashed giants the height of a house.

Out in the ocean, water is always on the move – an ever-changing pattern of small wave crests and troughs (Fig. 18.1). As the wind builds up so it catches these small waves in the same way as it catches the sail of a yacht. As a result the waves are pushed across the ocean, building ever higher as they trap more and more wind.

However, waves are not just surface features. The water movements on the surface are tied to movements below, so that water is stirred up to a depth about equal to half the distance between wave crests. Usually this is nothing more than a surface ripple on a deep ocean, but as the waves approach shallow seas so this stirred-up water starts to churn up sediment on the sea bed. As soon as this happens, frictional drag with the sea bed slows the waves down. At the same time they change shape, first becoming higher and then forming the **breakers** with which we are so familiar. Now as the waves complete their journey, all the power transferred from the wind is suddenly thrust upon the shore. The results can be dramatic, but they depend on the nature of the shore: there is a big contrast between the effects on cliffs and on sandy beaches.

ROGER JONES

Figure 18.2 Wave action on cliffs
As waves come onshore they increase in height and form breakers which then crash against cliffs. The pressure effect of such action can be dramatic.

Crumbling cliffs *Wave erosion*

When a wave crashes against a cliff, all the pent-up energy is spent in a fraction of a second, yet the cliff yields but little. Instead, a massive curtain of spray is flung up over the cliff face as the water rebounds back into the sea (Fig. 18.2). However, cliffs are mostly made from jointed rocks, and the impact of the waves on joints is very different. Air is often trapped in them by the inrush of water and this is driven like a hammer even deeper into the cliff (Fig. 18.3). The force of the wave is spent very quickly and, almost as soon as it has struck the cliff, water pours out of the joints and back into the sea. However, the sea is patient and can repeat its hammer blows for as long as required. And patience is rewarded. Every so often a block is dislodged and the water is presented with a new piece of rock to tackle.

The sea has other tools besides the force of water. Continual wetting of rock causes rotting (chemical weathering) in just the same way as it does on land. A few pebbles thrown against the cliff by the water everytime a wave breaks have an abrasive effect very similar to sandpaper and every bit as effective in erosion.

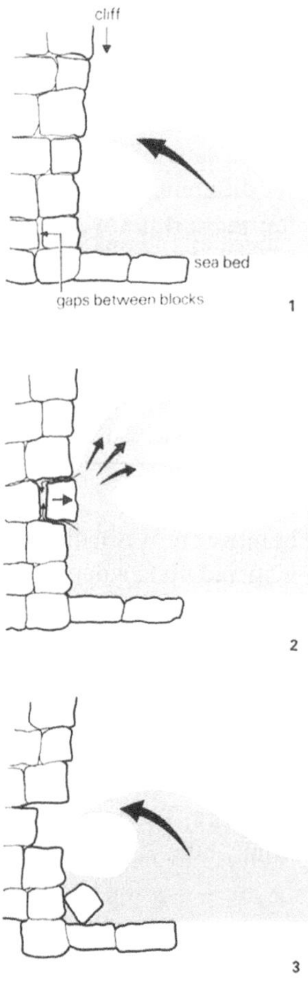

Figure 18.3 Wave action on cliffs
(1) A wave breaks against a cliff made from jointed rock. (2) The wave breaks and water is thrust into the gaps between blocks, gradually dislodging surface rock. (3) As the wave retreats the loose block is pulled out and it falls to the bottom of the cliff, leaving more rock exposed to the waves.

Figure 18.4 Cliff erosion
The process shown in Figure 18.3 leads to undercutting of the cliff base. (1) With a small overhang the cliff remains stable. (2) Undercutting has now proceeded so far that the cliff is ready to collapse. (3) The cliff has collapsed and the resultant rock debris must now be removed.

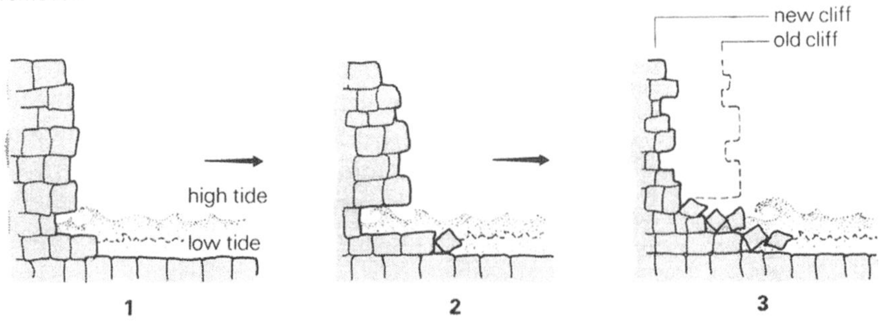

Cliffs are mostly attacked with all of these tools between high and low tide level and the result of this concentrated attack is **undercutting.** An undercut cliff rapidly becomes unstable because of the large weight of unsupported rock above the overhang. As a result, cliffs crumble, crash or collapse into the sea (Fig. 18.4). Whatever the speed of collapse, the debris scattered at the foot of the cliff protects the new cliff for a while. Now the waves have the task of breaking down the debris into pieces of moveable size. But as soon as the loose rock has been neatly cleared away, the waves resume their ceaseless pounding.

Caves, arches and stacks

Coastline under attack

Well jointed cliffs retreat by block removal and the cliff retreats in steps (Fig. 18.4). Notice that, in many ways, cliffs are very like hillsides. The angle of the cliff depends on the type of rock in which it is formed, the rate of erosion at its base and sometimes on the dip of the rock (Fig. 18.5). However, very few rocks have joints of just one size evenly spaced out across a cliff face. More commonly, rocks have concentrations of joints in some places, fewer well cemented particles somewhere else, and perhaps a fault or two. As a result cliffs are attacked most effectively in places where the weaknesses are greatest, and sections of cliff retreat at different rates.

The first sign of weakness may be localised undercutting and the formation of a **cave** (Fig. 18.6). A stretch of coast with many weaknesses may be full of small caves and collapsing cliffs. Where there are many such weaknesses the coast will retreat rapidly and a **bay** will begin to form. Any resistant stretch of coast will be eroded more slowly and will gradually form a **headland.** Bays and headlands are extremely common along the western shores of Britain and Ireland where there are many faults and much variation in rock resistance (Fig. 18.7).

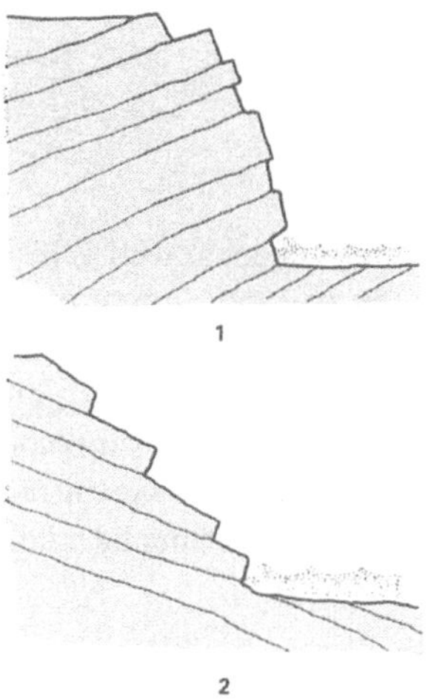

Figure 18.5 Cliff profiles The dip of rocks can affect the cliff steepness. (a) Steep cliffs with rocks that dip away from the coast. (b) More gentle cliffs with rocks that dip towards the sea. Blocks easily slide from the cliff in this case.

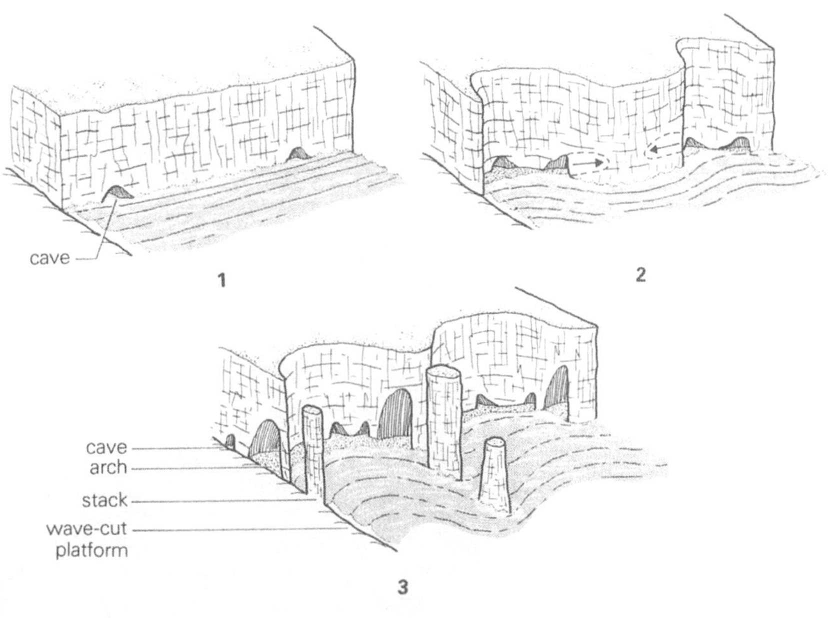

Figure 18.6 Summary of the stages of cliff retreat (1) Caves are produced as waves erode the base of the cliff. (2) Collapse of cliffs above caves leads to cliff retreat and bays and headlands are produced. (3) Headlands are attacked from the sides. Through caves produce arches; later cave roofs collapse to to produce stacks. The retreat of the cliffs leaves a wave cut platform behind.

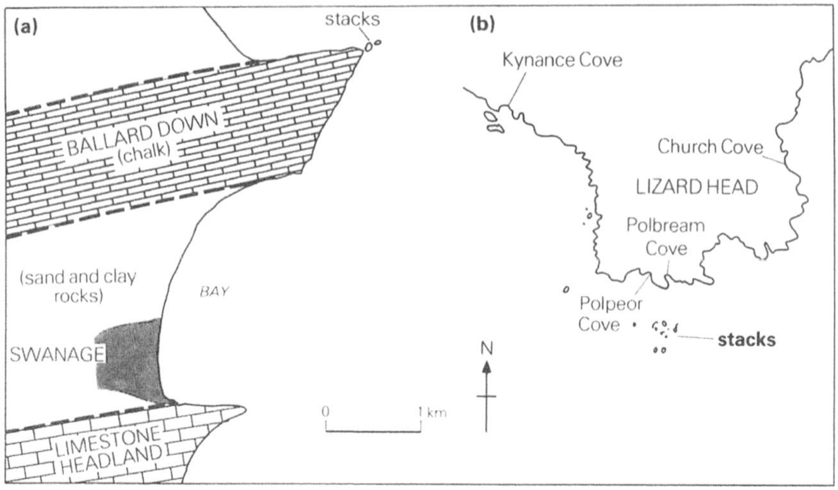

Figure 18.7 Bays and headlands
(a) Swanage bay in Dorset has been eroded from soft sands and clays whilst resistant limestone and chalk have been made into headlands. Notice the stacks at the end of the chalk show it is retreating by block collapse. (b) The Lizard, Cornwall, has many changes of rock, together with much faulting. This is reflected in small coves rather than large bays.

As soon as a headland is produced, it is attacked from the sides as well as head on. Enlargement of joints by waves may now lead to caves developing in the sides of the headland. If this happens and the caves meet, a tunnel is produced which, when enlarged further, leaves an **arch** standing out to sea (Fig. 18.8). The slow demolition of King Arthur's castle at Tintagel (Cornwall) shows this well (Fig. 18.9). Not surprisingly, such arches do not remain for long and soon the roof collapses, leaving a pillar of rock (called a **stack**) to indicate the former headland position (Fig. 18.7).

Figure 18.8 Durdle Door, an arch in limestone, Dorset, England

Figure 18.9 The disappearing castle of King Arthur, Tintagel (Cornwall)

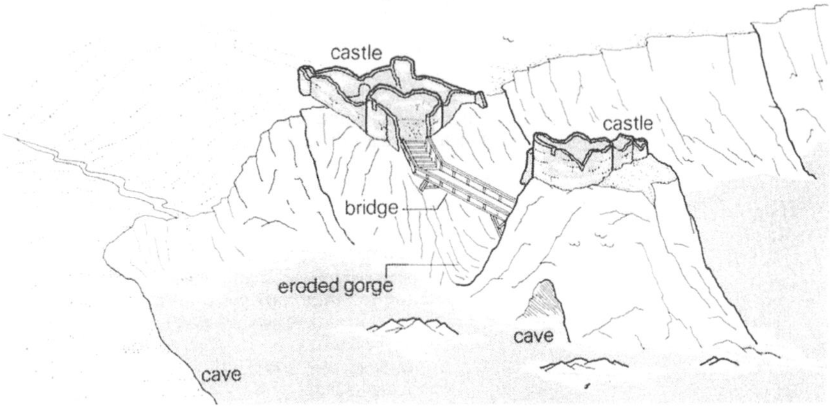

(a)

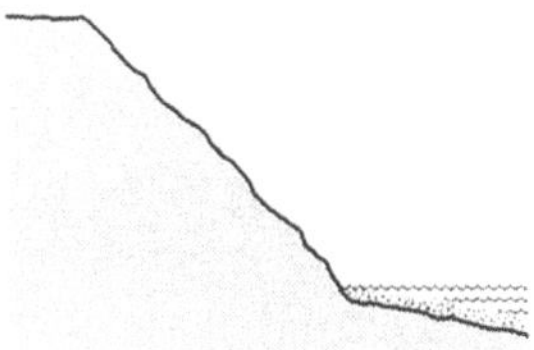

(b)

Figure 18.10 Cliff profiles

Exercise 18.1 Cliffs

Bays often have sandy beaches backed by gently sloping cliffs. When a wave breaks on a beach the energy given to it by the wind causes the sand and gravel to move. This uses up much of the wave energy.

1 What happens when a beach is not present to absorb the wave energy along a coast, say at a headland?
2 Figure 18.2 shows a wave breaking on a headland. Describe what is happening within the cliff. What is the result of prolonged wave action?
3 Match the two profiles of Figure 18.10 with the bay and headland of Figure 18.7a. Make a copy of these profiles and write 'bay' and 'headland' beneath the appropriate diagrams. Explain why you made your decision.
4 Figure 18.11 shows the remains of the village of Hallsands in Devon (England). Earlier this century gravel was taken from the beach below the cliffs. Describe and explain what has subsequently happened. Why might it have been a good idea to tip hundreds of tonnes of rock at the cliff foot at an early stage?

However, wave attack causes little erosion below low tide and cliff retreat leaves a wide rocky **wave-cut platform** behind. Stacks and arches usually project from wave-cut platforms. These platforms are very extensive, and beneath a metre or two of beach sand and gravel a wave-cut platform *always* hides.

Moving sand on beaches

Beside the seaside

Figure 18.11 Hallsands, Devon

Stand on any steeply sloping beach on a fine day and feel the water surging past your feet, pushing toward the land. Look down at the brown-coloured water of this **swash** containing thousands of grains of sand on the move. A few seconds later feel the water pouring back towards the sea, making wakes around your heels. Notice how the sand is pulled from beneath your heels by this **backwash.** Stand on a very steep beach made from pebbles and you can also hear the rush and watch the roll and tumble of the pebbles as they are carried back and forth with every wave. Stand on a very flat beach at the back of a bay and the waves come round your feet as thin sheets of placid water, each riding over the back of the one in front, but rarely appearing to return to the sea.

Do not stand on a beach in a storm. The rush of water would knock you off your feet and easily carry you out to sea. Each onrush of waves, surges boiling

Exercise 18.2 Coastal erosion

Most of the rocks of the Shetland Isles are hard, banded metamorphics which stand nearly vertically in the cliffs. In addition there are some limestone bands and prominent fault lines, all of which have a north–south 'grain' (Fig. 18.12).

1 Make a copy of the coast and mark in colour the inlets that have been eroded from limestone.
2 Using a different colour, mark on those that have been eroded along fault lines.
3 Describe how important you think faults and differences in rock resistance have been in shaping the coast.
4 Can you suggest reasons for the inlets that remain unshaded?
5 The Shetlands are now famous for their oil terminal at Sullom Voe, which is one of the largest in Europe. Explain what advantages Sullom Voe might have as an oil terminal (a) because of its coastal erosional history and (b) because of its position in the North Sea.
6 Why might Sullom Voe have been chosen in preference to other voes (inlets) on the map?

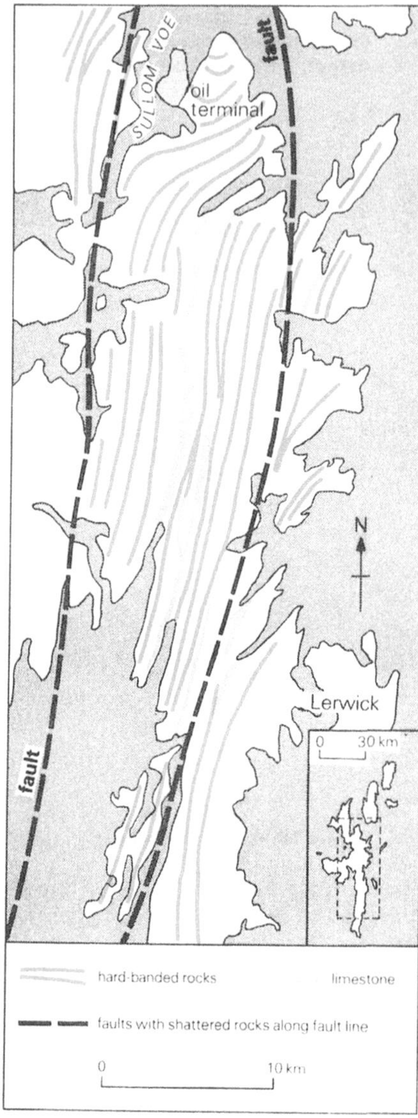

Figure 18.12 Part of the island of Zetland in the Shetlands, Scotland

water over the beach and throws salt spray high up into the air. Each curling wave thunders its way up the beach and hurls sand and pebbles before it. Only at the back of a bay are the waves more gentle and calm.

Beaches are clearly places of considerable contrast. Waves change shape and behave differently because of variations in wave energy, beach slope and beach sediment size. The same waves out at sea can behave very differently on a beach with a gentle slope from that with a steep slope (Fig. 18.14). Material of pebble size moves quite differently from sand grains. These variations are the keys to the wide range of beaches that we see.

SUMMARY: COASTAL EROSION

Cliffs are eroded by (a) the *force of waves* thrusting water and air into joints in the rock (b) *abrasion* by pebbles thrown against the cliff, and (c) *chemical weathering* and solution.

Jointed cliffs retreat mainly by block removal. The sequence of steps is (1) wave action dislodges blocks from the cliff causing an overhang, (2) the cliff collapses, (3) the accumulated blocks protect the cliff and are slowly broken down by the waves and removed, and (4) this leaves the cliff exposed once more.

Because coasts are made from a variety of rocks, the least resistant will be removed first, possibly by the formation of **caves** and the collapse of **undercut cliffs.** Slowly, bays form which leave **headlands** of more resistant rock open to attack from the sides. Cave formation in the sides of a headland can cause **tunnels** and **arches** to form. Eventually the headland's earlier positions are marked only by lines of **stacks.** The width of cliff retreat is shown by the width of the **wave-cut platform** (Fig. 18.13). Many wave-cut platforms are not very wide because the sea has not remained at a constant level for very long.

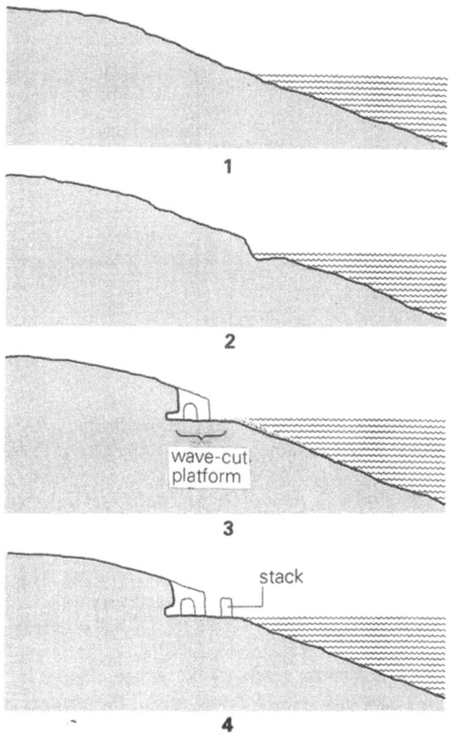

Figure 18.13 Stages in cliff formation

However, it is not just material in the surf zone that is on the move when waves break. The whole beach is being churned over to a region well outside where the breakers first start to form. And if the churning effect is noticeable on the edge of the surf zone, think how much more power is available to churn up and transport sediment beneath the place where the waves first break. We need to look closely at the form of the breaking wave at this critical point. Is it thrusting forwards and pushing material up the beach or is it plunging downwards and eroding the sediment from where it strikes?

Beaches are made from individual grains of rock which are easily moved by each breaking wave. This is the reason a beach never looks the same from one day to the next. But the effect of the waves on a beach is very closely involved with the tides. All the while the waves are breaking, the tide is moving in or out. **Spilling** waves that push material landwards will be helped by a rising tide and will leave material at the top of the beach, building it up. They are often called constructive waves. **Plunging waves** are erosive and they carry material seawards. If these waves occur on a falling tide, a metre's depth of sand can be stripped from a beach in just a few hours. These waves are often called destructive waves.

Topic Wave shape

Figure 18.14 Waves on the beach
(a) Swell waves, (b) storm waves, (c) gently sloping beach, (d) steeply sloping beach.

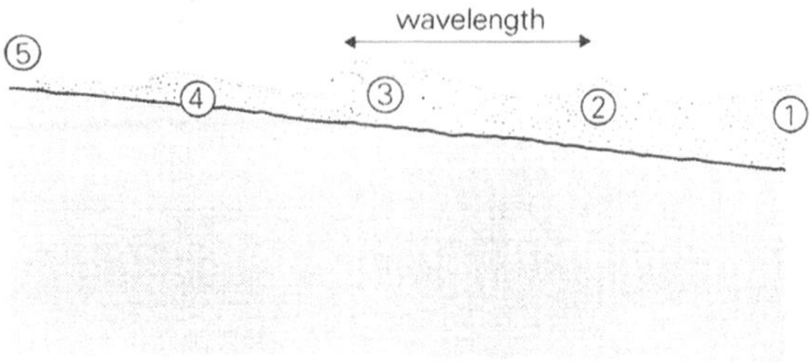

(a)

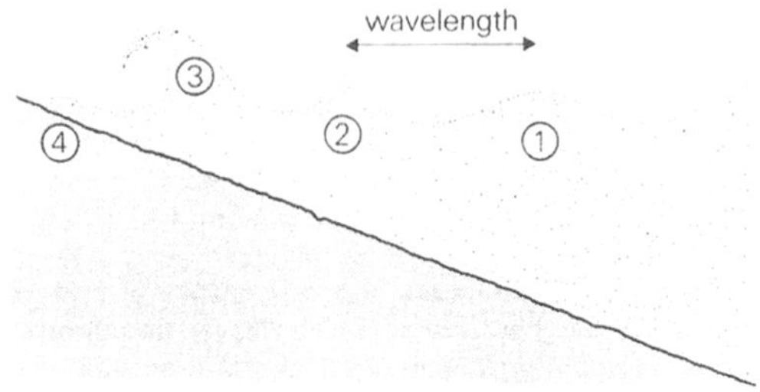

(b)

(a) **Gently sloping beach.** (1) Waves are affected by the sea floor some distance from the coast and they change from smooth rolling forms. The wave first develops a crest (2) which starts to break. As the waves move further inshore the crest water *spills* forward (3) as a breaking wave. The wave then rushes up the beach as 'swash' (4) finally reaching the upper beach as a shallow flat sheet of water. Much of this water sinks into the beach sand and very little returns as a backwash.

(b) **Steeply sloping beach.** (1) Waves are affected by the sea floor much closer to the shore. All the changes of shape happen closer in. Here the wave becomes crested at (2) but the steep slope causes water to be pushed up the beach and the wave *grows* in height as it breaks. These waves *plunge* vertically and erode the beach. Such waves occur mainly during a storm. The steep beach prevents much forward rush of water (swash). In fact it encourages a backwash and helps erosion. Small waves on steep beaches do not become high enough to plunge and they tend to spill forwards, helping to repair any storm erosion. Waves that spill forward are sometimes called **constructive** waves, those with a large backwash and plunging breaker are sometimes called **destructive** waves.

The size of a wave, its energy and its capacity to move sand onshore or offshore are all related. Large waves are produced by strong winds and are associated with storms. In many cases storm waves are plunging destructive waves. (Fig. 18.14b). But storm waves at the back of a bay are rarely plunging, because the beach is too flat for them to develop; so destructive waves are mostly found on steeply sloping exposed beaches. By contrast, for most of the time spilling waves occur on all beaches. They are not driven on by any strong local storm wind and they are called swell waves (Fig. 18.14a). Their effect is much more subtle than that of the stormy giants. Because they have less energy, they move only small amounts of sediment forwards at each tide, so it may take a couple of months to replace the material lost in a single tide of storm waves.

Figure 18.15 Stages forming longshore drift

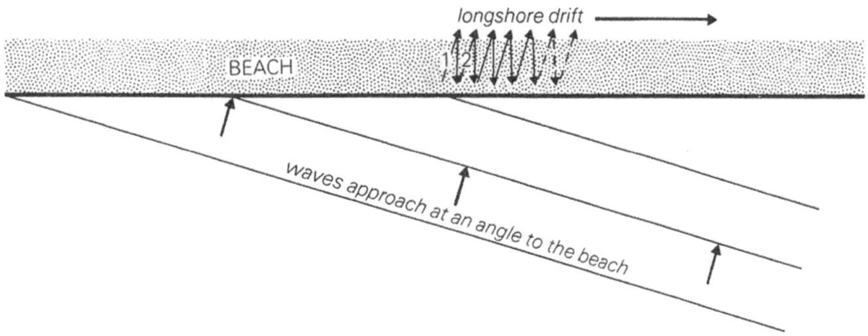

Drifting sands

Longshore drift

On an open coast, waves rarely approach parallel to the shore but are in long sweeping curves. There are no sharp bends in the plan of waves, and many coastal features reflect their shape. Bays, for instance, contain sand that forms long arcs of beach, widest at the centre, narrowest towards the headlands. Bays collect and retain much of the material eroded from the headlands. Some sand escapes, however, and there is a gradual movement of sand (and sometimes even large particles such as shingle) along the coast. This is why it is very common to find pebbles on a beach that do not look like the rock of nearby cliffs. The movement of material along a coast is called **longshore drift.**

When waves reach the coast they usually approach at an angle. As the waves break they push material ahead of them in the swash, again at an angle to the coast (Fig. 18.15(1)). However, although the swash direction is controlled by the wave approach, the backwash is always down the steepest line of the beach (2). As a result the beach material is moved along in the sawtooth motion called longshore drift. Because of this millions of tonnes of material are constantly moving along the beaches of the world every day. It is one of the main ways in which material is removed from the foot of cliffs, allowing further erosion to proceed (Fig. 18.16).

Figure 18.16 Longshore drift
(1) As waves break, material in the surf zone is pushed up the beach at an angle. (2) As the surf rushes back, it follows the line of steepest slope. The constant up and down motion of the beach material not only makes it rounded, but it also wears it into small fragments which can be more easily transported.

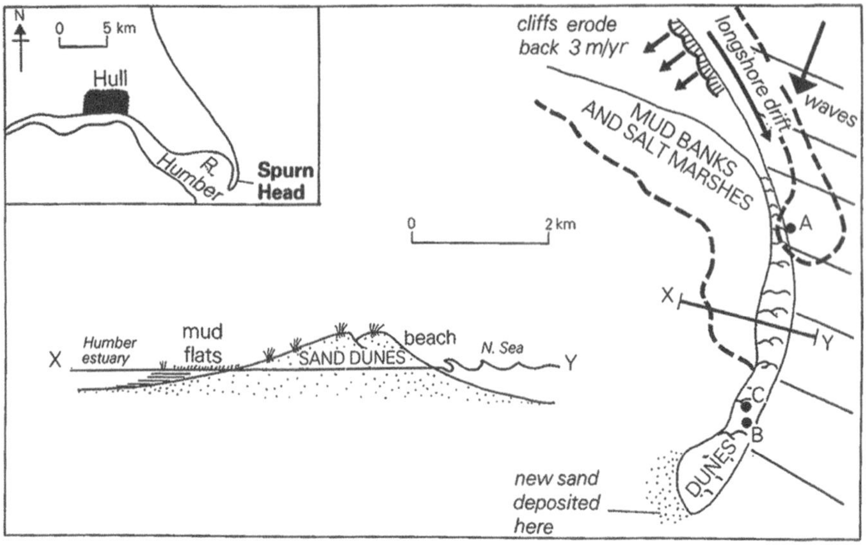

Figure 18.17 Spurn Head, a sand spit at the end of the River Humber

Spit formation

Sands of time

The drift of sand around coasts is a complex affair involving coastal current and the action of waves. But there can be no question that large amounts of sand, silt and clay are on the move. Nowhere is this clearer than at the mouth of the Humber on England's North Sea coast. Here we find a long, irregular, curving ridge of sand called Spurn Head (Fig. 18.17). There is no rocky core to this ridge even though it extends half way across the river estuary. This long ridge of sand is capped by sand dunes which have been colonised by marram grass, and on its more sheltered river side, mud flats with salt marsh vegetation survive. It clearly receives its material from the north, where the coast is eroding rapidly.

We know much about the way Spurn Head has grown because it has been such a major hazard to navigation into the River Humber estuary and has been charted for hundreds of years. Imagine yourself in a small sailing barge in the 17th century. It is dark and no light shows anywhere. There is a strong wind blowing and it is difficult to keep the boat running ahead of the wind. You are nearly home but you are straining your eyes for a sight of the low, dark silhouette of the dreaded sand bar that juts out into the estuary. Then, suddenly, the boat rears up and a grinding sound tells you the bottom is aground on the sand – you have struck Spurn Head and are now at the mercy of the pounding breakers.

Many deaths resulted from this unmarked **sand spit** and so it was decided to build a lighthouse at the end (A on Fig. 18.17). Soon after this lighthouse had been built the ease with which sand can be moved became apparent. The lighthouse was built on the edge of the spit, but as the years went by the spit grew longer and longer while the coast (and the spit) retreated further and further west. Finally 100 years after it had been built, the lighthouse was washed away. Clearly a new one was needed, so in 1776 it was built on what had then become the new tip (B). You can see that subsequently the spit has grown even longer and now the new 20th century lighthouse (C) is not at the end.

The disappearing sea

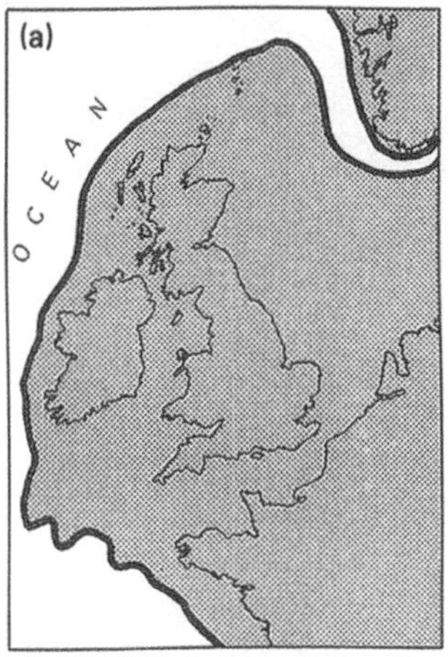

Figure 18.19 Britain's changing coastline
The coastline of Britain is almost entirely the result of changes in sea level. (a) Only 50 000 years ago the coast was up to 200 km out into the Atlantic. (b) If the ice now locked up on land were melted, much of our lowland would disappear. Clearly the effects of marine erosion on our present coastline are really quite trivial: they amount to no more than nibbling at the edges.

A man standing on the cliffs near Calais during the maximum of the last Ice Age would have viewed not the choppy waters of the Straits of Dover but a barren frozen waste of dry land. He might have walked across to Dover with nothing stopping him besides an occasional river swollen with the melting waters of the northern ice sheet. From Dover he could have walked to the Netherlands, probably skirting some large frozen lakes and other torrential meltwater streams.

A journey across such a cold wasteland with no shelter from the biting winds would have been extremely difficult for Stone Age man. But the journey to the sea was equally difficult. He would have had to walk many kilometres to the south-west of the Scilly Isles just to find a cold grey sea full of icebergs.

It is difficult to imagine such a scene occurring just a few thousand years ago, but when the ice locked up so much water that the water cycle could not be completed, the level of the oceans dropped about 80 m and the European coast probably looked somewhat like Figure 18.19.

Beyond the ice margin, rivers that carried away meltwater found themselves 'high and dry' as they reached the former coast. With their new sea level 80 m lower they had to extend their courses considerably and cut down deeply into their old channels. For year after year they cut deep channels into their previous floodplains until rivers such as the Thames were embedded in deep gorges. Then, when the climate eventually became warmer, the vast ice sheets melted

Exercise 18.3 Longshore drift

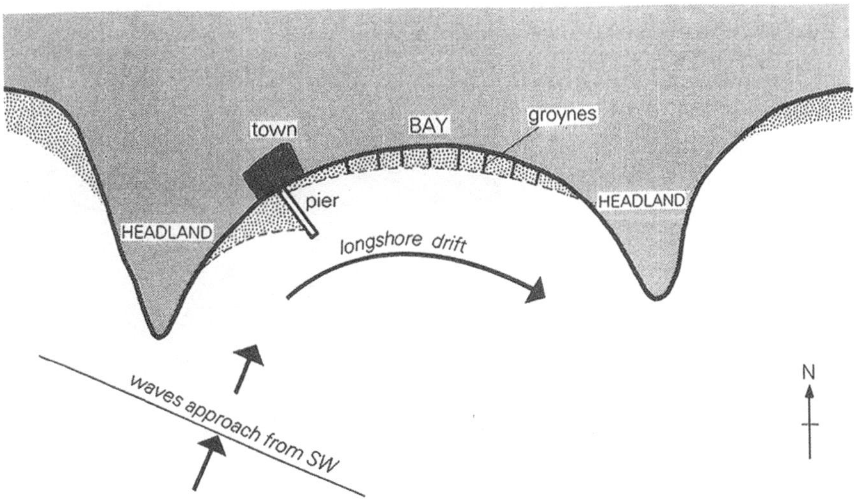

Figure 18.18 A sandy beach altered by man's construction

1. Notice that, because of longshore drift, sediment is being moved continuously along the coasts of Britain. Figure 18.18 shows a bay and headlands. What has happened following the construction of a stone pier 400 m out to sea, thereby interrupting the flow of material?
2. Why would you use the north side of the pier for tying up boats, rather than the south side?

You will have seen slatted 'fences' half buried in some beaches. These **groynes** are designed to prevent beaches being literally washed away by longshore drift. They are needed where, for some reason, the normal movement of beach material along the shore is interrupted so that sand is lost and not replaced. The bay in Figure 18.18 did not need groynes before the pier was built.

3. Copy Figure 18.18 and draw on the pattern of sand you think would be trapped by the groynes (use the pier as a guide).
4. Why should anyone be bothered about beach loss in any case?

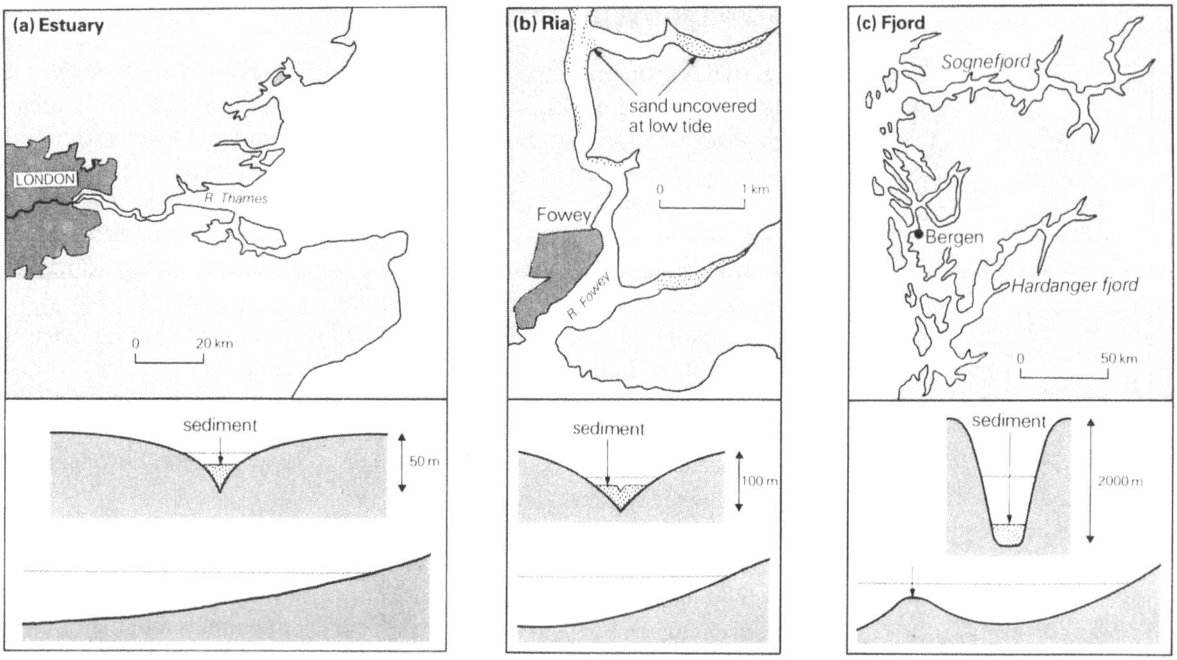

ROGER JONES

Figure 18.20 Drowned valleys
(a) Estuaries, (b) rias and (c) fjords all represent coastal valleys deepened during the Ice Age that have subsequently been flooded. Many are now refilled with sediment as the photograph shows. This is a sand-filled sea loch (fjord) in Sutherland (Scotland) at low tide.

quite quickly and the sea flooded back again, drowning the ends of all these new valleys. In only a few thousand years Ireland had once again become an island and soon after Britain was cut off from France. Now the rivers were obliged to adjust to these new circumstances by depositing their sediment in their channels. Some, like the Thames, have already mainly succeeded in filling in these gorged channels, but many drowned valleys still exist because the rivers flowing in them have not been able to provide enough sediment to complete their tasks. It is these drowned river valleys that we call **estuaries or rias** (Fig. 18.20).

SUMMARY: COASTAL DEPOSITION

Some of the most interesting coastal features are associated with sand movements along the shore. Waves causing longshore drift of sand do not readily bend to take in sharp changes of coastline because they are controlled by the smooth shape of the sea bed, *not* the coast of headlands and bays that form the edge of the land. So waves continue to sweep across the inlets of river valleys or bays until they are deflected by the force of river water, or they enter water too deep to cause breakers.

Because sand movement is related to breaking wave patterns, sand is dumped as soon as the water of the bay or inlet becomes too deep. Gradually the pile of dumped sand becomes large enough to reach the surface and form an extension to the land. This promontory of sand is called a **spit** (Fig. 18.17). The growth of a spit is controlled by a ready supply of sand, so spits are more common in places where the coast is easily eroded to provide the required material. Usually spits terminate a little way across an inlet, but occasionally the build up of material continues right across from one headland to the other. When a sand or shingle ridge connects two headlands it is called a **bar.** The Loe bar near Helston in Cornwall is a sand and shingle bar (Fig. 18.21).

Even more occasionally ridges of sand or shingle connect an island to the coast. In these cases the ridge is called a **tombolo.** The most famous tombolo in Britain is Chesil Beach which connects the Isle of Portland to the coast just west of Weymouth in Dorset. Here the shingle ridge is over 10 km long (Fig. 18.22). Many tombolos may not be entirely formed by longshore drift.

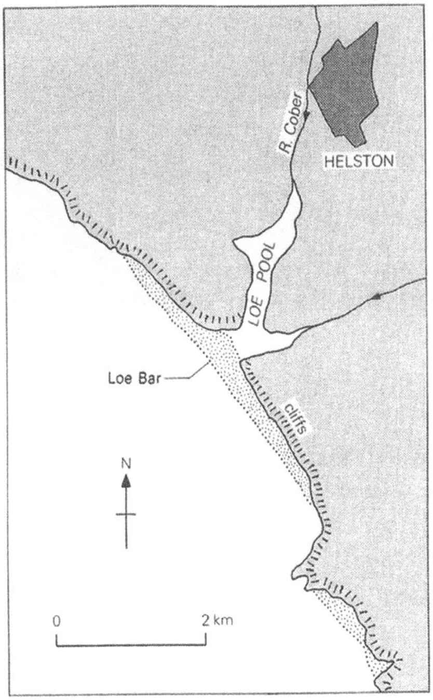

Figure 18.21 The Loe Bar, Helston, Cornwall

Figure 18.22 Chesil beach, Weymouth, Dorset

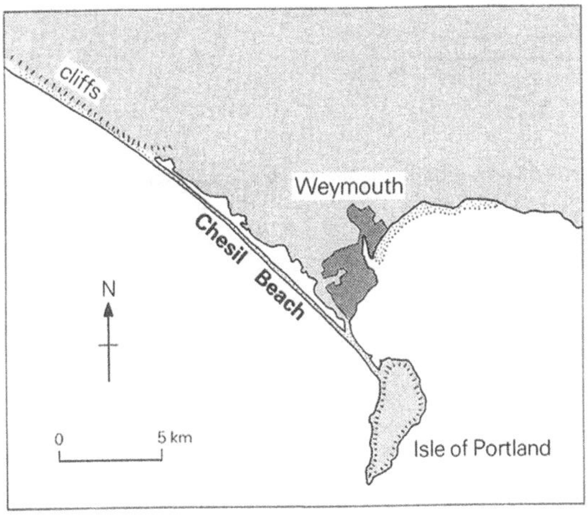

SUMMARY

Finally, it is worth noting the effect that the sea can have on the coast if waves nibble at the edges of the land for a long period without a change of sea level. We have already seen that wave-cut platforms will replace cliffs. You may already have noticed how flat cliff tops often seem to be (e.g. Fig. 18.23). This is not an accident. These cliffs are old wave-cut platforms raised high above sea level long ago by some movement of the Earth and are now themselves being consumed by the ceaseless waves.

Topic Estuaries, rias and fjords

The effects of recent changes of sea level have been dramatic (Fig. 18.20). Estuaries, rias and fjords are all the result of recent valley drowning.

Rias are drowned valleys in moderate relief. They occur in places such as the south-west peninsula of England and south-west Ireland. In lowland areas such as the south and east coasts of England the drowned valleys have much more subdued slopes and are called **estuaries.**

Fjords are a distinctive type of drowned valley, partly because they are found only in mountainous regions, but mostly because they were eroded by glaciers rather than rivers. Fjords occur in Scotland (where they are usually called **sea lochs**), New Zealand (where they are called **sounds**), Canada, Norway and some other coastal mountain areas that have been glaciated. It seems that glaciers were much more effective at eroding down to the low sea level of the Ice Age than were rivers. As a result some fjords, such as the famous Hardanger fjord in Norway, are very deep (up to 1000 m below present water level) and also very long (some are more than 200 km). Further, most fjords are shallower near the coast. In effect they have a lip, called a **threshold**, where the valley glacier ceased eroding as it met the sea.

Figure 18.23 The rising land
These flat-topped cliffs in Cornwall, (southwest England) are 70 m above sea level, but they are the remains of an old wave-cut platform.

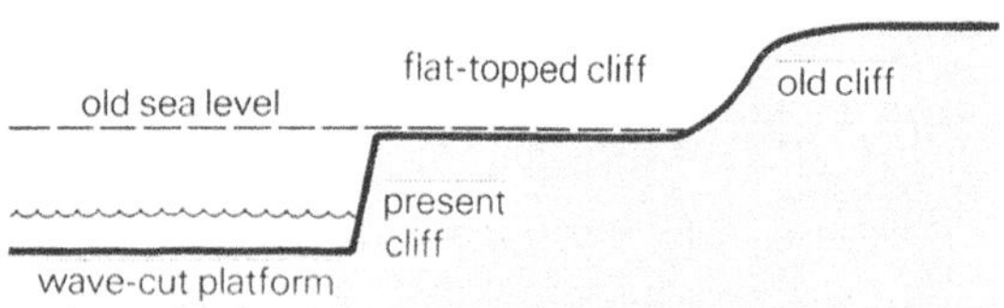

New land from the sea

The growing land

One of the most important jobs in a port is that of 'marine refuse collector'. But the sort of refuse that is collected by port authorities cannot go into a dustbin. Every year millions of tonnes of sediment have to be cleared out of our ports so that they will remain open for navigation. It is not a small or insignificant task and it must be continued year in year out in this area of constant battle between nature and man.

Ports are often built at the mouths of rivers because they provide access to inland areas and shelter from storms. But modern ships need access to deep water. Something of the order of 20 m is a commonly dredged depth for an international port – often more. If a five-storey block of flats stood in one of these channels, the top would be below water level.

Of course the river mouth is just where the flow of the river becomes slack and sediment is deposited. In fact, given time a delta would form. So a conflict arises wherever man tries to keep a port open. In the long term the river must win, but in the short term only continual dredging will keep the channels open. Nowadays there are large dredging boats to deal with the task, but in the Middle Ages there was no means of preventing silting, and history is full of ports that had to be abandoned when the river silted up. Famous examples include Brugge (Belgium) and the one-time port of Chester in north-west England.

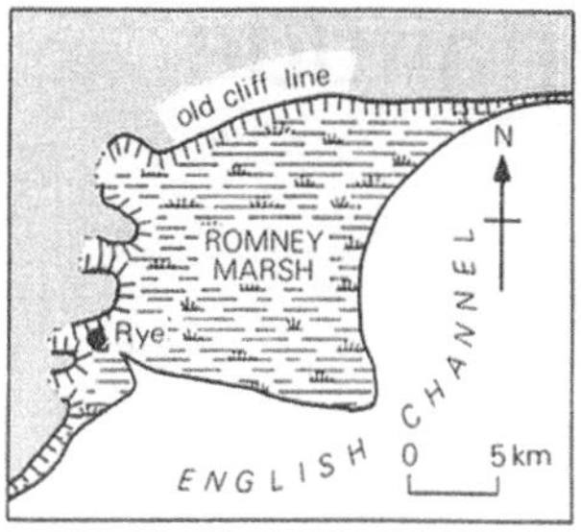

Figure 18.24 Romney marsh This is just one example of a part of Britain's coast that is growing.

Longshore drift can also cause silting in harbours. Indeed, many ports have to keep open not only river channels but also the coastal reaches. When nothing can be done the seashore gradually builds up new land, as has been the case of Romney Marsh in south-east England. Behind the now-considerable expanse of land in Kent lie the mediaeval ports of Rye and other towns. Some are now several kilometres from the open water (Fig. 18.24). It is as well to remember that, although rivers and the sea take away from some places, they are always giving back in others.

EXAMINATION-STYLE QUESTIONS

1 Name one example of a waterfall. By means of labelled diagrams explain the development of the named example.

2 Explain what is meant by the terms river meander, river cliff, and oxbow lake. Show how a river meander is responsible for a floodplain and sometimes for the formation of oxbow lakes.

3 Name one example of a delta. By means of labelled diagrams describe and explain the main features of a delta and how they are produced.

4 With the aid of diagrams explain the formation of a waterfall, river terraces, and a delta.

5 What is meant by the term *lake?* Explain how lakes may be formed by (a) earth movement, (b) glacial erosion, and (c) glacial deposition.

6 For a river floodplain that you have studied, describe the characteristic features of relief and drainage and show how the natural features of the floodplain have influenced settlement and economic activities within the area.

7 What is meant by a swallow hole, a stalactite, and a limestone pavement? By means of labelled diagrams describe how (a) a swallow hole and (b) a cave form in a limestone area.

8 Using labelled diagrams explain how an oasis might have formed. Describe and explain some of the major features of the desert region that involve running water.

9 Explain what is meant by a barchan and suggest how it may have been formed in a desert. Explain why sand dunes such as barchans are rare in a desert.

10 By means of a labelled diagram show the relationship between a mountain front, alluvial fans and playa lakes.

11 With the aid of diagrams explain briefly how each of the following features may have been formed: (a) a corrie, (b) a terminal moraine, and (c) a pyramidal peak.

12 With the aid of diagrams, explain how a valley headwater can be turned into a corrie.

13 Name four features of glacial erosion. For each feature describe and explain how it may have been formed.

14 Explain what is meant by a U-shaped valley and describe the main processes of ice action that change a river valley into a glaciated valley.

15 Explain what is meant by the terms *abrasion* and *quarrying*. By means of labelled diagrams explain how these are important in the formation of (a) a corrie and (b) truncated spurs.

16 Using labelled diagrams, describe and explain the formation of four features of lowland glaciation.

17 Name examples of a sea spit and a bar. Choose one of these and, by means of a labelled diagram, show its main characteristics and explain how it may have developed.

18 By means of labelled diagrams explain what is meant by the terms *ria, estuary* and *fjord*. Explain the main differences between a ria and a fjord.

19 Describe the main ways in which the sea erodes the land. Name two features that result from sea erosion and two features that result from the deposition of eroded material. For any one feature, name an example and describe how it was formed.

20 Explain why some waves tend to remove material from a beach and others tend to bring material back up onto a beach. What weather conditions are usually associated with each type of wave?

21 By means of labelled diagrams describe and explain how a coast develops cliffs and explain how the development of cliffs often results in the formation of stacks, arches, and wave-cut platforms.

22 Describe the formation of a bay and headland coast indicating particularly the features you would expect to find at the headland.

Appendix 1 Latitude and longitude

Where am I, where are you?

Geography is concerned with places on the Earth's surface. It tries to show the distribution of natural phenomena and those created by man. But to describe a distribution requires a map, and to construct a map some sort of frame of reference is needed. Communications and navigation would be difficult without a frame of reference; try explaining to somebody the location of New York and its position relative to London without a map!

There are some natural starting points for our frame of reference. Because the Earth rotates about an axis, the points of rotation on the surface can be useful. We call these the **North and South geographic poles.** They are nearly in the same places as the **magnetic North and South Poles** – a useful feature as most navigation over long distances is achieved with the aid of a magnetic compass. The **equator** is another natural reference because it is half way between the Poles and because places north and south of the Equator show many similarities of climate, soils and vegetation.

An easy start to a frame of reference is achieved by drawing circles on the Earth's surface parallel to the equator. These are **lines of latitude** (Fig. A.1). By convention the Equator is regarded as having a latitude of 0° and the circles parallel to it are designated in degrees north and south until the Poles are reached at 90° north and south respectively.

There is no natural starting point for a set of lines in any other direction. By agreement, therefore, a set of circles passing through both Poles are spaced evenly across the globe, starting with the circle that passes through Greenwich in London, which is designated 0°. All the other lines (called **lines of longitude** or **meridians**) are measured east and west from this, meeting in the Pacific ocean. The meridian of 180° east and west is chosen as the International Date Line, as described in Appendix 2.

Combining lines of latitude and longitude produces a grid system with which we can locate all places on the Earth's surface. Thus New York is at 41°N, 74°W, and London is at 51°N, 0°W to the nearest whole degree. We can also say that New York is at the same latitude as Naples, Italy; 1° farther north than Peking; and so on.

We are at grid reference . . .

Lines of latitude and longitude are useful on maps of the whole world or of continents and large countries, but they are not very convenient for use with maps of small countries such as Britain. So when the British Army Ordnance Corps was given the task of mapping the British Isles in the 18th century, they chose an artificial square grid whose starting point is just off the south-west of Cornwall. The grid is simply laid over the country and is not related to the lines of longitude in a strict way, although true north and the north of this grid are nearly the same. It is an example of a small-scale system chosen for convenience. Today the grid lines on an Ordnance Survey map are spaced 1 km apart. Most

Figure A.1

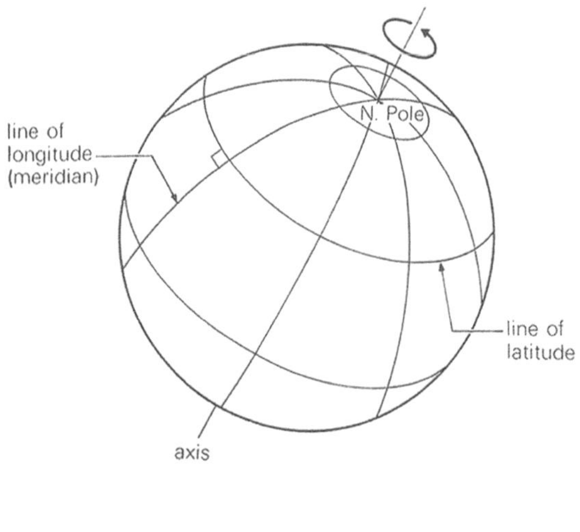

Figure A.2

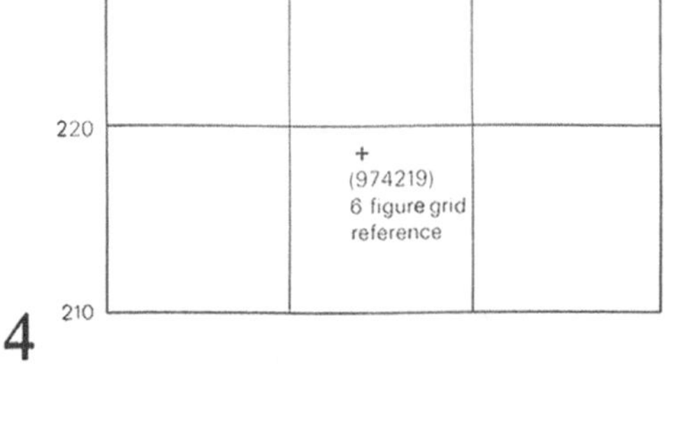

other national surveys use the same type of artificial grid.

The grid is used to locate places, just as lines of latitude and longitude are. Grid lines running from north to south are called **eastings** and those from east to west are **northings** (that is, east and west of the grid origin). Points on a map are then located by a six-figure reference as shown on Figure A.2. The first three figures are an easting (the vertical line); the last three are a northing (horizontal line). Some people find it easier to remember the order of a reference by saying to themselves 'along the corridor and up the stairs.'

Appendix 2 Time

Today is Tuesday . . . or is it Wednesday?

The **International Date Line,** fixed by international agreement, follows the 180° meridian, except where it deviates to avoid dividing land areas (see inside back cover). The international day begins and ends on this line and sweeps westward with the Sun (Fig. A3).

Exactly at midnight (on the date line) it may be Tuesday over the whole globe. Just west of the line it is Tuesday midnight; just east of the line Tuesday is beginning. At every 15°W of the line, it is one hour before midnight. This is because 24 h divide into the 360° of the globe 15 times. Therefore, for example, at 165°W it is 23.00 h Tuesday night and at Greenwich (0°) it is mid-day; a moment later Wednesday has started to the west of the line; an hour later Wednesday has begun 15°W; and so on . . .

Figure A.3

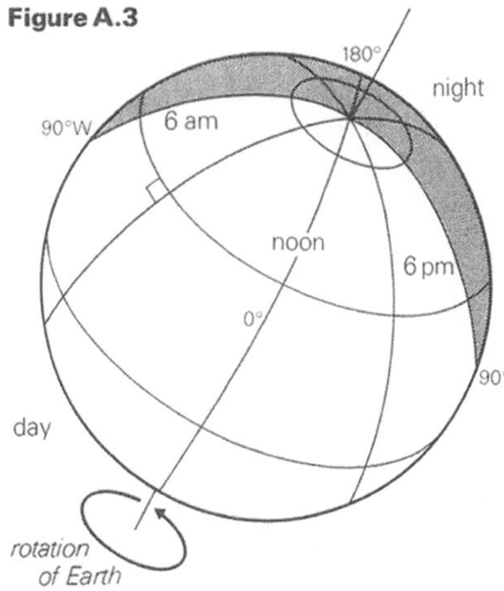

Appendix 3 Instruments to measure things

Much of the information we have used to build up a picture of the way the Earth's surface has been formed and the processes at work in its continued modification comes from direct measurement. Below are some of the pieces of equipment in common use.

Weather

The main elements of the weather that are often measured are air temperature, rainfall, wind speed and direction, duration of sunshine, and pressure. Many of the instruments required for such measurements are best kept in a special box called a **Stevenson's screen** (Fig. A.4). This has slatted sides designed to allow free air movement but prevent direct sunshine affecting the readings of the instruments. The box also ensures that the instruments are kept a standard 1.5 m off the ground to allow comparison between stations.

Air temperature is measured using a set of thermometers. These are read directly at regular intervals during the day. Usually one thermometer, called a **dry bulb thermometer,** is used for normal air temperature. Another has a wet muslin bag tied over the bulb (a **wet bulb thermometer**). The difference in temperature between them is a measure of the relative humidity of the air. When they read the same the air is saturated. These thermometers are usually mounted vertically within the Stevenson's screen.

In addition a **maximum and minimum** thermometer set are usually provided (Fig. A.5). These record the range of temperatures experienced over a 24-hour period. One of these thermometers is usually filled with mercury. A small

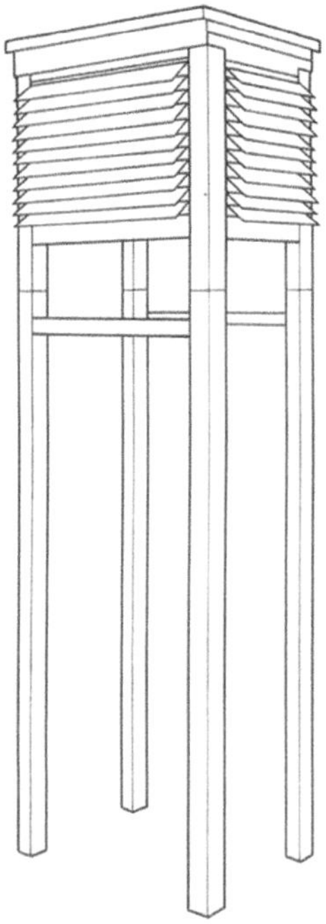

Figure A.4

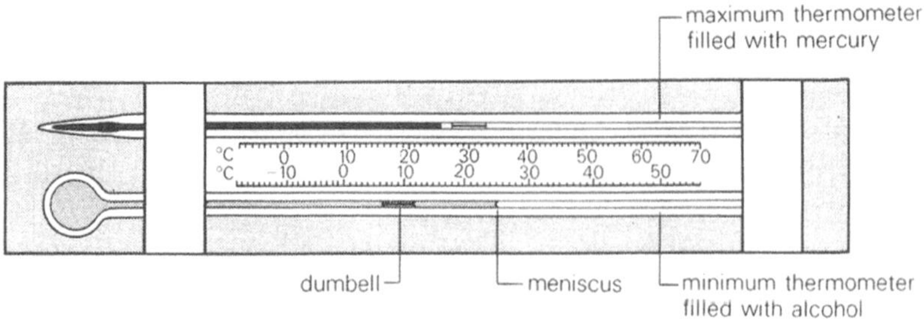

Figure A.5

wire dumbell in the thermometer tube is pushed forward as the mercury advances and is left behind as temperatures fall, thus recording the maximum value. The minimum temperature is also measured with a wire dumbell, but this time the thermometer is filled with alcohol and the dumbell is dragged along with the liquid as it contracts with lower temperatures. The dumbells can be seen clearly on the diagrams. Notice that these thermometers are kept horizontally and they need resetting each day by bringing the dumbell back against the meniscus in each thermometer.

Rainfall is measured with a standard raingauge whose upper rim is placed 30 cm above the ground (Fig. A.6). The receiving bottle is emptied and the contents measured daily. The gauge must not be sited near any sheltering trees or buildings.

Pressure is usually measured using an aneroid barometer which has a recording pen and drum mounted in its case (Fig. A.7). The pen records expansion and contraction of the partially evacuated container X with changes of air pressure.

The duration of continuous sunshine is measured with a recording device comprising a glass sphere which concentrates the Sun's rays and burns a trace on a piece of sensitised paper placed in the metal holder (Fig. A.8).

Wind speed is measured with a 3-cup rotating device on a spindle called an anemometer (Fig. A.9). As the wind blows, it turns the cups and the number of revolutions of the spindle are converted into wind speeds in m/s or km/hr.

A good exposed position away from any buildings or trees is essential.

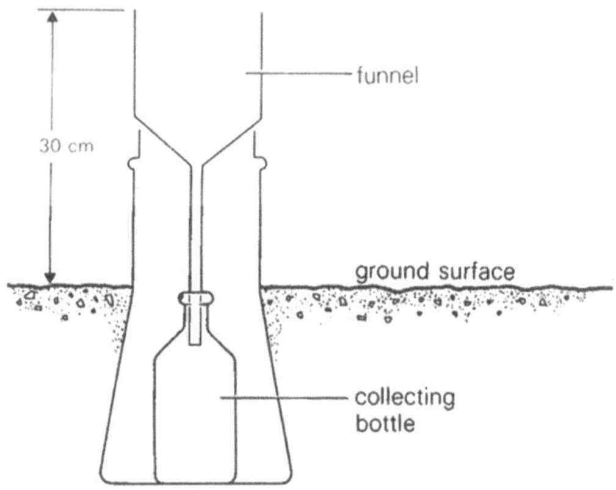

Figure A.6

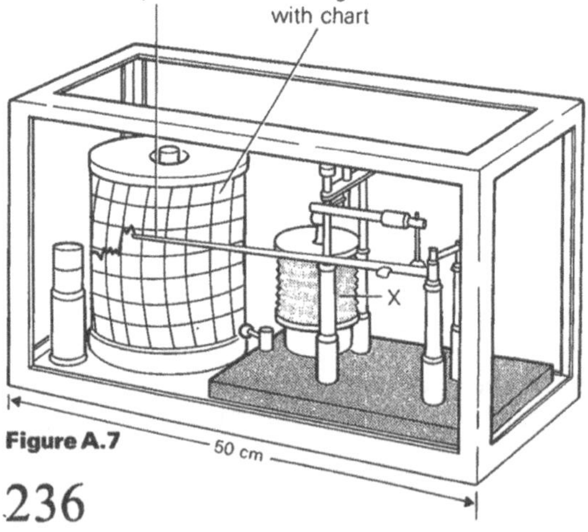

Figure A.7

Water cycle

In addition to weather elements the water cycle requires the measurement of evaporation and streamflow.

Evaporation is measured by a large tank which is daily refilled with water to a marker (Fig. A.10). The amount of water added is equal to the amount lost by evaporation over the previous day.

Streamflow is usually measured with a V-notch weir (Fig. A.11). There is a known relationship between the height of water ponded up behind the V-notch construction (h) and the flow of water. The height is recorded daily or continuously on a drum.

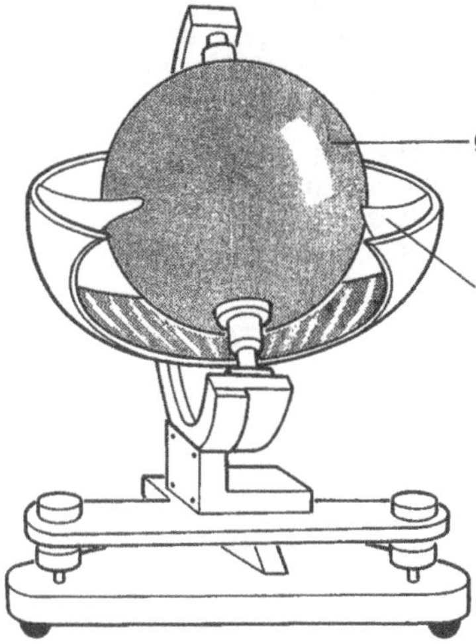

Figure A.8

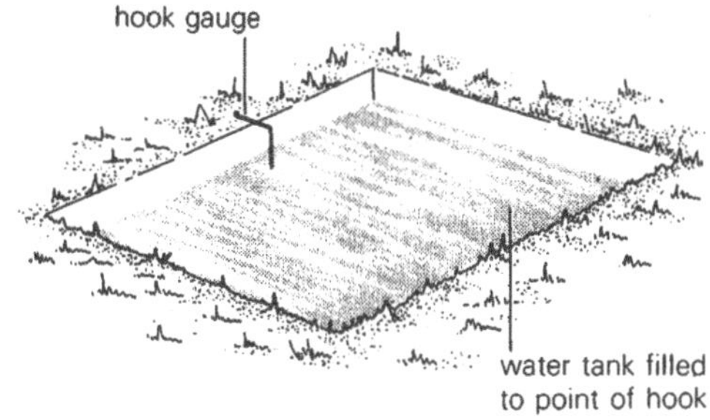

Figure A.10

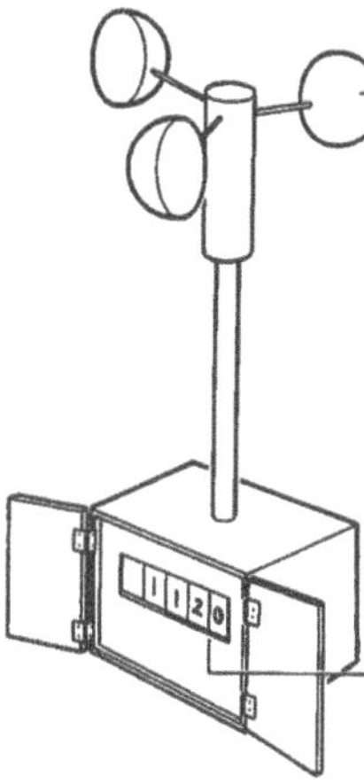

Figure A.9

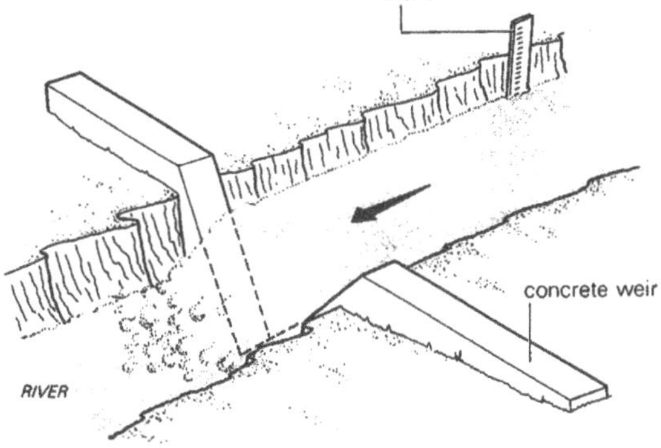

Figure A.11

Index

abrasion, by
  ice 207, 213
  water 174
  wind 200
acacia tree 130
accelerated soil erosion 143
adaptation by vegetation 125, 126
advection fog 65
aeroplane routes 70
air photos 2
air pressure 57, 63
air temperature 57
alluvial fan 196
alluvium 116, 181, 187
Alpine mountain building 40
altitude 205
Andes mountains 41
andesite 11, 41
anemometer 237
anticline 36, 37
anticyclone 61, 68
Appalachians 2
aquifer 93, 99
arch 221, 222, 225
arete 210, 211, 213
ash 12, 13, 16, 19
aspect 52
atmosphere 50
atoll 27, 28
available water 98
avalanche 164

backwash 223
bar, sand 231
barchans 201
barometer 63
barrier reef 27, 28
basalt 9
batholith 11, 15, 16, 41
bay 221
beach 224–6
bedding plane 112
black soil, *see* chernozem
bloom, of plankton 87
bore 85
braided stream 180
breakers 219
British climate 73
brown earth soil 117, 120
building materials 29
Bumpass Hell, USA 21
bush fallow system 140–2

cactus 125
caldera 17, 18
Caledonian mountain building 40, 46
California 60
canyon 199
carbon dioxide 51, 113
Cascade mountains, USA 16, 17
catastrophe 161

cave, in
  ice 207
  limestone 192
  sea cliff 221, 225
chalk 194
chemical weathering 108, 111, 113
chernozem soil 118, 120
cinder cone 9
circulation, of atmosphere 54, 56
cirque, *see* corrie
cirrus cloud 58, 59, 61
clay 113, 114
cliff 220, 221, 223
climate 71
  cold 78, 79
  cool temperate 71, 73, 79
  desert 76, 78, 79
  hot 71, 79
  monsoon 74, 76, 78, 79
climatic change 80
climax vegetation 127
clint 191
cloud 57–9, 61
coal
  field 24, 30
  formation 30
  mining 32, 39
coastal breeze 55
cold climate 78, 79
cold front 68
cold ocean current 83
cold period 204
coffee trees 150
colonisation by plants 125, 128
Colorado river, USA 23, 98
concealed coalfield 39
condensation 53
conduction of heat 53, 54
coniferous forest 131
conservation, soil 146
continental climate 73
constructive wave 226
continental drift 40, 45, 133
contour ploughing 147
convection
  in the atmosphere 53–5, 67
  in the Earth 7
  rainfall 67
cool temperate climate 73
coral reef 26, 28
core of the Earth 8
corrie 208, 209, 213
cove 222
crag and tail 208
Crater Lake, USA 18
crevasse 209, 211
crust, of Earth 8, 40
cuesta 171
cumulonimbus cloud 58, 59, 61
cumulus cloud 58, 59, 61, 62, 66
curtain stalactites 192, 193
cwm, *see* corrie

date line (*see* international date line)
  134
Death Valley 60
deciduous forest 131
decomposers 115
delta 188, 189
dendritic drainage 171
depression, in atmosphere 61, 64, 68
desert 110, 195
  climate 76, 78, 79
  farming 149
  vegetation 125, 127, 197
desertification 153, 155
destructive wave 226
dew 57
dip slope 171
disaster, from
  drought 90
  earthquakes 33
  flood 172
  landslides 166
  marine erosion 223
  soil erosion 154
  volcanoes 19
Doldrums 56
drainage basin 91, 170
drizzle 59
drought
  effect on vegetation 133, 126, 125
  effect on water cycle 90
drowned valleys 230
drumlins 214, 217
dry-bulb thermometer 235
Dust Bowl, USA 143
dykes
  in igneous rock 9, 15, 16
  in river 184

Earth
  origin 6
  structure 8
earthquakes 10, 33
easting 235
ecosystem 107, 121, 122, 138
effluent 101
energy, solar 51
equatorial climate 76, 79
equinox 72
erg 200
erosion
  by ice 207
  by river 175
  by sea 220
  by wind 200
  of soil 143
erratic 206, 214
eruption, of volcano, *see* volcano
esker 216, 217
estuary 230
evaporation measurement 237
European climate 73
exfoliation 109, 110, 113

exposed coalfield 39
extrusive rock 16

fan, *see* alluvial fan
farming
by high technology 149
by primitive methods 138, 151
in the past 139
fault 33, 36, 37
fauna 134
fertility, of soil 142
fertiliser 142, 143
fire 138
firn 206
fishing grounds 86
fjord 230, 232
flash flood 195, 198, 199
flooding 13, 187
floodplain 181
flora 134
flowstone 192, 193
fog 65
fold 35–7
fold mountain 35
food chain 123
fossil 24
freeze-thaw, *see* frost shatter
fringing reef 27, 28
front, in the atmosphere 62
frontal rain 62
frost shatter 108, 109, 113, 210
frost-shattered peak 210

gas 24, 30, 32
geothermal power 22
geyser 21
glacial erosion 207
glacier ice 206
corrie 208
valley 210
global winds 56
gorge 178, 179
granite 11, 112
Grand Canyon, USA 23
grassland 130
gravel 114
Great Barrier Reef, Australia 27
Great Glen Fault, Scotland 34
Great Whin Sill, England 15
greenhouse effect 51
Greenwich 235
grid reference 234
ground moraine 207, 212
ground water 93, 99
groynes 229
gryke 191
Gulf Stream, Atlantic ocean 54, 82, 84, 88
gullies 145, 163, 167, 171

habitat 122
hanging valley 177, 210, 211, 213
hamada 201
headland 221, 225
heating power of Sun 51
Helgafell volcano, Iceland 9, 14
Hercynian mountain building 40, 46
high pressure region 55, 61, 63, 68
high tide 85
horse latitudes 56
horst 37
hot climates 76
hot spring 21, 22
humus 115
hurricane 54, 69, 78, 79
hydro-electric power (HEP) 183, 208

Ice Age 204
ice
bergs 83
cave 207
crystals, *see* snowflakes
fall, 209
sheet 204, 214
igneous rock 10, 16
Indian monsoon 77
infiltration 91
insolation 51
interglacial period 204
international date line 235
intrusive rock 16
iron ore 31
iron pan 120
irrigation 101–3
island arc 42
isobar 63
isotherm 72, 73

jet stream 54, 56, 70, 77
joints 37, 112
jungle 129

lagoon 27, 28, 31
lake 185
landslide 157, 159, 160, 161, 167
lateral moraine 210, 212
laterite, *see* tropical red soil
latitude 234
lava 7, 8, 9, 16
leaching 119, 120, 164, 167
levée 187
lightning 67
limestone 191
pavement 191
lip, of corrie 208
loch 230
lode, of metal ore 22
loess 116, 202
longitude 234
longshore drift 227
long profile 178, 179, 180
lowland
glaciation 214
soils 117
low pressure region 55, 56, 61, 63, 64, 68

magma 7, 16
mantle 8
man-made ecosystems 138
maritime climate 73
mass movement of soil 158
maximum thermometer 235
meandering 180–2
medial moraine 210, 212
Mediterranean climate 80
meltwater 212
meridian 234
mesquite bush 125
metal ores 22
metamorphic rock 10, 44
mid-latitude gears 56
migration of species 134
millibar 63
mineral 10
minimum thermometer 235
Mississippi delta, USA 190
monoculture 148
monsoon 74, 76, 77, 79
moraine 206, 207, 212
mountain
building 40, 42, 43, 45
climates 136
vegetation 136
Mt Pelee, Caribbean 19
Mt St Helens volcano, USA 5, 11–13, 17
mudflats 189
mudflow 13, 167
mudpool 21
mulch 148

natural selection 122, 123
neap tide 85
névé 206, 213
Nile, R., Egypt 103
nimbostratus cloud 58, 61
normal fault 37
nomadic agriculture 153, 154
northing 235
north
geographic pole 234
magnetic pole 234
nutrient cycle 142
nutrients 114

oasis 202
occluded front 68
ocean
currents 54, 82
heating 81
trench 41, 42
waves 219
oil 24, 29, 31, 32
ores 22
outwash plain 216, 217
overfold 36
oxbow lake 181

Pacific Ring of Fire 10, 11
pack ice 83
palm oil plantation 151
parent material of soil 116
peat 117, 119
pebbles 114
permafrost 167, 168
physical weathering 108
plankton 87
plantation farming 149, 150–2
plate, crustal 40, 45

playa lake 198, 199
plunge pool 177
plunging wave 225
podzol soil 119
Polar easterly winds 56
pollution
  of atmosphere 113
  of rivers 100, 104
potholes 178
prairie 130
pressure, of air 55
pyramidal peak 208, 212, 213

quarrying 207, 213

radiation
  fog 65, 68
  solar 51, 53, 54
rain 61
  -drops 58
  forest 130
  gauge 236
  shadow 60
  splash 163, 167
range of species 124
rapids 176, 177
redwood tree 126
reef, coral 28
relief
  cloud 60
  rain 60
rendzina soil 120
reservoir 99, 100
reversed fault 37
Rhine rift valley, Germany 37
ria 230, 232
ribbon lake 212
rift valley 37
rill 145
river
  channel 90
  control 103, 174
  flow 94, 172
  levées 187
  pollution 100
  terrace 183, 184, 186
roche moutonnée 208, 213
rock basin lake 210, 213
  fall 167
rotation, of Earth 54
rubber tree plantation 152

St Vincent 19
Sahara desert, Africa 199
Sahel, Africa 145, 153
salt
  lake 198
  marsh 228
  rock 31
  water 81
San Andreas fault, USA 34
sand 114
  bar 231
  dune 199, 228
  sea 199
  spit 228
saturated air 57
savanna 130
scarp 2, 171
scouring by river 174
scree 109, 117, 158, 167
seasons 72
sea breeze 55
sediment 23, 123
sedimentary rock 10, 23, 25
seepage 92
sill 9, 15, 16
silt 114
smog 65
snow 205
snowflakes 57, 206
soil 107
  animals 115
  components 114
  conservation 146, 148
  creep 162, 167
  erosion 143, 148
  fertility 142
  formation 114, 116, 120
  groups, *see* under individual soil name
  humus 114, 115, 118
  layers 116
  lowland 117
  moisture 91
  nutrients 114, 115
  organisms 114, 116
  parent material 116
  pores 115
  solution 119, 164
  upland 118
  water 114
solar
  energy 51
  radiation 53
solstice 72
solution 164, 167, 175, 191, 192
south
  geographic pole 234
  magnetic pole 234
species 122
spilling wave 225
spinning Earth 56
spit, *see* sand spit
spring 92, 93, 171
  hot 21
  limestone 171
  tide 85
stack 221, 222, 225
stalactites 192, 193
stalagmites 192, 193
steppe 130
Stevenson screen 235
stone pavements 201
storm waves 219
stratus cloud 58, 59, 61
streamflow 94
striations 207, 213
Sudan, Africa 1, 154
Suez, Egypt 196
sunshine recorder 237
surf 227
surface wash 163, 167
survival of species 127
suspension 173, 175
swallow hole 192
swash 223
swell wave 226
syncline 36, 37

tarn 211
tear fault 37
temperate climate 73
  deciduous forests 131
terminal moraine 212, 215, 217
terracettes 160, 167
thermometer 235
thunder 67
tides 85
till 212–15, 217
tin mine 22
tombolo 231
tornado, *see* hurricane
trade winds 56
transpiration 124
trellis pattern 171
tropical
  climate 78, 79
  rainforest 127, 130
  red soil 118, 120
Tropics 54, 58, 76
truncated spur 210–13
tundra
  climate 79
  vegetation 132
tunnel, in sea cliff 221, 225
turbulence in atmosphere 53
typhoon, *see* hurricane

undercut cliff 221, 225
undercutting 221
upland
  glaciation 203
  soil 118
urban water movement 95
U-shaped valley 210, 211

vale 2, 171
valley
  glacier 210, 212, 213
  river 188
  shape 188
vegetation 107, 121, 197
vein, metal ore 22
Vesuvius, Italy 14
viticulture 52, 53
volcano
  benefits 20–1
  disaster 11–13, 19
  distribution 10, 19
  eruption 5, 7, 11, 14, 16, 18, 41
  materials from 16

wadi 195, 196
warm front 68
  ocean currents 83
  sector 68
  temperate climate 79
waste disposal 168

water
  cycle 89, 95, 170
  demand 97
  droplets 58
  vapour 51
waterfall 175, 178, 183
  power 172
wave 219, 220, 226
wave-cut platform 221, 223, 225
weather 50, 71
weather chart 63, 68
  map symbols 64
weathering 108, 113, 175
weir 237
westerly winds 56
wet-bulb thermometer 235
wind 56
wind erosion
  accelerated 144, 145

Yosemite valley, USA 38

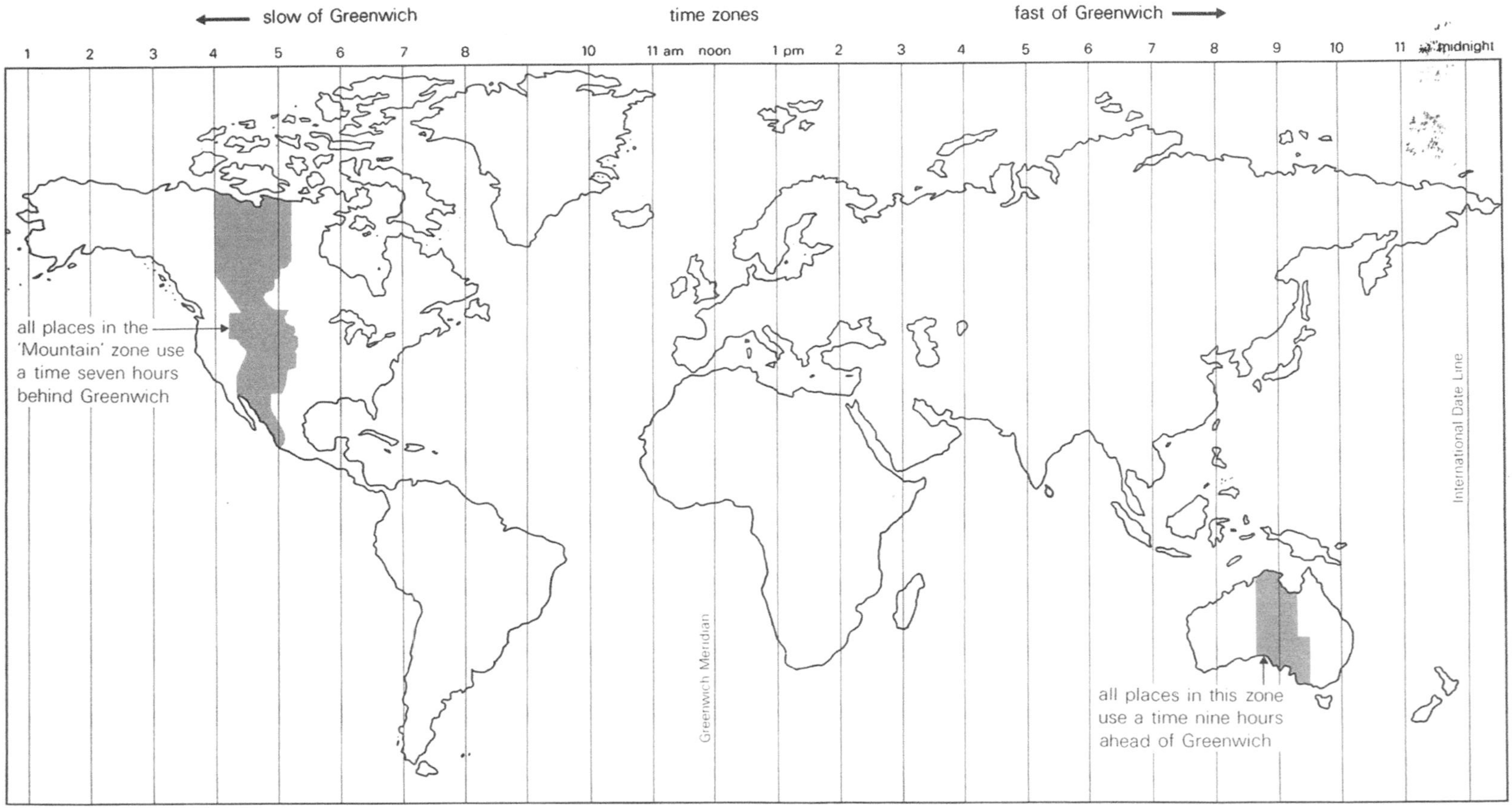

These are time zones around the world related to Greenwich mean time. The time zones used by many countries do not correspond exactly with this scheme. Some examples are shown. Can you suggest practical reasons for these broad time bands?